METHODS IN MOLECULAR BIOLOGY

Series Editor
John M. Walker
School of Life and Medical Sciences
University of Hertfordshire
Hatfield, Hertfordshire, UK

For further volumes:
http://www.springer.com/series/7651

For over 35 years, biological scientists have come to rely on the research protocols and methodologies in the critically acclaimed *Methods in Molecular Biology* series. The series was the first to introduce the step-by-step protocols approach that has become the standard in all biomedical protocol publishing. Each protocol is provided in readily-reproducible step-by step fashion, opening with an introductory overview, a list of the materials and reagents needed to complete the experiment, and followed by a detailed procedure that is supported with a helpful notes section offering tips and tricks of the trade as well as troubleshooting advice. These hallmark features were introduced by series editor Dr. John Walker and constitute the key ingredient in each and every volume of the *Methods in Molecular Biology* series. Tested and trusted, comprehensive and reliable, all protocols from the series are indexed in PubMed.

Lipid Signalling

Methods and Protocols

Edited by

Marco Brotto

Bone-Muscle Research Center, College of Nursing and Health Innovations, Department of Graduate Nursing, The University of Texas at Arlington, Arlington, TX, USA

Sanjoy K. Bhattacharya

Bascom Palmer Eye Institute, Miami Integrative Metabolomics Research Center, University of Miami, Miami, FL, USA

Humana Press

Editors
Marco Brotto
Bone-Muscle Research Center
College of Nursing and Health
Innovations, Department of Graduate
Nursing
The University of Texas at Arlington
Arlington, TX, USA

Sanjoy K. Bhattacharya
Bascom Palmer Eye Institute, Miami Integrative
Metabolomics Research Center
University of Miami
Miami, FL, USA

ISSN 1064-3745 ISSN 1940-6029 (electronic)
Methods in Molecular Biology
ISBN 978-1-0716-3901-6 ISBN 978-1-0716-3902-3 (eBook)
https://doi.org/10.1007/978-1-0716-3902-3

Cover Caption: Cover image designed by Merina Daniel, Kamal Awad, Sanjoy Bhattacharya, and Marco Brotto.

This Humana imprint is published by the registered company Springer Science+Business Media, LLC, part of Springer Nature.
The registered company address is: 1 New York Plaza, New York, NY 10004, U.S.A.

Preface

The history of lipids spans almost four centuries, starting with Techenius Otto in 1673. Otto described acidic compounds in fat by noting that the alkali is neutralized by animal fat in the process of soap manufacturing. Lipids, in addition to forming cellular boundary, are also highly efficient signaling molecules. Each fragment of a lipid can potentially generate a different signal in living cells.

Our protocol chapters encompass many areas of cell biology and several tissues. The protocols encompass a wide range of cell types, including neuronal cells like retinal ganglion cells and other generic neurons, microglial cells, muscle cells, epithelial cells, and serum from human samples. It encompasses protocols for optimized extraction for signaling studies, including clinical studies in humans. There are protocols that involve recent tagging methods such as azo analogs for probing phosphatidic acid mediated signaling, imaging phospholipase D activity with clickable alcohols via transphosphatidylation and lipid GPCR molecular interactions by proximity labeling, and a new TRAP-method for targeted lipid signaling mediators. The protocols encompass diverse disease spectrum from Alzheimer's, musculoskeletal diseases, eye diseases, to metabolic diseases. This book therefore provides a diversity of protocols encompassing different cells, tissues, and diseases.

Lipid signaling mediators play a crucial role in various biological processes by relaying signals within cells and between cells. These signaling molecules are derived from lipids, which are a diverse group of molecules that serve as structural components of cell membranes and as energy sources.

In the 1960s and 1970s, major discoveries would forever reshape the field of lipid signaling and cell biology, physiology, and medicine. First, the pioneering work of Konrad Bloch and Feodor Lynen "for their discoveries concerning the mechanism and regulation of the cholesterol and fatty acid metabolism," earning them the Nobel Prize in 1964. Second, prostaglandins, leukotrienes, and thromboxanes were discovered through the work of Bergström, Bengt I. Samuelsson, and John R. Vane, and these molecules were initially recognized for their roles in inflammation, blood clotting, and smooth muscle contraction. These scientists were honored with the Nobel Prize in 1982. In 1984, Michael Brown and Sir Joseph Goldstein were selected for the Nobel Prize for their discoveries concerning the regulation of cholesterol metabolism. Goldstein also had over the years begin to guide the field into the regenerative role of PGE_2, a critical recognition of the modulatory roles of lipid mediators.

Later, in the 1990s, researchers made significant advancements in understanding more about lipid signaling mediators. They identified other important lipid molecules, including sphingolipids and endocannabinoids, which are known to modulate numerous physiological functions.

Since then, the field of lipid signaling has continued to expand with the discovery of additional lipid mediators, such as lysophospholipids, ceramides, and specialized pro-resolving lipid mediators (SPMs). These newly discovered lipid signaling molecules have been found to participate in various biological processes, including immune response regulation, cell growth and differentiation, neuroinflammation, and resolution of inflammation, with the entire family of Resolvins lipid mediators, and the major implication that failure of resolution of inflammation is a culprit leading to fibrosis.

Overall, the history of lipid signaling mediators has shown that these molecules are involved in a wide range of physiological and pathological conditions. Their discovery and study have provided valuable insights into cellular communication and have paved the way for the development of new therapeutic approaches targeting lipid signaling pathways.

The protocols assembled here are detailed and precise. We hope that they will be useful in the deeper and integrative investigation of biological questions in biomedical research laboratories worldwide.

Arlington, TX, USA *Marco Brotto*
Miami, FL, USA *Sanjoy K. Bhattacharya*

Contents

Contributors

AHMED AL SAEDI • *Division of Endocrinology, Boston Children's Hospital, Boston, MA, USA; Department of Pediatrics, Harvard Medical School, Boston, MA, USA*

LOUAY ALDAHER • *Bascom Palmer Eye Institute, Miami Integrative Metabolomics Research Center, University of Miami, Miami, FL, USA*

AMR ALMOBAYED • *Bascom Palmer Eye Institute, Miller School of Medicine at University of Miami, Miami, FL, USA; Miami Integrative Metabolomics Research Center, Miami, FL, USA*

KANNADASAN ANAND BABU • *R.S. Mehta Jain Department of Biochemistry and Cell Biology, KNBIRVO, Vision Research Foundation, Chennai, India; Anderson Clinical Genetics, Anderson Diagnostic Services Private Limited, Chennai, India*

NARAYANASAMY ANGAYARKANNI • *Department of Biochemistry, Medical Research Foundation, Sankara Nethralaya, Chennai, India; R.S. Mehta Jain Department of Biochemistry and Cell Biology, KNBIRVO, Vision Research Foundation, Chennai, India*

ZEINAB FOTOUHI ASHIN • *Bone Muscle Research Center, The University of Texas at Arlington, Arlington, TX, USA; Department of Materials Science and Engineering, The University of Texas at Arlington, Arlington, TX, USA*

PRANESH ASWATH • *Department of Materials Science and Engineering, The University of Texas at Arlington, Arlington, TX, USA*

KAMAL AWAD • *Department of Materials Science and Engineering, The University of Texas at Arlington, Arlington, TX, USA; Bone-Muscle Research Center, College of Nursing and Health Innovations, Department of Graduate Nursing, The University of Texas at Arlington, Arlington, TX, USA*

MARIAN N. AZIZ • *Cyclotron and Radiochemistry Program, Radiology Department, University of Texas Southwestern Medical Center, Dallas, TX, USA*

OMAR BADLA • *Bascom Palmer Eye Institute, University of Miami Miller School of Medicine, Miami, FL, USA; Miami Integrative Metabolomics Research Center, Miami, FL, USA*

SANJOY K. BHATTACHARYA • *Miller School of Medicine, University of Miami, Miami, FL, USA; Bascom Palmer Eye Institute, Miami Integrative Metabolomics Research Center, University of Miami, Miami, FL, USA*

K'MANI BLYDEN • *Bascom Palmer Eye Institute, Miami Integrative Metabolomics Research Center, University of Miami, Miami, FL, USA; Medical College of Georgia, Augusta University, Augusta, GA, USA*

LETICIA BROTTO • *Bone-Muscle Research Center, College of Nursing and Health Innovation, The University of Texas at Arlington, Arlington, TX, USA*

MARCO BROTTO • *Bone-Muscle Research Center, College of Nursing and Health Innovations, Department of Graduate Nursing, The University of Texas at Arlington, Arlington, TX, USA*

MARIA D. CABRERA GONZALEZ • *Department of Ophthalmology, Bascom Palmer Eye Institute, Miller School of Medicine at University of Miami, Miami, FL, USA; Miami Integrative Metabolomics Research Center, Miami, FL, USA*

CHRISTOPHER P. CARDOZO • *Spinal Cord Damage Research Center, James J. Peters VA Medical Center, Bronx, NY, USA; Departments of Medicine and Rehabilitation Medicine and Human Performance, Icahn School of Medicine at Mount Sinai, New York, NY, USA*

Yan Chang • *Department of Graduate Nursing, College of Nursing and Health Innovation, Bone and Muscle Research Center, The University of Texas at Arlington, Arlington, TX, USA*

Ishan Chaplot • *Bascom Palmer Eye Institute, Miami Integrative Metabolomics Research Center, University of Miami, Miami, FL, USA*

Subashchandrabose Chinnathambi • *Department of Neurochemistry, National Institute of Mental Health and Neuro Sciences (NIMHANS), Institute of National Importance, Bangalore, Karnataka, India*

Carolina Cruz-Wegener • *Bascom Palmer Eye Institute, Miami Integrative Metabolomics Research Center, University of Miami, Miami, FL, USA*

Merina Daniel • *Bone Muscle Research Center, The University of Texas at Arlington, Arlington, TX, USA*

Noa deHaseth • *Bascom Palmer Eye Institute, Miami Integrative Metabolomics Research Center, University of Miami, Miami, FL, USA; Department of Biomedical Engineering, University of Miami, Coral Gables, USA*

Li Dong • *Bone-Muscle Research Center, Department of Kinesiology, College of Nursing and Health Innovation, University of Texas at Arlington, Arlington, TX, USA*

Gustavo Duque • *Research Institute of McGill University Health Center, Department of Medicine, McGill University, Montreal, Québec, Canada*

Sindhu Guntur • *Department of Materials Science and Engineering, The University of Texas at Arlington, Arlington, TX, USA*

Lauren Harlow • *Spinal Cord Damage Research Center, James J. Peters VA Medical Center, Bronx, NY, USA*

Jian Huang • *Bone-Muscle Research Center, College of Nursing and Health Innovations, Department of Graduate Nursing, The University of Texas at Arlington, Arlington, TX, USA*

Samyak Jain • *Miller School of Medicine, University of Miami, Miami, FL, USA; Bascom Palmer Eye Institute, Miami Integrative Metabolomics Research Center, University of Miami, Miami, FL, USA*

Alexa Jauregui • *Bascom Palmer Eye Institute, Miller School of Medicine at University of Miami, Miami, FL, USA; Miami Integrative Metabolomics Research Center, Miami, FL, USA*

David Karasik • *The Musculoskeletal Genetics Laboratory, The Azrieli Faculty of Medicine, Bar-Ilan University, Safed, Israel*

Molly Kennedy • *Miller School of Medicine, University of Miami, Miami, FL, USA; Bascom Palmer Eye Institute, Miami Integrative Metabolomics Research Center, University of Miami, Miami, FL, USA*

Jonah Kutell • *Bascom Palmer Eye Institute, Miami Integrative Metabolomics Research Center, University of Miami, Miami, FL, USA*

Alexandra Lens • *Bascom Palmer Eye Institute, Miller School of Medicine at University of Miami, Miami, FL, USA; Nova Southeastern University, Fort Lauderdale, FL, USA*

Ang Li • *Bone-Muscle Research Center, Department of Kinesiology, College of Nursing and Health Innovation, University of Texas at Arlington, Arlington, TX, USA*

Xuejun Li • *Bone-Muscle Research Center, Department of Kinesiology, College of Nursing and Health Innovation, University of Texas at Arlington, Arlington, TX, USA*

Isabella Moceri • *Bascom Palmer Eye Institute, Miami Integrative Metabolomics Research Center, Miller School of Medicine, University of Miami, Miami, FL, USA*

Jawad Muayad • *Texas A & M University School of Medicine, Houston, TX, USA*

EMILY J. NEAG • *Bascom Palmer Eye Institute, University of Miami Miller School of Medicine, Miami, FL, USA; Miami Integrative Metabolomics Research Center, Miami, FL, USA; Michigan State University College of Osteopathic Medicine, East Lansing, MI, USA*

STEPHEN LEROY NEWHART • *Department of Kinesiology, The College of Nursing and Health Innovation, The University of Texas at Arlington, Arlington, TX, USA*

TAYLOR NGO • *Bascom Palmer Eye Institute, Miami Integrative Metabolomics Research Center, University of Miami Miller School of Medicine, Miami, FL, USA; California Northstate University College of Medicine, Elk Grove, CA, USA*

HIDERU OBINATA • *Education and Research Support Center, Gunma University Graduate School of Medicine, Maebashi, Gunma, Japan*

ZUI PAN • *Department of Graduate Nursing, College of Nursing and Health Innovation, The University of Texas at Arlington, Arlington, TX, USA; Bone and Muscle Research Center, College of Nursing and Health Innovation, The University of Texas at Arlington, Arlington, TX, USA; Department of Kinesiology, College of Nursing and Health Innovation, The University of Texas at Arlington, Arlington, TX, USA*

PADMANABHAN PARANJI PATTABIRAMAN • *Department of Ophthalmology, Glick Eye Institute, Indiana University School of Medicine, Indianapolis, IN, USA; Stark Neuroscience Research Institute, Indiana University School of Medicine, Indianapolis, IN, USA*

DONAKA RAJASHEKAR • *The Musculoskeletal Genetics Laboratory, The Azrieli Faculty of Medicine, Bar-Ilan University, Safed, Israel*

MARWA SAADY • *Chemistry Department, School of Sciences & Engineering, The American University in Cairo, Cairo, Egypt*

MEHER A. SALEEM • *Bascom Palmer Eye Institute, University of Miami Miller School of Medicine, Miami, FL, USA; Miami Integrative Metabolomics Research Center, Miami, FL, USA; Georgetown University School of Medicine, Washington, DC, USA*

AVINASH SOUNDARARAJAN • *Department of Ophthalmology, Glick Eye Institute, Indiana University School of Medicine, Indianapolis, IN, USA*

EVAN B. STUBBS JR. • *Research Service, Department of Veterans Affairs, Edward Hines Jr. VA Hospital, Hines, IL, USA; Department of Ophthalmology, Stritch School of Medicine, Loyola University Chicago, Maywood, IL, USA*

SOPHIA E. THOMPSON • *Bone Muscle Research Center, The University of Texas at Arlington, Arlington, TX, USA*

CARLOS A. TORO • *Spinal Cord Damage Research Center, James J. Peters VA Medical Center, Bronx, NY, USA; Departments of Medicine and Rehabilitation Medicine and Human Performance, Icahn School of Medicine at Mount Sinai, New York, NY, USA*

VENU VARANASI • *Department of Materials Science and Engineering, The University of Texas at Arlington, Arlington, TX, USA; Bone Muscle Research Center, The University of Texas at Arlington, Arlington, TX, USA*

AINSELY VERKAIK • *Bascom Palmer Eye Institute, Miami Integrative Metabolomics Research Center, University of Miami Miller School of Medicine, Miami, FL, USA; College of Optometry, Nova Southeastern University, Fort Lauderdale, FL, USA*

SHYAMAL P. WAGHWALA • *Bascom Palmer Eye Institute, Miami Integrative Metabolomics Research Center, University of Miami Miller School of Medicine, Miami, FL, USA; The University of Texas at Southwestern Medical School, Dallas, TX, USA*

MICHAEL WILSON • *Department of Materials Science and Engineering, The University of Texas at Arlington, Arlington, TX, USA*

AHMED S. YACOUB • *Bone-Muscle Research Center, College of Nursing and Health Innovation, The University of Texas at Arlington, Arlington, TX, USA; Department of Kinesiology, College of Nursing and Health Innovation, The University of Texas at Arlington, Arlington, TX, USA*

JAMES Y. YI • *Department of Kinesiology, College of Nursing and Health Innovation, The University of Texas at Arlington, Arlington, TX, USA*

JIANXUN YI • *Bone-Muscle Research Center, Department of Kinesiology, College of Nursing and Health Innovation, University of Texas at Arlington, Arlington, TX, USA*

XIAN ZHAO • *Department of Biochemistry, Gunma University Graduate School of Medicine, Maebashi, Gunma, Japan; Department of Pharmacy, Nanjing Drum Tower Hospital, The Affiliated Hospital of Nanjing University Medical School, Nanjing University, Nanjing, China*

JINGSONG ZHOU • *Bone-Muscle Research Center, Department of Kinesiology, College of Nursing and Health Innovation, University of Texas at Arlington, Arlington, TX, USA*

NOEL ZIEBARTH • *Bascom Palmer Eye Institute, Miami Integrative Metabolomics Research Center, University of Miami, Miami, FL, USA; Department of Biomedical Engineering, University of Miami, Coral Gables, USA*

Chapter 1

The Interplay of Lipid Signaling in Musculoskeletal Cross Talk: Implications for Health and Disease

Ahmed Al Saedi, Ahmed S. Yacoub, Kamal Awad, David Karasik, Marco Brotto, and Gustavo Duque

Abstract

The intricate interplay between the muscle and bone tissues is a fundamental aspect of musculoskeletal physiology. Over the past decades, emerging research has highlighted the pivotal role of lipid signaling in mediating communication between these tissues. This chapter delves into the multifaceted mechanisms through which lipids, particularly phospholipids, sphingolipids, and eicosanoids, participate in orchestrating cellular responses and metabolic pathways in both muscle and bone. Additionally, we examine the clinical implications of disrupted lipid signaling in musculoskeletal disorders, offering insights into potential therapeutic avenues. This chapter aims to shed light on the complex lipid-driven interactions between the muscle and bone tissues, paving the way for a deeper understanding of musculoskeletal health and disease.

Key words Cross talk, Lipid signaling, Lipids, Musculoskeletal cross talk, Fat

1 Introduction

1.1 Background

The musculoskeletal system, comprising muscles and bones, plays a crucial role in locomotion, structural support, and metabolic regulation. The communication between these two tissues is vital for maintaining musculoskeletal homeostasis [1, 2]. Recent research has unveiled the indispensable role of lipids in this interplay, revealing a network of signaling pathways that govern cellular processes [3]. There is accumulating evidence to suggest that certain types of lipids, specifically sphingolipids and ceramides, may play a role in the development and progression of sarcopenia and osteoporosis [4, 5].

Lipid signaling mediators, also known as lipid mediators or bioactive lipids, are a diverse group of molecules that play important roles in cellular signaling and the regulation of physiological processes in the body. These molecules are derived from various types of lipids, including phospholipids, sphingolipids, and fatty

Marco Brotto and Sanjoy K. Bhattacharya (eds.), *Lipid Signalling: Methods and Protocols*, Methods in Molecular Biology, vol. 2816, https://doi.org/10.1007/978-1-0716-3902-3_1,

acids. Some examples of lipid signaling mediators include Prostaglandins, Leukotrienes, Platelet-activating factor (PAF), Sphingosine-1-phosphate (S1P), Endocannabinoids.

Lipid signaling mediators are involved in a wide range of physiological processes and can have both beneficial and detrimental effects on health, depending on the context. Dysregulation of lipid signaling can contribute to the development of various diseases, including cancer, cardiovascular disease, sarcopenia, osteoporosis, musculoskeletal and neurodegenerative disorders [6].

Chromatographic methods have been developed to determine lipid signaling mediators in different biological matrixes, such as the muscle, bone marrow, urine, female reproductive tissues, liver tissue, milk, and cerebrospinal fluid [7, 8]. Over the past years, liquid chromatography coupled with tandem mass spectrometry (LC-MS/MS) has become the reference technique to determine ECs in biological samples, because it is highly sensitive and selective, and it suits the physicochemical properties of lipidomics [9]. With the advent of effective tools to study lipids, including mass spectrometry-based lipidomics, lipids are emerging as an important player in musculoskeletal tissues.

This chapter aims to provide a comprehensive overview of the lipid signaling mechanisms involved in the interaction between muscle and bone tissues. We will explore the diverse classes of lipids, their receptors, and downstream effectors, and how they contribute to cellular responses. Additionally, we will delve into the clinical implications of perturbed lipid signaling in musculoskeletal disorders.

2 Lipid Classes and Signaling in Musculoskeletal Cross Talk

2.1 Phospholipids

Phospholipids, integral components of cell membranes, serve as substrates for numerous signaling molecules [10]. These molecules constitute the lipid bilayer defining the outer confines of a cell, but also serve as the structural entities, which confine subcellular components. Mass spectrometry has emerged as a powerful tool useful for the qualitative and quantitative analysis of complex phospholipids, including glycerophospholipids and the sphingolipid, sphingomyelin. Collision-induced decomposition of both positive and negative molecular ion species yields rich information as to the polar head group of the phospholipid and the fatty-acyl substituents esterified to the glycerophospholipid backbone. In the context of muscle-bone interaction, phospholipids play pivotal roles in cell membrane integrity, ion channel function, and intracellular signaling cascades [8].

2.1.1 Phosphatidylcholine (PC) and Phosphatidylethanolamine (PE)

The maintenance of phospholipid homeostasis is increasingly being implicated in metabolic health. Phosphatidylethanolamine (PE) is the most abundant phospholipid on the inner leaflet of cellular membranes [11]. PC and PE, major phospholipids, have been shown to influence osteogenic differentiation and myogenic differentiation, respectively. PC-derived lipids modulate Wnt/β-catenin signaling in bone [12]; this study showed that PC ameliorated the microstructure of the trabecular bone and accelerated bone mineral apposition rate, additionally increasing bone mineral density and biomechanical properties of the bone. Furthermore, gene and protein expression levels suggest that PC inhibited overactive osteogenesis via downregulation of the expression of the osteogenesis-related Wnt/β-catenin signaling pathway [12]. In muscle, PE influences myogenic regulatory factors in muscle cells, and aberrant lipid metabolism promotes the development of skeletal muscle insulin resistance [13]. Furthermore, a study showed that PC induces osteogenic morphology, physiology, gene expression, and phenotype in vascular smooth muscle cells [14].

2.1.2 Phosphatidylinositol Phosphates (PIPs)

Phosphatidylinositol phosphates (PIPs) are low-abundance phospholipids that participate in a range of cellular processes, including cell migration and membrane traffic [15]. PIP levels and subcellular distribution are regulated by a series of lipid kinases and phosphatases. PIPs act as secondary messengers in signaling cascades, regulating processes such as cell proliferation, cytoskeletal dynamics, and vesicle trafficking [16]. They play essential roles in the regulation of osteoblast differentiation and muscle contraction through distinct PIP isoforms [17]. Furthmore, PIPs and their regulatory enzymes are key players in many cellular processes and are required for aspects of vertebrate development and skeletal muscle formation [18]. Dysregulated PIP metabolism has been implicated in skeletal myopathies that feature structural defects in the triad [18].

2.2 Sphingolipids

Sphingolipids, a diverse class of lipids, have garnered attention for their roles in apoptosis, cell proliferation, and differentiation. Emerging evidence suggests their involvement in musculoskeletal health [19]. Emerging evidence based on genetically manipulated mouse models and molecular biology allows us to obtain new insights into the role sphingolipid played on skeletal remodeling [20]. There is a cross talk between sphingomyelin, ceramide, and sphingosine-1-phosphate (S1P) of sphingolipids family and the cells, especially osteoblasts and osteoclasts of the bone through which bone is remodeled during life constantly.

2.2.1 Sphingosine-1-Phosphate (S1P)

Given the significant progress made in understanding the sphingolipid pathobiology of several diseases, sphingolipid-related checkpoints emerge as attractive targets. Recent data indicate the multifaceted contribution of the sphingolipid machinery to osteoclast-osteoblast cross talk, representing one of the pivotal interactions underlying bone homeostasis [19, 21]. Furthermore, recent studies on S1P showed a critical role as autocrine and paracrine mediator of bone remodeling and its potential interaction with immune cells that have emerged as important player in skeletal biology [22]. In muscle tissues, S1P has been implicated in muscle regeneration, S1P accumulates in mouse skeletal muscle upon aging, and both genetic and pharmacological inhibition of sphingolipid synthesis prevents age-related decline in muscle mass while enhancing strength and exercise capacity [23]. Inhibition of sphingolipid synthesis confers increased myogenic potential and promotes protein synthesis [24]. Therefore, S1P receptors in both tissues, bone and muscle, serve as key mediators of these effects.

2.2.2 Ceramides

Ceramides, precursors of bioactive sphingolipids, have been linked to apoptosis and insulin resistance. Ceramide, due to the differing acyl CoAs that can be used to produce it, is technically a class of molecules rather than a single molecule and, therefore, may have different biological functions depending on the acyl chain it is composed of [25]. At the apex of complexity is the group of lipids known as glycosphingolipids (GSL), which contain dozens of different sphingolipid species differing by both the order and type of sugar residues attached to their headgroups [26].

Recent studies indicate their potential influence on muscle and bone cross talk. Hamrick et al. have shown that the association of increased ceramides, especially C18:0 and C24:1, with adverse bone phenotypes in elderly people could be explained mainly by the increase in osteoclastogenesis and bone resorption [27]. Furthermore, long-chain ceramides found to be strongly associated with low muscle function cardiovascular fitness in older adults and may be implicated in the pathogenesis of low fitness with aging [28]. However, longitudinal studies are needed to further validate these associations and investigate the relationship between ceramides and health outcomes.

3 Eicosanoids

Eicosanoids are bioactive lipids derived from arachidonic acid, exerting a wide array of physiological effects [29]. They play crucial roles in inflammation, pain perception, and immune response regulation, with emerging implications for musculoskeletal physiology [30]. Eicosanoids have multiple groups such as Prostaglandins and Leukotrienes.

3.1 Prostaglandins

Prostaglandins, a subgroup of eicosanoids, have garnered attention for their roles in bone remodeling and muscle contractility. Prostaglandins are potent regulators of bone formation and bone resorption that can have both stimulatory and inhibitory effects [31]. They are likely to play a central role in the ability of the skeleton to respond to mechanical forces as well as to humoral mediators. The stimulatory and inhibitory responses appear to have different structure-activity relations and signal transduction pathways.

Prostaglandin E2, for instance, has been implicated in osteoblast differentiation and osteoclast activity regulation [32]. Chen et al. showed that prostaglandin E2 (PGE2) secreted by osteoblastic cells activates PGE2 receptor 4 in sensory nerves to regulate bone formation by inhibiting sympathetic activity through the central nervous system [33]. Therefore, PGE2 mediates sensory nerve to control bone homeostasis and promote regeneration. Furthmore, Brotto's lab has showed that PGE [2] induced acceleration of myogenesis of C2C12 myoblasts that was linked with significant modifications in intracellular calcium homeostasis and presented PGE [2] as new concept of bone–muscle cross talk for the development and application of inventions designed to treat muscle diseases characterized by enhanced muscle wasting, such as sarcopenia [34]. Furthermore, the activity of prostaglandins in bone tissue is defined by maintaining bone remodeling balance and its reactions to humoral mediators and mechanical stress. Most data confirm that among prostaglandins, PGE2 takes part in all processes of trauma response, including homeostasis, inflammation, and healing, and plays a key role in bone physiology [35]. However, further analysis of PG receptors and their downstream G-protein signaling in bone could provide important clues to the regulation of skeletal cell growth in both health and disease.

3.2 Leukotrienes

Leukotrienes, another subgroup of eicosanoids, are known for their pro-inflammatory properties [36]. Leukotrienes are biologically active 5-lipoxygenase products of arachidonic acid metabolism that are involved in the mediation of various inflammatory disorders. Recent studies suggest their involvement in muscle inflammation and potentially in bone resorption, highlighting their dual role in musculoskeletal health [37]. Leukotrienes play significant role in the muscle atrophy and weakness that occur in inflammatory diseases and are the major clinical features of the autoimmune myopathies [38]. Furthermore, leukotriene type B4 has been proved to play an essential role in the induction of pain and bone damage. Skeletal diseases promote the production of leukotriene type B4, which acts on their receptors, leading to the increased release of pro-inflammatory cytokines and ROS to reduce neuron viability and pain threshold [39]. Recently, Oliveira et al. showed leukotrienes' role in bone turnover, particularly the suppression of bone formation by modulating the function of osteoclasts and osteoblasts [40]. The study investigated the effects of Leukotrienes on bone metabolism and their impact on osteogenic differentiation and osteoclastogenesis using a 5-LO knockout mouse mode and demonstrated that the absence of 5-LO is related to the greater osteogenic profile [40]. Therefore, the comprehension of the mechanisms in which Leukotrienes regulate osteoblastogenesis and osteoclastogenesis is critical to understanding the pathophysiology of inflammatory disorders and its impact on skeletal tissue to develop novel therapies for bone-related diseases.

4 Clinical Implications and Therapeutic Perspectives

Disrupted lipid signaling pathways have been implicated in various musculoskeletal disorders, including osteoporosis, sarcopenia, and muscular dystrophies. Understanding these pathways could lead to targeted therapeutic interventions.

Targeting lipid signaling pathways holds promise for therapeutic interventions in musculoskeletal disorders. Modulating specific lipid receptors or enzymes involved in lipid metabolism may offer novel strategies for preventing and treating musculoskeletal conditions. For instance, phosphoinositide provides a key implication in regulating integral membrane signaling such that unlike in peripheral membrane recruitment, this interaction will almost certainly be driven by both interactions between the phosphoinositide's headgroup and the acyl chains [41, 42]. Recent structures of phosphoinositide kinases have revealed unique molecular insight into their regulation [42].

Novel insight into targeting pathological phosphoinositide metabolism in disease has been investigated, the most clinically advanced strategy targeting pathologies in phosphoinositide signaling have been targeting the PI3K pathway in cancer, immunodeficiencies, and growth disorders. Currently, there are four clinically approved PI3K inhibitors [43]. However, similar approaches are needed in the musculoskeletal field particularly when inhibitors specific to the PIK3CA PI3K isoform led to a large increase in insulin signaling in muscle tissues and play a regulating role in osteoblast-osteoclast interaction [44, 45]. These approaches show promising results as a negative regulator of bone homeostasis by regulating osteogenesis and osteoclastogenesis and provide a potential therapeutic target for bone and muscle degenerative diseases.

Importantly, advances in native mass spectrometry [46], cryo-electron microscopy [47], and molecular dynamic simulations [48] have been particularly powerful in providing insight into the molecular mechanisms that mediate how phosphoinositide can bind and regulate integral membrane protein function [49]. In summary, knowledge of the classical lipid-driven pathways is now maturing into a state where pharmacologic inactivation of their metabolism can alleviate a variety of malignant, infectious, and inflammatory diseases and partially in musculoskeletal diseases. As our basic mechanistic knowledge of lipid mediators signaling expands, we expect therapeutic opportunities derived from this knowledge to expand even further.

5 Methods of Detection of the Lipid Signaling Mediators

5.1 Mass Spectrometry (MS)

Mass spectrometry stands as a cornerstone in lipidomics, offering unparalleled sensitivity and specificity in detecting and quantifying lipid signaling molecules. This technique involves ionizing molecules, separating them based on their mass-to-charge ratio, and then measuring the abundance of each ion. In the field of lipid signaling, MS enables the identification of a diverse range of mediators, including prostaglandins, leukotrienes, and sphingolipids, providing invaluable insights into the lipidome [50].

5.2 Liquid Chromatography Mass Spectrometry (LCMS)

Liquid chromatography, coupled with mass spectrometry (LC-MS), has emerged as a powerful tool for the separation and quantification of lipid signaling molecules. This method enables the precise separation of complex lipid mixtures before subjecting them to mass spectrometric analysis. LC-MS facilitates the accurate identification of individual lipid species, allowing researchers to discern sensitive variations in the lipid composition that may hold key biological significance [51].

5.3 Enzyme-Linked Immunosorbent Assay (ELISA)

ELISA, a widely used immunoassay, plays a fundamental role in the targeted detection of specific lipid mediators. By utilizing antibodies that selectively bind to a particular lipid species, ELISA enables the quantification of these molecules with high sensitivity. This approach proves particularly valuable when investigating the involvement of specific lipid signaling pathways in various physiological and pathological conditions [52].

5.4 Nuclear Magnetic Resonance (NMR) Spectroscopy

NMR spectroscopy provides a nondestructive means of analyzing lipid-signaling mediators by exploiting the magnetic properties of atomic nuclei. This technique allows researchers to elucidate the structural characteristics of lipids and monitor dynamic changes in lipid composition. While not as widely employed as MS, NMR spectroscopy contributes unique insights into the conformation and interactions of lipid molecules [53].

5.5 Fluorescence-Based Methods

Fluorescence-based techniques offer a flexible approach for visualizing and quantifying lipid signaling mediators in live cells. By employing fluorescent probes with specificity lipid classes, researchers can monitor the real-time dynamics of lipid signaling events. These methods are particularly valuable for studying localized lipid changes and can be employed in conjunction with advanced imaging technologies [54].

In navigating the complex landscape of lipid signaling, researchers control a various range of techniques, each with its own set of advantages and limitations. The integration of these methods not only expands our understanding of lipid mediators but also unveils the details that underlie cellular communication and homeostasis. As we delve further into the subsequent chapters, we will explore the application of these methodologies in unraveling the complexities of lipid signaling pathways and their implications for human health and disease.

6 Conclusion

The burgeoning field of lipid signaling in musculoskeletal cross talk provides a rich landscape for exploration. This chapter has highlighted the diverse classes of lipids and their multifaceted roles in mediating communication between the muscle and bone tissues. Further research in this area promises to unveil novel therapeutic strategies for managing musculoskeletal disorders and ultimately enhancing overall health and well-being.

References

1. Dalle Carbonare L, Minoia A, Zouari S, Piritore FC, Vareschi A, Romanelli MG et al. (2023) Crosstalk between Bone and Muscles during Physical Activity Cells 12(16): 2088. https://doi.org/10.3390/cells12162088. PMID: 37626898; PMCID: PMC10453939
2. Brotto M, Bonewald L (2015) Bone and muscle: interactions beyond mechanical. Bone 80: 109–114
3. Kirk B, Feehan J, Lombardi G, Duque G (2020) Muscle, bone, and fat crosstalk: the biological role of myokines, osteokines, and adipokines. Curr Osteoporos Rep 18:388–400
4. Hue I, Capilla E, Rosell-Moll E, Balbuena-Pecino S, Goffette V, Gabillard JC et al. (2023) Recent advances in the crosstalk between adipose, muscle and bone tissues in fish. Front Endocrinol (Lausanne) 14: 1155202. https://doi.org/10.3389/fendo.2023.1155202. PMID: 36998471; PMCID: PMC10043431
5. Brotto M, Johnson ML (2014) Endocrine crosstalk between muscle and bone. Curr Osteoporos Rep 12(2):135–141. https://doi.org/10.1007/s11914-014-0209-0. PMID: 24667990; PMCID: PMC4374433
6. Al Saedi A, Debruin DA, Hayes A, Hamrick M (2022) Lipid metabolism in sarcopenia. Bone 164:116539
7. Wang Z, Mo C, Awad K, Bonewald L, Brotto M (2023) Mass spectrometry approaches for detection and determination of prostaglandins from biological samples. Methods Mol Biol 2625:299–311
8. Kirkwood JS, Broeckling CD, Donahue S, Prenni JE (2016) A novel microflow LC-MS method for the quantitation of endocannabinoids in serum. J Chromatogr B Analyt Technol Biomed Life Sci 1033–1034:271–277

9. Ottria R, Ravelli A, Gigli F, Ciuffreda P (2014) Simultaneous ultra-high performance liquid chromathograpy-electrospray ionization-quadrupole-time of flight mass spectrometry quantification of endogenous anandamide and related N-acylethanolamides in bio-matrices. J Chromatogr B Analyt Technol Biomed Life Sci 958:83–89
10. Nuhn P, Kertscher HP, Rüger HJ (1982) Phospholipide und biologische Aktivität [Phospholipids and biologic activity]. Pharmazie 37(10): 681–701. German. PMID: 6757975
11. Grapentine S, Singh RK, Bakovic M (2023) Skeletal muscle consequences of phosphatidylethanolamine synthesis deficiency. Function (Oxf) 4:zqad020
12. Fu M, Tian Y, Zhang T, Zhan Q, Zhang L, Wang J (2020) Comparative study of DHA-enriched phosphatidylcholine and EPA-enriched phosphatidylcholine on ameliorating high bone turnover via regulation of the osteogenesis-related Wnt/beta-catenin pathway in ovariectomized mice. Food Funct 11: 10094–10104
13. Ferrara PJ, Rong X, Maschek JA, Verkerke AR, Siripoksup P, Song H et al. (2021) Lysophospholipid acylation modulates plasma membrane lipid organization and insulin sensitivity in skeletal muscle. J Clin Invest 131(8): e135963. https://doi.org/10.1172/JCI135963. PMID: 33591957; PMCID: PMC8262507
14. Vickers KC, Castro-Chavez F, Morrisett JD (2010) Lyso-phosphatidylcholine induces osteogenic gene expression and phenotype in vascular smooth muscle cells. Atherosclerosis 211:122–129
15. De Camilli P, Emr SD, McPherson PS, Novick P (1996) Phosphoinositides as regulators in membrane traffic. Science 271(5255): 1533–1539. https://doi.org/10.1126/science.271.5255.1533. PMID: 8599109
16. Zhang X, Xie H, Iaea D, Khelashvili G, Weinstein H, Maxfield FR (2022) Phosphatidylinositol phosphates modulate interactions between the StarD4 sterol trafficking protein and lipid membranes. J Biol Chem 298(7): 102058. https://doi.org/10.1016/j.jbc.2022.102058. Epub 2022 May 20. PMID: 35605664; PMCID: PMC9207681
17. Gibbs EM, Feldman EL, Dowling JJ (2010) The role of MTMR14 in autophagy and in muscle disease. Autophagy 6:819–820
18. Smith L, Fabian L, Al-Maawali A, Noche RR, Dowling JJ (2020) De novo phosphoinositide synthesis in zebrafish is required for triad formation but not essential for myogenesis. PLoS One. 15(8):e0231364. https://doi.org/10.1371/journal.pone.0231364. PMID: 32804943; PMCID: PMC7430711
19. Higashi K, Matsuzaki E, Hashimoto Y, Takahashi-Yanaga F, Takano A, Anan H et al (2016) Sphingosine-1-phosphate/S1PR2-mediated signaling triggers Smad1/5/8 phosphorylation and thereby induces Runx2 expression in osteoblasts. Bone 93:1–11
20. Qi T, Li L, Weidong T (2021) The Role of Sphingolipid Metabolism in Bone Remodeling. Front Cell Dev Biol. 2021 Nov 29;9: 752540. https://doi.org/10.3389/fcell.2021.752540. PMID: 34912800; PMCID: PMC8666436
21. Meshcheryakova A, Mechtcheriakova D, Pietschmann P (2017) Sphingosine 1-phosphate signaling in bone remodeling: multifaceted roles and therapeutic potential. Expert Opin Ther Targets 21:725–737
22. Dziak R (2013) The role of sphingosine-1-phosphate (S1P) and lysophosphatidic acid (LPA) in regulation of osteoclastic and osteoblastic cells. Immunol Investig 42:510–518
23. Laurila PP, Wohlwend M, Imamura de Lima T, Luan P, Herzig S, Zanou N et al. (2022) Sphingolipids accumulate in aged muscle, and their reduction counteracts sarcopenia. Nat Aging 2 (12):1159–1175. https://doi.org/10.1038/s43587-022-00309-6. Epub 2022 Dec 16. PMID: 37118545
24. Laurila PP, Luan P, Wohlwend M, Zanou N, Crisol B, Imamura de Lima T et al. (2022) Inhibition of sphingolipid de novo synthesis counteracts muscular dystrophy. Sci Adv 8: eabh4423
25. Gault CR, Obeid LM, Hannun YA (2010) An overview of sphingolipid metabolism: from synthesis to breakdown. Adv Exp Med Biol 688:1–23
26. Yamaji T, Horie A, Tachida Y, Sakuma C, Suzuki Y, Kushi Y et al. (2016) Role of Intracellular Lipid Logistics in the Preferential Usage of Very Long Chain-Ceramides in Glucosylceramide. Int J Mol Sci 17(10): 1761. https://doi.org/10.3390/ijms17101761. PMID: 27775668; PMCID: PMC5085785
27. Kim BJ, Lee JY, Park SJ, Lee SH, Kim SJ, Yoo HJ et al. (2019) Elevated ceramides 18:0 and 24:1 with aging are associated with hip fracture risk through increased bone resorption. Aging (Albany NY). 11(21):9388–9404. https://doi.org/10.18632/aging.102389. Epub 2019 Nov 1. PMID: 31675352; PMCID: PMC6874435
28. Fabbri E, Yang A, Simonsick EM, Chia CW, Zoli M, Haughey NJ et al (2016) Circulating

ceramides are inversely associated with cardiorespiratory fitness in participants aged 54-96 years from the Baltimore longitudinal study of aging. Aging Cell 15:825–831

29. Baker RR (1990) The eicosanoids: a historical overview. Clin Biochem 23:455–458
30. Plotkin LI, Davis HM, Cisterna BA, Saez JC (2017) Connexins and Pannexins in bone and skeletal muscle. Curr Osteoporos Rep 15:326–334
31. Raisz LG (1995) Physiologic and pathologic roles of prostaglandins and other eicosanoids in bone metabolism. J Nutr 125:2024S–2027S
32. Feigenson M, Jonason JH, Shen J, Loiselle AE, Awad HA, O'Keefe RJ (2020) Inhibition of the prostaglandin EP-1 receptor in periosteum progenitor cells enhances osteoblast differentiation and fracture repair. Ann Biomed Eng 48: 927–939
33. Chen H, Hu B, Lv X, Zhu S, Zhen G, Wan M et al (2019) Prostaglandin E2 mediates sensory nerve regulation of bone homeostasis. Nat Commun 10:181
34. Mo C, Romero-Suarez S, Bonewald L, Johnson M, Brotto M (2012) Prostaglandin E2: from clinical applications to its potential role in bone- muscle crosstalk and myogenic differentiation. Recent Pat Biotechnol 6:223–229
35. Lisowska B, Kosson D, Domaracka K (2018) Lights and shadows of NSAIDs in bone healing: the role of prostaglandins in bone metabolism. Drug Des Devel Ther 12:1753–1758
36. Ogawa Y, Calhoun WJ (2006) The role of leukotrienes in airway inflammation. J Allergy Clin Immunol 118:789–98; quiz 99–800
37. Hikiji H, Takato T, Shimizu T, Ishii S (2008) The roles of prostanoids, leukotrienes, and platelet-activating factor in bone metabolism and disease. Prog Lipid Res 47:107–126
38. Korotkova M, Lundberg IE (2014) The skeletal muscle arachidonic acid cascade in health and inflammatory disease. Nat Rev Rheumatol 10:295–303
39. Zheng LX, Li KX, Hong FF, Yang SL (2019) Pain and bone damage in rheumatoid arthritis: role of leukotriene B4. Clin Exp Rheumatol 37:872–878
40. Amadeu de Oliveira F, Tokuhara CK, Veeriah V, Domezi JP, Santesso MR, Cestari TM et al (2023) The multifarious functions of leukotrienes in bone metabolism. J Bone Miner Res 38:1135–1153
41. Hammond GRV, Burke JE (2020) Novel roles of phosphoinositides in signaling, lipid transport, and disease. Curr Opin Cell Biol 63:57–67
42. Burke JE (2018) Structural Basis for Regulation of Phosphoinositide Kinases and Their Involvement in Human Disease. Mol Cell. 6;71(5):653–673. https://doi.org/10.1016/j.molcel.2018.08.005. PMID: 30193094
43. Goncalves MD, Hopkins BD, Cantley LC (2018) Phosphatidylinositol 3-kinase, growth disorders, and cancer. N Engl J Med 379: 2052–2062
44. Hopkins BD, Pauli C, Du X, Wang DG, Li X, Wu D et al. (2018) Suppression of insulin feedback enhances the efficacy of PI3K inhibitors. Nature. 560(7719):499–503. https://doi.org/10.1038/s41586-018-0343-4. Epub 2018 Jul 4. Erratum in: Nature 2018 Aug 29. PMID: 30051890; PMCID: PMC6197057
45. Tang L, Wu M, Lu S, Zhang H, Shen Y, Shen C et al (2021) Fgf9 negatively regulates bone mass by inhibiting Osteogenesis and promoting Osteoclastogenesis via MAPK and PI3K/AKT signaling. J Bone Miner Res 36:779–791
46. Liu Y, LoCaste CE, Liu W, Poltash ML, Russell DH, Laganowsky A (2019) Selective binding of a toxin and phosphatidylinositides to a mammalian potassium channel. Nat Commun 10: 1352
47. Timcenko M, Lyons JA, Januliene D, Ulstrup JJ, Dieudonne T, Montigny C et al (2019) Structure and autoregulation of a P4-ATPase lipid flippase. Nature 571:366–370
48. Song W, Yen HY, Robinson CV, Sansom MSP (2018) State-dependent Lipid Interactions with the A2a Receptor Revealed by MD Simulations Using In Vivo-Mimetic Membranes. Structure 27(2):392–403.e3. https://doi.org/10.1016/j.str.2018.10.024. Epub 20. PMID: 30581046; PMCID: PMC7031699
49. Robinson CV, Rohacs T, Hansen SB (2019) Tools for understanding nanoscale lipid regulation of ion channels. Trends Biochem Sci 44: 795–806
50. Ross DA-O, Guo J, Bilbao AA-O, Huan TA-O, Smith RA-O, Zheng XA-O (2023) Evaluating software tools for lipid identification from ion mobility spectrometry-mass spectrometry Lipidomics data. LID - 10.3390/molecules28083483 [doi] LID - 3483.
51. Yang K, Han X (2016) Lipidomics: techniques, applications, and outcomes related to biomedical sciences. Trends Biochem Sci 41:954–969
52. Aroonrerk N, Suksamrarn A, Kirtikara K (2007) A sensitive direct ELISA for detection of prostaglandin E2. J Immunoassay Immunochem 28:319–330

53. Shimamoto S, Nakahata Y, Hidaka Y, Yoshida T, Ohkubo T (2022) NMR resonance assignments of mouse lipocalin-type prostaglandin D synthase/prostaglandin J(2) complex. Biomol NMR Assign 16:225–229

54. Turk J, Weiss SJ, Davis JE, Needleman P (1978) Fluorescent derivatives of prostaglandins and thromboxanes for liquid chromatography. Prostaglandins 16(2):291–309. https://doi.org/10.1016/0090-6980(78)90031-x. PMID: 704934

Chapter 2

Zebrafish as a Model for Lipidomics and Similar Investigations

Kamal Awad, Jian Huang, Donaka Rajashekar, Gustavo Duque, Marco Brotto, and David Karasik

Abstract

Zebrafish (*Danio rerio*) has emerged as a pivotal model organism in vertebrate development research over several decades. Beyond its contributions to developmental biology, zebrafish have increasingly played a crucial role in the field of lipidomics. Lipidomics, a comprehensive analysis of lipids within biological systems, offers profound insights into lipid metabolism and signaling pathways. This chapter explores the zebrafish's unique attributes that make it an ideal candidate for lipidomics studies. With a genome sharing numerous genetic similarities with humans, zebrafish serve as a powerful model for dissecting lipid metabolism and unraveling the complexities of lipid mediator-related diseases. In this chapter, we delve into specific protocols tailored for utilizing zebrafish in lipidomics research and similar investigations. Through a comprehensive exploration of zebrafish as a model organism, this chapter aims to provide researchers with valuable insights and methodologies for advancing lipidomics studies using zebrafish.

Key words Zebrafish, Lipidomics, Lipid mediators, Arachidonic acid, Muscle

1 Introduction

In recent years, the small teleost, zebrafish, and medaka have significantly grown as animal models for the study of vertebrate development, homeostasis, and disease [1–3]. Zebrafish is an attractive model animal for studying metabolic diseases because of the functional conservation in lipid metabolism, adipose tissue biology, pancreas and liver structure, skeletal muscle, and glucose homeostasis; it is also suitable for the identification of novel targets associated with the risk and treatment of obesity, dyslipidemia, and diabetes [4]. Zebrafish are vertebrates possessing notable genetic, anatomical, and physiological resemblances to humans, particularly in their lipid storage mechanism and metabolic pathways, which play pivotal roles in adipocyte differentiation, energy regulation, and cholesterol metabolism [5, 6].

Marco Brotto and Sanjoy K. Bhattacharya (eds.), *Lipid Signalling: Methods and Protocols*, Methods in Molecular Biology, vol. 2816, https://doi.org/10.1007/978-1-0716-3902-3_2,

The effects of nutritionally induced obesity on metabolic pathways are well-documented [7] since the nutritional requirements of zebrafish are known. Several researchers established various models by changing the standard diet (such as overfeeding with usual chow or providing a high-fat diet to develop obesity, hyperglycemia, and dyslipidemia) to induce metabolic stress [7]. Formulation of a high-cholesterol diet is also crucial for the study of dyslipidemia [5], and the histopathological changes in zebrafish fed a high level of cholesterol are very similar to human atherosclerosis. Like humans, excessive nutrient intake causes elevated triglyceride levels and hepatic steatosis in zebrafish. Lipids accumulate in visceral, intramuscular, and subcutaneous adipose tissue, facilitating a comprehensive assessment of the body's fat distribution. The conservation, distribution, and formation of fat in the adipose tissue of zebrafish are comparable to those of mammals, and the optical transparency of larvae enables in vivo monitoring of adipocyte formation and fatty acid uptake in a temporal manner [8].

Advanced approaches in the fields of microscopy, transcriptomics, and metabolomics have succeeded in providing fundamental insights into the mechanisms of diseases using a zebrafish experimental model [9]. Applications of studying the metabolic disturbances include different techniques such as lipidomics using liquid chromatography–tandem mass spectrometry (LCMS/MS) [6, 10], targeted transcriptomics [6], and transcriptomic sequencing [7]. SREBP1 is a known adipocyte differentiation factor that directly regulates the transcription of over 200 genes involved in the de novo synthesis of fatty acids, triglycerides, and cholesterol [11]. Thus, *SREBF1* gene (coding SREBP1) is responsible for the regulation of lipid homeostasis [12]. In a study by Shochat C. et al., using LCMS/MS for lipidomics profiling, they reported that specific lipid mediators such as 11,12 epoxyeicosatrienoic acid (11, 2-EET) levels are significantly affected by the *SREBF1* alleles [12].

In this chapter, we will report step by step the methods and protocols to be used to perform lipidomics analysis using zebrafish model. We will also explore in detail how to dissect the organs and back muscle from zebrafish. We will mainly focus on the detection of the arachidonic acid (AA) lipid metabolites. AA derived lipid mediators, eicosanoids, have widespread effects on physiological pathways. They are synthesized through the metabolism of arachidonic acid by cyclooxygenase and lipoxygenase. Eicosanoids, diverse in structure and function, act directly on receptors in target cells. Eicosanoids can include prostaglandins (PGs), leukotrienes (LTs), lipoxins (LXs), and thromboxanes (TXs) [13, 14].

2 Materials

1. Adult zebrafish (age 3 months and older)
2. Tricaine – dissolve 0.030 g Tricaine in 100 mL nanopure water
3. Phosphate-buffered saline (PBS) stock solution
4. Dissecting tray (Styrofoam)
5. Tools: Scalpel, scissors, tweezers, ruler
6. Disinfectant: Ethanol
7. Petri plates, Weighing paper
8. Eppendorf tubes
9. Scales for weighting
10. Needles or pins
11. PTFE/silicon screw cap
12. Glass vial for LCMS
13. Clear autosampler glass vials

3 Methods

3.1 Dissection of Organs and Back Muscle Dissection

Carry out all procedures at room temperature unless otherwise instructed. Dissection of the organs from the adult zebrafish is performed according to modified method from Gupta and Mullins' report; see also open-access online video at (http://www.jove.com/index/Details.stp?ID=1717).

1. Euthanize the fish according to your IACUC-approved protocol. To euthanize the adult fish, first anesthetize it in 0.2% Tricaine, then transfer with a spoon to ice-cold water for 15 min. Rinse the fish in fish water to remove the Tricaine. Lightly pat the fish dry on a paper towel so that it is damp, not slippery.
2. Put the fish on weighing paper, place a ruler next to the fish, and take a picture with the ruler next to the fish number.
3. Weigh the fish.
4. Pin the fish onto the dissecting tray with the left side of the fish facing up.
5. Cut the skin from the anal fin along the belly of the fish to the operculum.
6. Remove the operculum and the pectoral fin, including the pectoral girdle.

7. Cut the skin beginning from above the now exposed gills posteriorly along the side of the fish and then down to the anal fin.
8. Remove the skin from the side of the fish. Identify the gonads, the anterior and posterior chambers of the swim bladder, the liver, and the intestine. Remove the testis or ovaries and place them in the PBS.
9. Carefully remove the liver and intestine. You should be able to identify the spleen and the gall bladder within the viscera.
10. Remove and discard the swim bladder. Unpin the fish and re-pin it ventral side up. Identify the head, body, and tail regions of the kidney.
11. Find the heart, located posterior and ventral to the gills. The heart is delicate, so it is helpful to remove most of the surrounding tissue before trying to remove it from the fish.

To collect the muscles:

1. Cut the tail and pectoral fins of the fish
2. Re-pin the fish on a dissecting tray with the pins/needles
3. Cut the fish's head off with a scalpel blade
4. Use micro-scissors to cut open the skin of the fish along the side facing you. Remove the skin using fine forceps being careful not to remove dorsal axial muscle. This region was chosen since it contains mostly myotome and avoids the intestine and anal regions, which correspond to hollow structures that are often difficult to section.
5. Use a scalpel or micro-scissors to take off the trunk/tail muscle between the regions of the pelvic fins to the caudal part. Directly place the dorsal trunk/tail muscles in liquid nitrogen.
6. For myosin extraction, put the fish muscle in an Eppendorf tube with PBS and immediately fast freeze in liquid nitrogen. Transfer the tube into the −80 °C freezer until the time of analysis.

3.2 Lipidomics Protocol

Solid-phase extraction (SPE) pretreatment for muscle samples followed by a LC/ESI-MS/MS method using the Shimadzu LCMS-8060 system (Shimadzu Scientific Instruments, Inc., Columbia, MD, USA) is described for the profiling and quantification of AA derived LMs. This method provides a simultaneous quantification for AA derived LMs with on-column detection at femto-molar level without the necessity of chemical derivatization. Arachidonic acid has several metabolites that have significant roles as LMs (see Table 1).

Table 1
Arachidonic acid metabolites/lipid mediators

NO.	Compounds	NO.	Compounds	NO.	Compounds
1	Tetranor-PGFM	34	15-keto Prostaglandin F2α	66	12-oxo LTB4
2	Tetranor-PGEM	35	11β-Prostaglandin E2	67	tetranor-12(S)-HETE
3	Tetranor-PGDM	36	5(S),14(R)-LXB4	68	N-acetyl LTE4
4	Tetranor-PGJM	37	Prostaglandin K2	69	(±)14(15)-DiHET
5	Tetranor-PGAM	38	Prostaglandin D2	70	12(S)-HHTrE
6	20-hydroxy Prostaglandin F2α	39	11β-13,14-dihydro-15-keto Prostaglandin F2α	71	(±)11(12)-DiHET
7	20-hydroxy Prostaglandin E2	40	15-keto Prostaglandin E2	72	(±)8(9)-DiHET
8	18-carboxy dinor LTB4	41	14,15-LTC4	73	20-carboxy arachidonic acid
9	13,14-dihydro-15-keto-tetranor Prostaglandin F1β	42	13,14-dihydro-15-keto Prostaglandin F2α	74	(±)5(6)-DiHET
10	2,3-dinor-8-iso Prostaglandin F2α	43	5(S),6(R)-Lipoxin A4	75	19(S)-HETE
11	2,3-dinor Thromboxane B2	44	13,14-dihydro-15-keto Prostaglandin E2	76	15-deoxy-delta12,14-PGJ2
12	13,14-dihydro-15-keto-tetranor Prostaglandin F1α	45	5(S),6(S)-Lipoxin A4	77	20-HETE
13	2,3-dinor-11β-Prostaglandin F2α	46	14,15-LTE4, Eoxin E4	78	(±)18-HETE
14	6-keto-Prostaglandin F1α	47	13,14-dihydro-15-keto Prostaglandin D2	79	(±)17-HETE
15	13,14-dihydro-15-keto-tetranor Prostaglandin D2	48	Leukotriene C4	80	(±)16-HETE
16	20-carboxy leukotriene B4	49	11-trans LTC4	81	15(S)-HETE
17	20-hydroxy leukotriene B4	50	Leukotriene D4	82	11(S)-HETE
18	11-dehydro-2,3-dinor Thromboxane B2	51	Leukotriene E4	83	8(S)-HETE
19	13,14-dihydro-15-keto-tetranor Prostaglandin E2	52	Leukotriene F4	84	12(S)-HETE
20	6,15-diketo-13,14-dihydro Prostaglandin F1α	53	8-iso Prostaglandin A2	85	15-OxoETE
21	iPF2α-IV	54	11-trans LTD4	86	15(S)-HpETE
22	8-iso-15(R)-Prostaglandin F2α	55	Prostaglandin A2	87	(±)9-HETE
23	8-iso Prostaglandin F2α	56	11-trans LTE4	88	5(S)-HETE

(continued)

Table 1
(continued)

NO.	Compounds	NO.	Compounds	NO.	Compounds
24	Thromboxane B2	57	Prostaglandin J2	89	12(S)-HpETE
25	11β-Prostaglandin F2α	58	Prostaglandin B2	90	12-OxoETE
26	(±)5-iPF2α-VI	59	8,12-iso-iPF2α-VI 1,5- lactone	91	(±)5,6-DHET-lactone
27	8-iso-15-keto Prostaglandin F2α	60	8(S),15(S)-DiHETE	92	5(S)-HpETE
28	Prostaglandin F2α	61	6-trans LTB4	93	(±)14(15)-EET
29	8-iso-13,14-dihydro-15-keto Prostaglandin F2α	62	5(S),15(S)-DiHETE	94	5-OxoETE
30	8-iso Prostaglandin E2	63	13,14-dihydro-15-keto Prostaglandin A2	95	(±)11(12)-EET
31	Prostaglandin E2	64	Leukotriene B4	96	(±)8(9)-EET
32	11-dehydro Thromboxane B2	65	13,14-dihydro-15-keto Prostaglandin J2	97	(±)5(6)-EET

3.2.1 Solid-Phase Extraction

SPE is an analytical sample preparation technique that is used to isolate, concentrate, and purify specific compounds for further analysis. In this method, we have used our developed and detailed SPE protocol published in this book in the chapter entitled "Detailed Protocol for Solid Phase Extraction for Lipidomic Analysis." This protocol is adapted for different sample types (such as cells, serum, plasma, body fluids, soft tissues, and bone marrow); please see the mentioned chapter for full SPE details.

3.2.2 Standards for the AA Lipid Mediators

Standards of the targeted LMs and their related deuterated Internal Standards (ISs) can be purchased from any company, but the purity must exceed 98%. Methanol, acetonitrile, water, and other solvents used for the LC mobile phase must be HPLC-MS analytical grade. Reagents like formic acid used during the protocol could be a reagent grade (>95%). The following steps are taken to prepare the IS stock solution and the standards working solution:

1. *Internal Standards (ISs) stock solution*: The IS stock solution is prepared by mixing the specific volume of each isotope labeled internal standard with methanol for a final IS stock according to the concentrations mentioned in Table 2. After preparing the IS stock, it is stored as aliquots in small tubes and frozen at −80 °C for future use.
2. *Standards Stock Solution (Std)*: The standard compounds are mixed in methanol to prepare the stock solution of standards mixture with final concentration of 2.5 μg/mL of each compound in the final stock volume of 1 mL in methanol as in Table 3.

Table 2
Preparation of IS stock solution

Internal Standards (IS)	Stock conc. (μg/μL)	Stock volume (μL)	Final conc. in mixture (μg/mL)	Volume (μL) for 500 μL IS stock
Tetranor-Prostaglandin E Metabolite-d6 (tetranor-PGEM-d6)	0.1	250	0.5	2.5
6-keto Prostaglandin F1α-d4 (6-keto-PGF1α-d4)	0.1	250	0.5	2.5
Thromboxane B2-d4	0.1	250	0.5	2.5
Prostaglandin F2α-d4 (PGF2α-d4)	0.1	250	0.5	2.5
Prostaglandin E2-d4 (PGE2-d4)	0.5	100	0.5	0.5
Prostaglandin D2-d4 (PGD2-d4)	0.1	250	0.5	2.5
Leukotriene C4-d5 (LTC4-d5)	0.1	250	0.5	2.5
Leukotriene B4-d4 (LTB4-d4)	0.1	250	0.5	2.5
5(S)HETE-d8	0.1	250	0.5	2.5
12(S)HETE-d8	0.1	250	0.5	2.5
15(S)HETE-d8	0.1	250	0.5	2.5
Arachidonic Acid-d8 (AA-d8)	10	50	10	0.5
Methanol				474
Final IS stock volume (μL)				**500**

Table 3
Preparation of standards stock solutions (Std)

Standards stock solution (Std)	Stock conc. (μg/μL)	Final conc. in mixture (μg/mL)	Volume (μL) for 1 mL stock
6-keto Prostaglandin F1α (6-keto-PGF1α)	5	2.5	50
Prostaglandin F2α (PGF2α)	10	2.5	25
Prostaglandin E2 (PGE2)	8.8	2.5	28.4
Arachidonic Acid (AA)	76	2.5	3.29
Methanol			893.31
Final volume of the prepared standards stock solution			1000

3. *Working Solution and Calibration Curve:* Using the prepared Std stock, the working solution can be prepared by serial dilutions of eight concentrations in the range from 1 μg/mL to 1 ng/mL in methanol. At least 50 μL of the highest concentration (1 μg/mL) is prepared to be used for the lower concentrations and the calibration curve as well. To prepare the final mixture for the calibration curve, 5 μL of each working solution +5 μL of the IS solution +90 μL methanol is mixed in 300 μL of the target polyspring insert, placed in clear autosampler glass vials, then closed using the PTEF/silicon screw cap, and vortexed for 10 s to ensure proper mixing.

3.2.3 LCMS/MS Analysis

1. *Instrument:* Shimadzu LCMS-8060 system with the UF-Qarray ion guide that increases LC/MS/MS sensitivity by enhancing ion signal intensity and reducing noise. By improving the ion sampling device, the ion guide, and the vacuum efficiency, the LCMS-8060 delivers a new vision of sensitivity and makes a real difference to working better and faster. Multiple Reaction Monitoring (MRM) is successfully employed to provide sensitive and specific analysis of lipidomics, metabolomics, and proteomics.
2. *Mobile phase:* Pump A: 0.1% Formic acid in water, and Pump B: 100% Acetonitrile (See Table 4).
3. The LC system is equipped with system controller: CBM-20A, 3 LC-30 AD pumps, an SIL-30 AC autosampler, and a CTO-30A column oven. The LC separation is conducted on a C8 column (Ultra C8, 150 × 2.1 mm, 3 μm, RESTEK, Manchaca, TX, USA). The oven temperature for the analytical column is 40 °C; the specific LC program parameters are shown in Table 4.

Table 4
Mobile phase and LC program parameters

Time (min)	Mobile phase A	Mobile phase B	Flow rate (mL/min)
0.00–5.00	75	25	0.4
5.00–10.00	65	35	0.4
10.00–20.00	25	75	0.4
20.00–20.10	5	95	0.4
20.10–25.00	5	95	0.4
25.00–25.10	90	10	0.4
25.10–30.00	90	10	0.4

4. The triple quadrupole mass spectrometer (LCMS-8060 system) is operated and optimized under electrospray ionization in negative mode. MS parameters are optimized as follows: interface: Dual ion source (DUIS) electrospray ionization (ESI) and atmospheric pressure chemical ionization (APCI), voltage, 4.0 kV; interface temperature, 280 °C; desolvation line (DL) temperature, 275 °C; heating block temperature, 400 °C; drying gas (N_2), 10 L/min; nebulizing gas (N_2), 3 L/min; heating gas (air), 10 L/min; and collision-induced dissociation (CID) gas (Ar), 230 kPa.
5. Considering the MRM technique, the acquisition is divided into multiple segments. Start with MRM optimization method for lipidomics to obtain the m/z transitions and their MS parameters (Q1, CE, and Q3 voltages) based on the best MRM responses.
6. Once all parameters have been optimized, the optimized method is run with full range to scan and monitor all IS and Std LMs in a concentrated sample of IS + Std mixture.
7. Once all LMs are identified and detected, the exact retention time in the method is recorded and updated if needed.
8. The optimized protocol is saved, and the calibration curve samples are running; chromatograms are obtained from the calibration curve samples (Example is shown in Fig. 1).
9. The browser tab in the LabSolutions software is used to analyze and check the calibration curve accuracy with at least $R^2 > 99.99\%$.
10. Once all parameters are optimized and the calibration curve results met the standards, samples will begin preparation. The samples prepared by SPE are reconstituted with the dried extracts in the 1.5 mL tubes by adding 50 μL methanol, vertexing, and then transferring 20 μL of each sample directly into autosampler glass vials for auto injection of 10 μL.
11. After all samples are ran successfully, the raw data are uploaded into the browser with the calibration curve results to quantify the detected LMs in the tested samples.
12. An example of the quantified LMs from the zebrafish muscle samples[12] is presented in Fig. 2.
13. External calibration curves should be performed daily, with the limits of quantitation (LOQ) of approximately 0.1 nM for each analyte under the present LCMS/MS conditions.
14. All analyses and data processing are performed using the Shimadzu LabSolutions software V5.99 SP2.

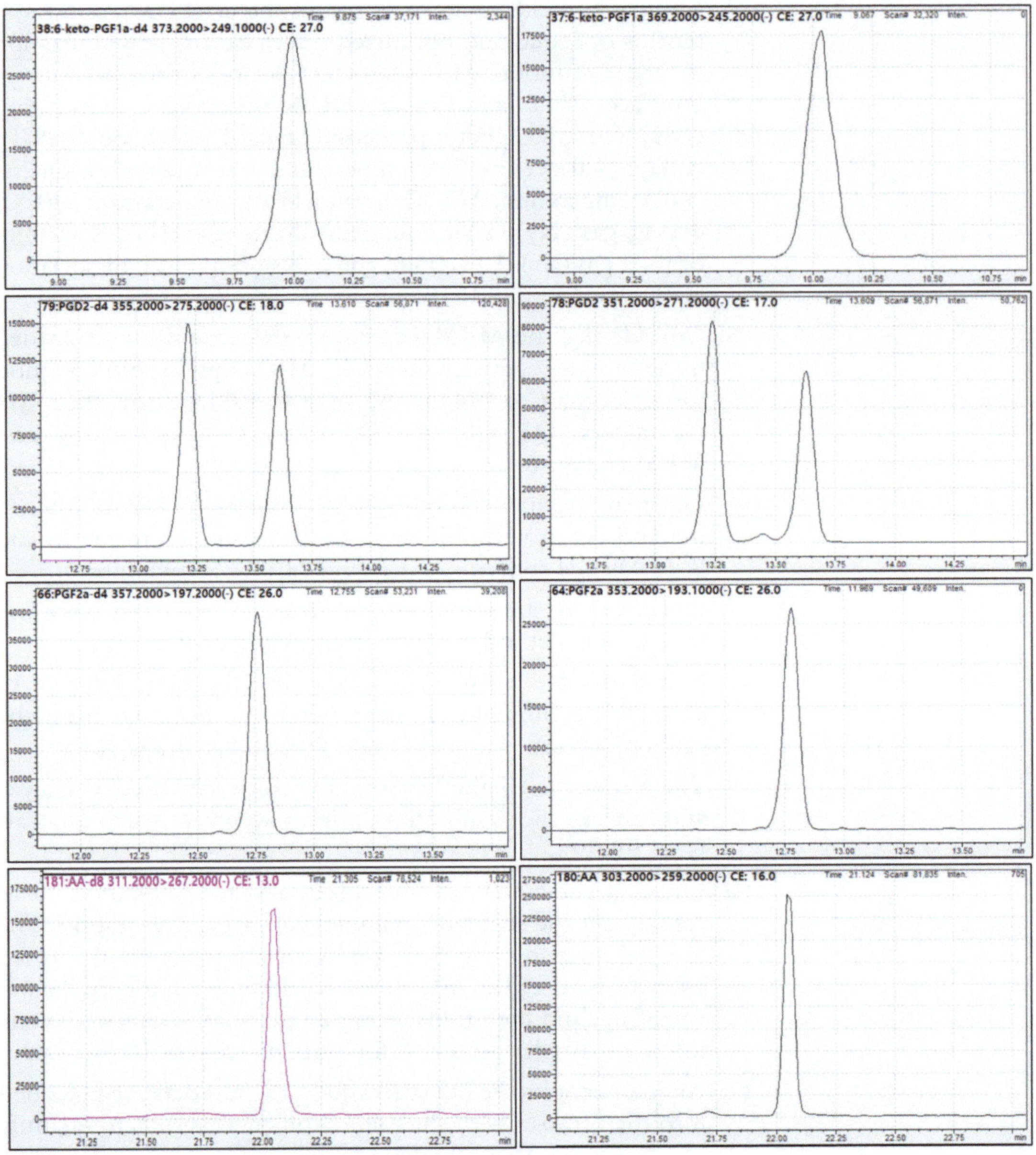

Fig. 1 MS/MS chromatograms of some of the detected arachidonic acid metabolites and their deuterated standards

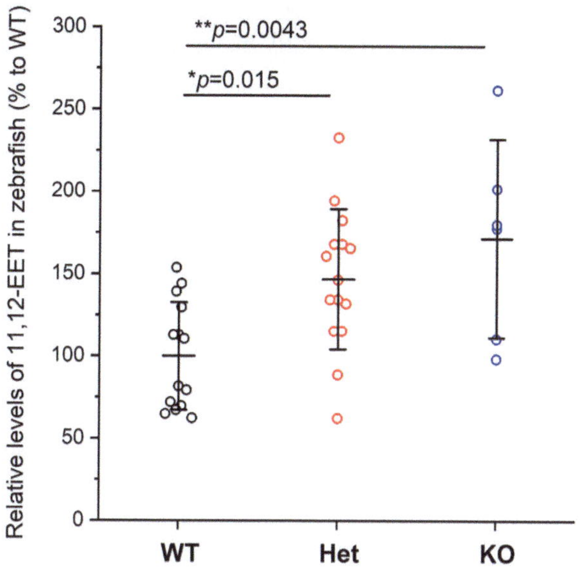

Fig. 2 Relative levels of 11,12-EET in dorsal muscles from adult zebrafish of different *srebf1* genotype. Wild type (WT, srebf1+/+); heterozygote (Het, srebf1+/−), and homozygote (knockout [KO], srebf1−/−). Mean ± SD, $n = 15$ for WT, $n = 15$ for Het, and $n = 6$ for KO. One-way analysis of variance (ANOVA) with Tukey post hoc test ($\alpha = 0.05$) was applied for statistical analysis. (Modified from Shochat C. et al[12])

Acknowledgment

This project was funded by the National Institutes of Health Grants: NIA 2-PO1AG039355, NIA-R01AG056504; NIDDK R01DK119066 and National Institutes of Neurological Disorders and Stroke (NINDS) 2-R01NS105621. The authors are thankful for the generous support from the George W. and Hazel M. Jay and Evanston Research Endowments and the UTA Shimadzu Institute for Research Technologies (https://www.uta.edu/sirt/). DK was funded by ISF grant #1121/19.

References

1. Kwon RY, Watson CJ, Karasik D (2019) Using zebrafish to study skeletal genomics. Bone. https://doi.org/10.1016/j.bone.2019.02.009
2. Dietrich K et al (2021) Skeletal biology and disease modeling in zebrafish. J Bone Miner Res 36:436–458. https://doi.org/10.1002/jbmr.4256
3. Witten PE, Harris MP, Huysseune A, Winkler C (2017) Small teleost fish provide new insights into human skeletal diseases. Methods Cell Biol 138:321–346. https://doi.org/10.1016/bs.mcb.2016.09.001
4. Ka J, Jin S-W (2021) Zebrafish as an emerging model for dyslipidemia and associated diseases. J Lipid Atheroscler 10:42–56
5. Oka T et al (2010) Diet-induced obesity in zebrafish shares common pathophysiological pathways with mammalian obesity. BMC Physiol 10:21. https://doi.org/10.1186/1472-6793-10-21

6. Luo L, Zhou J, Zhao H, Fan M, Gao W (2019) The anti-inflammatory effects of formononetin and ononin on lipopolysaccharide-induced zebrafish models based on lipidomics and targeted transcriptomics. Metabolomics 15:153. https://doi.org/10.1007/s11306-019-1614-2
7. Li X, Ge G, Song G, Li Q, Cui Z (2023) Effects of nutritionally induced obesity on metabolic pathways of zebrafish. Int J Mol Sci 24:1850
8. Hölttä-Vuori M et al (2010) Zebrafish: gaining popularity in lipid research. Biochem J 429: 235–242
9. Russo C et al (2023) Inflammation and obesity: the pharmacological role of flavonoids in the zebrafish model. Int J Mol Sci 24:2899
10. Monroe JD et al (2022) Identification of novel lipid biomarkers in xmrk- and Myc-induced models of hepatocellular carcinoma in zebrafish. Cancer Metab 10:7. https://doi.org/10.1186/s40170-022-00283-y
11. Eberlé D, Hegarty B, Bossard P, Ferré P, Foufelle F (2004) SREBP transcription factors: master regulators of lipid homeostasis. Biochimie 86:839–848. https://doi.org/10.1016/j.biochi.2004.09.018
12. Shochat C et al (2021) Deletion of SREBF1, a functional bone-muscle pleiotropic gene, alters bone density and lipid signaling in zebrafish. Endocrinology 162:bqaa189. https://doi.org/10.1210/endocr/bqaa189
13. Calder PC (2020) Eicosanoids. Essays Biochem 64:423–441. https://doi.org/10.1042/EBC20190083
14. Harizi H, Corcuff J-B, Gualde N (2008) Arachidonic-acid-derived eicosanoids: roles in biology and immunopathology. Trends Mol Med 14:461–469. https://doi.org/10.1016/j.molmed.2008.08.005

Chapter 3

Optimized Lipidomics Extraction of Sphingosine and Sphinganine from Optic Nerve for Signaling Studies

Alexa Jauregui, Emily J. Neag, Amr Almobayed, Alexandra Lens, and Sanjoy K. Bhattacharya

Abstract

Interconvertible sphingolipid metabolites represent germane constituents of eukaryotic membranes and are vital in the regulation of cellular homeostasis, proliferation, survival, and induction of autophagy. This protocol describes a step-by-step method for extractions of sphingosine and sphinganine from mammalian tissue samples, particularly from the murine optic nerve. These lipids are partitioned into a binary mixture of chloroform and methanol in a modified Bligh and Dyer method. This is followed with reverse phase ultrahigh-performance liquid chromatography fractionation with a C18+ column and subsequent tandem mass spectrometry (UHPLC-MS-MS) analysis of the biological abundance. These free sphingoid bases dissociate to form structurally distinctive carbocation product ions that can be confirmed with annotations of lipidomic databases or in-house fragmentation software.

Key words Lipidomics, Mass spectrometry, Imaging mass spectrometry, Bligh and Dyer, Optic nerve, Regeneration

1 Introduction

Progressive degenerative neuropathies, including glaucoma, entail the failure of adult mammalian retinal ganglion cells (RGCs) to regenerate, resulting in irreversible damage. Heretofore, few studies have explained the neurodegenerative nature of RGCs and the mechanisms leading to glaucomatous neurodegeneration [1–3]. The burgeoning growth of MS-omics over the past decade, including the subclass of lipidomics [4], has vastly increased the quantity and quality of biomarker prediction and systems approaches of CNS dysregulation. The complex systems-level analysis and characterization of lipids, known as lipidomics, have spurred advancements during recent years with consideration of the isobaric overlap and range of concentrations among species

Marco Brotto and Sanjoy K. Bhattacharya (eds.), *Lipid Signalling: Methods and Protocols*, Methods in Molecular Biology, vol. 2816, https://doi.org/10.1007/978-1-0716-3902-3_3,

[5, 6]. Despite this, much remains to be discovered of their role in axon injury and repair.

Bioactive sphingolipids are germane constituents of systemic and cellular events. First introduced by Thudichum as the abstruse compound "sphingosin" [7], sphingosine and sphinganine are the metabolomic precursors of complex sphingolipids, which are intimately ensheathed in both retinal physiology and pathology [8]. Beginning in the endoplasmic reticulum, they are synthesized de novo from the condensation of serine and a fatty acyl-CoA, commonly palmityl CoA, to form sphinganine, containing an 18-carbon long-chain base (LCB) backbone characteristic of typical sphingoid bases [9]. Sphinganine (also named dihydrosphingosine or d18:0), represents the first detectable sphingolipid metabolite [10] and serves as the sphingoid base for synthesis of ceramides and potential alteration with diverse headgroups to downstream complex sphingolipids. Sphingolipids are also found in high quantities within membrane rafts of the cells in the nervous system that control resident proteins and maintenance of cellular signaling functions [11, 12]. Alterations in the length and desaturation of the fatty acid linked to sphingosine play critical roles in the function of these dynamic membrane domains. The initial product of sphingosine catabolism, sphingosine-1-phosphate, and its ratio to SPH and sphinganine may affect CNS cellular survival [13]. The former sphingolipid mediator is an anti-apoptotic molecule that participates in the retention of photoreceptor and ganglion cell degeneration in juxtaposition with its sphingolipid relatives [14, 15]. Aberrant sphingolipid metabolism may also hold an association with primary open-angle glaucoma (POAG) [16]. Untargeted lipidomic profiling has recently revealed the total amount of these species within glaucomatous human optic nerves to be less than the control [16–18]. This indicates the utmost importance of sphingolipid metabolism regulation in relation to cell fate, which in turn highlights the importance of predicting and quantifying biological perturbations with respect to glaucoma and other age-related and progressive neurological diseases.

Because of the vast range of chemical, physical, and concentration distinctions between tissue samples, sample preparation workflows are not standardized in protocol literature. In the case of minute tissue samples (i.e., optic nerve), the differences in reported lipid compositions may result from the variety of solvent systems that have been used. In our wet-lab pipeline of lipidomic identification, we describe a modified extraction procedure according to Bligh and Dyer [19], as previously performed [20, 21]. This involves the partition of the lipid species into a binary ratio of chloroform and methanol (MeOH) at room temperature, where MeOH is added to the aqueous phase in a geometry-dependent rearrangement to disrupt the hydrogen bonding networks or electrostatic forces between lipids and proteins. This is followed with subsequent addition of the nonpolar solvent chloroform to allow the formation of a homogenate in a two-phase system from which the purified lipids are isolated.

After the preliminary separation technique, the tissue sample is introduced to an ionization mechanism, where particles receive a charge, enabling the detector to estimate the mass-to-charge ratio (m/z) of detected particles and intensity of each analyte. It has been well established that each class of sphingolipids move as a single head-group with normal chromatography [22]; thus, reverse-phase (RP) chromatography is preferred for separation based on carbon number, double bond position, fatty acid residues, and isomers with moieties at the sn-1 or sn-2 position [23]. With tandem mass spectroscopy capabilities, key precursor ions (MS1) are selected for fragmentation (MS/MS). Recommended data processing techniques and interpretation strategies are outlined further to prepare the data for quantification with statistical methods and identification via matches to *in-house* or open-source identification software. Using commercial sphingoid-type base standards, the optimal collision energy and tune conditions required to achieve the fragmentation patterns appropriate to the structural annotation of these species have been determined (Fig. 1). Of important note, abundant background ions will obscure lipid precursors and densely populate MS/MS spectra with abundant chemical noise [24]. The following parameters balance the targeted sphingolipid abundance with high sample throughput to provide semiquantitative lipid concentrations for endogenous samples of interest.

2 Materials

2.1 Chemicals

1. Liquid Nitrogen (N_2).
2. UHPLC-MS Grade Solvents: Chloroform:Methanol (1:1), Chloroform:Methanol (2:1).
3. Reconstitution Solution: IPA/ACN (1:1).
4. Ultrapure Argon gas.
5. UHPLC-MS-Grade Mobile-Phase Solvents:
   - Mobile Phase A: 50:50 Water:Acetonitrile, 5 mM Ammonium Formate, 0.1% Formic Acid.
   - Mobile Phase B: 88:10:2 Isopropanol:Acetonitrile:Water, 5 mM Ammonium Formate, 0.1% Formic Acid.
6. Lipid Standards: spingosine-d7 (Avanti® Polar Lipids, 860,657), sphinganine-d7 (Avanti® Polar Lipids, 860,658).

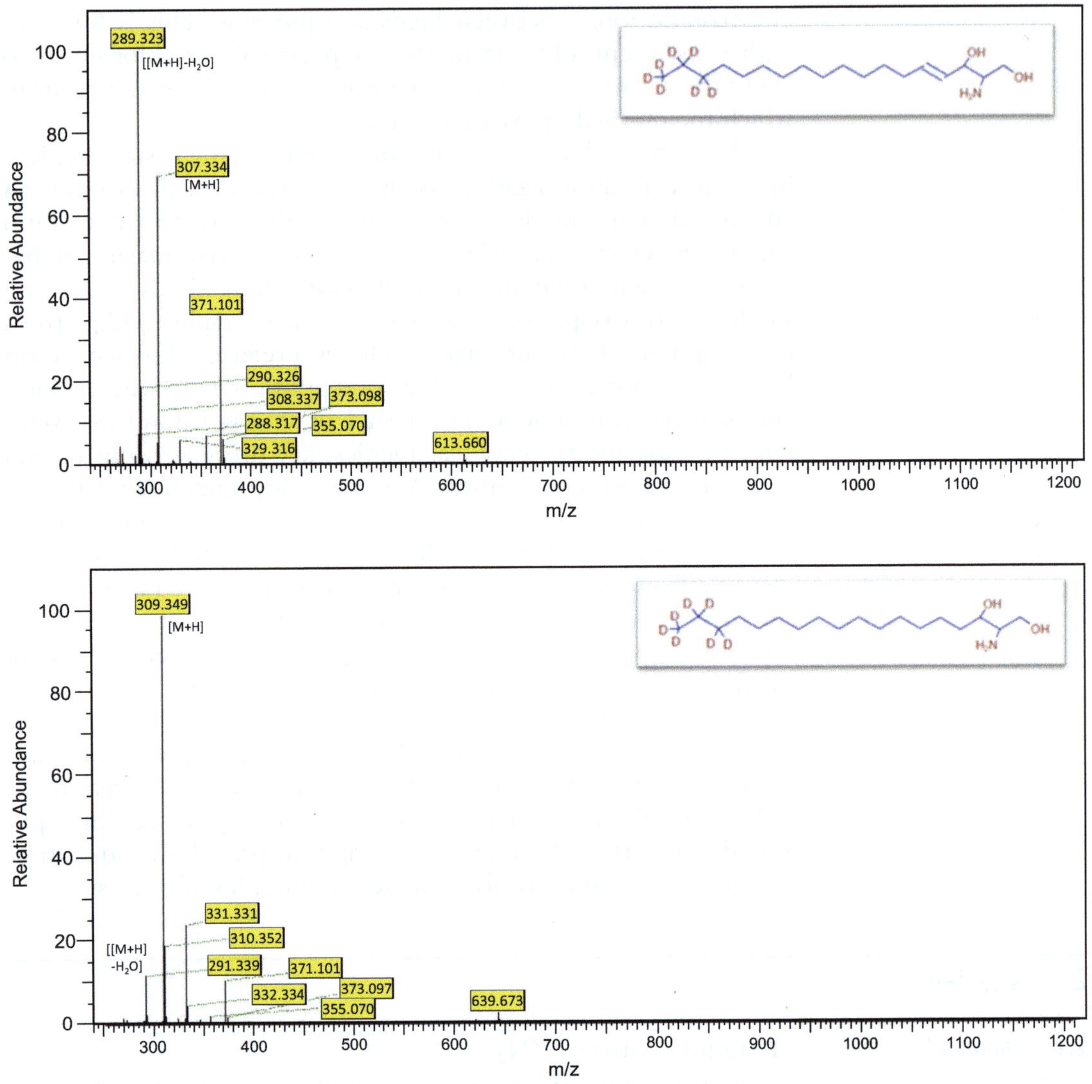

Fig. 1 Mass spectrum of sphingosine-d7 (Avanti® Polar Lipids, 860657) obtained in parent ion scan (PIS) positive ion mode using a ThermoScientificTMQ-Exactive. Positive acquisition mode of the [M+H] reveals additional fragmentation via single dehydration to a product ion of m/z 289.323

2.2 Instrumentation

2.2.1 Equipment

1. Dissection Tool Kit.
2. Calibrated pipette tips: 10–200 μL and 100–1000 μL.
3. Autoclaved 1.5 mL or 2 mL Eppendorf tubes.
4. Vortex Mixer.
5. Centrifuge (Beckman Microfuge 18) in 4 °C (refrigerated cold room).
6. Speed Vacuum Concentrator.
7. Sonicator/Ultrasonic Water Bath.

2.2.2 Ultrahigh-Performance Liquid Chromatography and Mass Spectrometer

1. HPLC vials w/ Clear Insert (12 × 32 DP Vial Clear ID w/insert 100 Vials/Package).
2. Accucore™ Vanquish™ C18+ Column (100 mm × 2.1 mm, 1.5 μm, Thermo Scientific™).
3. Thermo Scientific™ Vanquish™ Horizon UHPLC.
4. Thermo Scientific™ Q Exactive™ Orbitrap™ Mass Spectrometer.
5. +/− Polarity (Positive/Negative Acquisition Mode)
6. Thermo Scientific™ Xcalibur™ Software.

3 Methods

Carry out the extraction procedure in a 4 °C refrigerated room (*see* **Note 1**).

3.1 Tissue Preparation

1. Pre-label and weigh the empty Eppendorf tubes. Weigh the tubes again after tissue dissection (*see* **Note 2**).
2. Break the tissue membrane using five freeze/thaw cycles alternating between liquid N_2 and sonicator bath. Each step should be carried out for 5–10 min.
3. Mince the tissue for 60 s or until sample has homogenized repeatedly (*see* **Note 3**).
4. Flush with Argon gas (*see* **Note 4**).
5. Add 200 μL of Chloroform:Methanol (2:1).
6. Repeat **step 3**.
7. Add 150 μL of Chloroform:Methanol (1:1).
8. Repeat **step 3**.
9. Vortex for 20 s.

3.2 Lipid Isolation and Dessication

1. Centrifuge at 4 °C at 13,000 rpm for 20 min (*see* **Note 5**).
2. Once separation is observed, check for any bubbles on top of the aqueous layer and remove them prior to removing the lipids (next step).
3. Remove the bottom organic layer (contains lipids) with a pipette (only go in with syringe once and draw up all lipids) and place in a separate tube (*see* **Note 6–7**).
4. Transfer the bottom organic layer into a newly labeled Eppendorf tube.
5. Place the tubes into the speed vacuum (*see* **Note 8**).
6. Flush the samples with Argon gas and store at −80 °C until ready for Mass Spectrometry analysis (*see* **Note 9**).

3.3 Preparation of Lipid Samples

1. Reconstitute the samples after speed vac with 45 μL of IPA/ACN (1:1).
2. Sonicate the samples in a sonicator (ultrasonic water bath) for 20 min.
3. Centrifuge the samples.
4. Spike 5 μL of 10× IS into each sample.
5. Label the sample vials for both positive and negative modes.
6. Transfer 25 μL of each sample to its respective LC-MS vial.

3.4 Preparation of Blank Samples (See Note 10)

1. Label the blank vials for both positive and negative modes.
2. Transfer 25 μL of the reconstitution solution: IPA/ACN (1:1) to its respective LC-MS vial.

3.5 Ultrahigh-Performance Liquid Chromatography-Tandem Mass Spectrometry (UHPLC-MS)

3.5.1 Preparation for Ultrahigh Performance Liquid Chromatography

1. Prepare the following solvents for UHPLC (*see* **Note 11**):
   - Mobile Phase A: 50:50 Water:Acetonitrile, 5 mM Ammonium Formate, 0.1% Formic Acid.
   - Mobile Phase B: 88:10:2 Isopropanol:Acetonitrile:Water, 5 mM Ammonium Formate, 0.1% Formic Acid.
2. Set the Column Compartment and Pre-Heater temperature to 55 °C.
3. Set the flow rate to 260 μL/min.
4. Set the flow gradient as follows:
   - Increase Solvent B from 10% to 100% over the course of 30 min (Fig. 2).
5. Injection volume should be set to 8 μL during sequence setup using Xcalibur.

3.5.2 Q-Exactive and HESI-II Ion Source Settings

HESI-II Ion Source Settings

1. Set the spray voltage to 4.0 kV for positive acquisition mode and 2.5 kV for negative acquisition mode.
2. Set the capillary temperature to 325 °C for both acquisition modes.
3. Set the sheath gas flow rate to 35 for positive mode and 30 for negative mode.
4. Set the auxiliary gas to 15 units for both acquisition modes.
5. Probe heater temperature is the same for both modes and should be set to 275 °C.
6. Set the S-lens radio frequency (RF) level to 70.

   Q-Exactive Mass Spectrometry—Full MS/ddMS^2 (TopN).
   Full MS

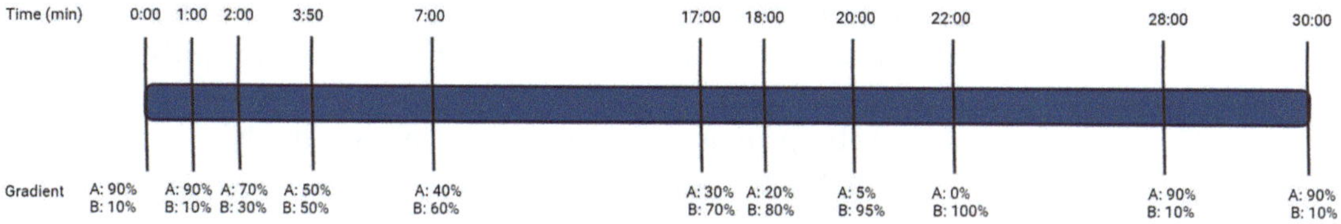

Fig. 2 UHPLC Vanquish schematic of gradient between mobile phase A: 50:50 water: acetonitrile, 5 mM ammonium formate, 0.1% formic acid and mobile phase B: 88:10:2 isopropanal: acentonitrile: water, 5 mM ammonium formate, 0.1% formic acid

- Scan range: 250–1200 *m/z*.
- Microscans: 1.
- Resolution: 70,000.
- AGC target: 1e6.
- Max. Ion injection time (IT): 250 ms.

dd-MS^2

- Microscans: 1.
- Resolution: 17,500.
- AGC target: 1e5.
- Max. Ion injection time (IT): 50 ms.
- Loop count/TopN: 10.
- Isolation window: 1.0 m/z.
- Dynamic exclusion: 8.0 s.
- Normalized Collision Energy (NCE): 24.

4 Notes

1. If unable to do so, pre-chill solutions at −20 °C for at least 1 h before use and keep all samples and solutions on ice during sample preparation. Wear lab coat throughout the entire procedure and change the gloves as needed.
2. Subtract the mass of the tissue and tube minus the weight of the tube prior to adding the sample. This will give the weight of the tissue sample.
3. Clean the scissors with EtOH.
4. Only perform this step if planning long-term storage at −20 °C. Samples stored at −20 °C are good for up to 1 week. If samples need to be stored overnight, a 4 °C refrigerator can be used.
5. An upper (aqueous) and a lower (lipid) fraction should form. If separation is not visible, vortex for 1 min and repeat **step 1**.

6. Keep the protein layer to use to determine the protein concentration in your sample (You can normalize the lipids using this concentration). If accidently went in with syringe twice, there may be protein in your organic phase (lipids), so centrifuge again to get separation again, then use a new syringe, and repeat the process.
7. Save and separate the original Eppendorf tube and place it on ice. This portion contains the protein pellet.
8. Make sure the heat setting is turned off. Samples may take up to 90 min to become bone dry.
9. Perform this step only if planning long-term storage of sample prior to running samples.
10. While the column should be conditioned and equilibrated before each run, it is strongly encouraged to run blank samples prior to analysis and between every 8 and 10 samples to optimize flow and temperature, as well as consistently test for instrument and column performance.
11. Dissolve Ammonium Formate in water first, then dilute up with the rest of the solvent.

References

1. Danesh-Meyer HV, Levin LA (2015) Glaucoma as a neurodegenerative disease. J Neuroophthalmol 35(Suppl 1):S22–S28. https://doi.org/10.1097/WNO.0000000000000293
2. Levin LA, Crowe ME, Quigley HA (2017) Lasker/IRRF initiative on astrocytes and glaucomatous neurodegeneration participants. Neuroprotection for glaucoma: requirements for clinical translation. Exp Eye Res 157:34–37. https://doi.org/10.1016/j.exer.2016.12.005. Epub 2016 Dec 9. PMID: 27955999; PMCID: PMC5413382
3. Lawlor M, Danesh-Meyer H, Levin LA, Davagnanam I, De Vita E, Plant GT (2018) Glaucoma and the brain: trans-synaptic degeneration, structural change, and implications for neuroprotection. Surv Ophthalmol 63(3): 296–306. https://doi.org/10.1016/j.survophthal.2017.09.010. Epub 2017 Oct 3
4. Castellanos DB, Martín-Jiménez CA, Rojas-Rodríguez F, Barreto GE, González J (2021) Brain lipidomics as a rising field in neurodegenerative contexts: perspectives with machine learning approaches. Front Neuroendocrinol 61:100899. https://doi.org/10.1016/j.yfrne.2021.100899. Epub 2021 Jan 12
5. Falomir-Lockhart LJ, Cavazzutti GF, Giménez E, Toscani AM (2019) Fatty acid signaling mechanisms in neural cells: fatty acid receptors. Front Cell Neurosci 24(13):162. https://doi.org/10.3389/fncel.2019.00162. PMID: 31105530; PMCID: PMC6491900
6. Schmitt S, Castelvetri LC, Simons M (2015) Metabolism and functions of lipids in myelin. Biochim Biophys Acta 1851(8):999–1005. https://doi.org/10.1016/j.bbalip.2014.12.016. Epub 2014 Dec 24
7. Thudichum JLW (1884) A treatise on the chemical constitution of brain. Bailliere and Cox, London
8. Simon MV, Basu SK, Qaladize B, Grambergs R, Rotstein NP, Mandal N (2021) Sphingolipids as critical players in retinal physiology and pathology. J Lipid Res 62:100037. https://doi.org/10.1194/jlr.TR120000972
9. Wigger D, Gulbins E, Kleuser B, Schumacher F (2019) Monitoring the sphingolipid de novo synthesis by stable-isotope labeling and liquid chromatography-mass spectrometry. Front Cell Dev Biol 1(7):210. https://doi.org/10.3389/fcell.2019.00210. PMID: 31632963; PMCID: PMC6779703
10. Gonzalez-Cabrera PJ, Hla T, Rosen H (2007) Mapping pathways downstream of sphingosine 1-phosphate subtype 1 by differential chemical perturbation and proteomics. J Biol Chem

282(10):7254–7264. https://doi.org/10.1074/jbc.M610581200. Epub 2007 Jan 10

11. Sezgin E, Levental I, Mayor S, Eggeling C (2017) The mystery of membrane organization: composition, regulation and roles of lipid rafts. Nat Rev Mol Cell Biol 18(6): 361–374. https://doi.org/10.1038/nrm.2017.16. Epub 2017 Mar 30. PMID: 28356571; PMCID: PMC5500228
12. Bieberich E (2018) Sphingolipids and lipid rafts: Novel concepts and methods of analysis. Chem Phys Lipids 216:114–131. https://doi.org/10.1016/j.chemphyslip.2018.08.003. Epub 2018 Sep 5. PMID: 30194926; PMCID: PMC6196108
13. Maceyka M, Spiegel S (2014) Sphingolipid metabolites in inflammatory disease. Nature 510(7503):58–67. https://doi.org/10.1038/nature13475. PMID: 24899305; PMCID: PMC4320971
14. Maceyka M, Payne SG, Milstien S, Spiegel S (2002) Sphingosine kinase, sphingosine-1-phosphate, and apoptosis. Biochim Biophys Acta 1585(2–3):193–201. https://doi.org/10.1016/s1388-1981(02)00341-4
15. Simon MV, Basu SK, Qaladize B, Grambergs R, Rotstein NP, Mandal N (2021) Sphingolipids as critical players in retinal physiology and pathology. J Lipid Res 62:100037. https://doi.org/10.1194/jlr.TR120000972. Epub 2021 Feb 6. PMID: 32948663; PMCID: PMC7933806
16. Chauhan MZ, Valencia AK, Piqueras MC, Enriquez-Algeciras M, Bhattacharya SK (2019) Optic nerve Lipidomics reveal impaired Glucosylsphingosine lipids pathway in glaucoma. Invest Ophthalmol Vis Sci 60(5): 1789–1798. https://doi.org/10.1167/iovs.18-25802. PMID: 31022733; PMCID: PMC6485987
17. Untargeted lipidomic profiling has of recent revealed the total amount of these species within glaucomatous human optic nerves to be less than control
18. Gutner UA, Shupik MA, Maloshitskaya OA, Sokolov SA, Rezvykh AP, Funikov SY, Lebedev AT, Ustyugov AA, Alessenko AV (2019) Changes in the metabolism of Sphingoid bases in the brain and spinal cord of transgenic FUS(1-359) mice, a model of amyotrophic lateral sclerosis. Biochemistry (Mosc) 84(10): 1166–1176. https://doi.org/10.1134/S0006297919100055
19. Bligh EG, Dyer WJ (1959) A rapid method of total lipid extraction and purification. Can J Biochem Physiol 37(8):911–917. https://doi.org/10.1139/o59-099
20. Edwards G, Aribindi K, Guerra Y, Bhattacharya SK (2014) Sphingolipids and ceramides of mouse aqueous humor: Comparative profiles from normotensive and hypertensive DBA/2J mice. Biochimie 105:99–109. https://doi.org/10.1016/j.biochi.2014.06.019. Epub 2014 Jul 9. PMID: 25014247; PMCID: PMC4163522
21. Milbeck SM, Bhattacharya SK (2020) Alteration in lysophospholipids and converting enzymes in glaucomatous optic nerves. Invest Ophthalmol Vis Sci 61(6):60. https://doi.org/10.1167/iovs.61.6.60. PMID: 32602905; PMCID: PMC7415893
22. Masood MA, Rao RP, Acharya JK, Blonder J, Veenstra TD (2012) Quantitation of multiple sphingolipid classes using normal and reversed-phase LC-ESI-MS/MS: comparative profiling of two cell lines. Lipids 47(2):209–226. https://doi.org/10.1007/s11745-011-3633-2. Epub 2011 Nov 29. PMID: 22124806; PMCID: PMC7480952
23. Lin J, Snyder L, McKeon TA (1998) Prediction of relative retention times of traicylglycerols in non-aqeous reversed-phase high-performance liquid chromatography. J Chromatgr 808(1): 43–49. https://doi.org/10.1016/S0021-9673(98)00134-4
24. Matyash V, Liebisch G, Kurzchalia TV, Shevchenko A, Schwudke D (2008) Lipid extraction by methyl-tert-butyl ether for high-throughput lipidomics. J Lipid Res 49(5): 1137–1146. https://doi.org/10.1194/jlr.D700041-JLR200. Epub 2008 Feb 16. PMID: 18281723; PMCID: PMC2311442

Chapter 4

Analysis of Sphingosine and Sphinganine from the Aqueous Humor for Signaling Studies Using Ultrahigh-Performance Liquid Chromatography-Mass Spectrometry

Meher A. Saleem, Molly Kennedy, Omar Badla, Emily J. Neag, and Sanjoy K. Bhattacharya

Abstract

Sphingolipids, including sphingosine and sphinganine, are one of the major classes of lipids. They serve as constituents of cell membranes and lipid rafts and aid in the performance of cell-cell communication and adhesion. Abnormal levels of sphingolipids in the aqueous humor can indicate impaired sphingolipid metabolism and associated ocular pathologies. Sphingolipids can be extracted from the aqueous humor by the methyl-*tert*-butyl ether (MTBE) lipid extraction method and subsequently analyzed by liquid chromatography-mass spectrometry (LC-MS). This chapter describes a modified protocol for an MTBE lipid extraction from the aqueous humor, followed by analysis with ultrahigh-performance liquid chromatography-mass spectrometry (UHPLC-MS).

Key words Sphingosine, Sphinganine, Sphingolipids, Aqueous humor, Lipidomics, Sphingolipidomics, Liquid chromatography, Ultrahigh-performance liquid chromatography, Mass spectrometry, LC, MS, LC-MS, UHPLC, UHPLC-MS, MTBE, Lipid extraction

1 Introduction

Sphingosine and sphinganine are both categorized as sphingolipids, one of major classes of lipids in eukaryotes [1]. Sphingolipids perform several essential cellular functions. They serve as components of cell membranes and lipid rafts and aid in the performance of cell-cell communication and adhesion [2]. Sphinganine and sphingosine perform independent functions in the cell and also serve as a sphingoid bases, founding blocks for the development of more complex sphingolipids [3, 1]. A broad range of sphingolipids can be formed by the multitude of potential combinations of sphingoid bases with fatty acids and head groups in an organism [3].

Marco Brotto and Sanjoy K. Bhattacharya (eds.), *Lipid Signalling: Methods and Protocols*, Methods in Molecular Biology, vol. 2816, https://doi.org/10.1007/978-1-0716-3902-3_4,

Sphingolipids, including sphingosine and sphinganine, can be found in aqueous humor [4]. The amount of resident sphingosine and sphinganine in aqueous humor can provide a quantitative assessment of multiple cellular responses including autophagy [5] and mitochondrial dysfunction [6]. Impairments in sphingolipid metabolism have been found to be associated with ocular hypertension [4] and primary open-angle glaucoma [7].

Lipidomics is the study of lipids using analytical chemistry. It allows the quantification of changes in lipid classes, subclasses, and molecular species that indicate metabolic differences. There are multiple mass spectrometry (MS) techniques used in lipidomics including liquid chromatography-mass spectrometry (LC-MS), shotgun lipidomics, matrix-assisted laser desorption/ionization-mass spectrometry (MALDI-MS), MS imaging, and ion-mobility MS.

LC-MS is a powerful technique that allows for the comprehensive analysis of lipid classes and molecular species [2]. LC separates compounds using a polar mobile phase and non-polar stationary phase in a column. The column contains a porous solid material (the stationary phase) in which a sample is injected with the solvent (the mobile phase). As the sample and solvent travel through the column, the components of the sample are separated and absorbed on the stationary phase based on their affinity to the stationary and mobile phase. High-performance liquid chromatography (HPLC) is an advanced type of LC. In LC, the solvent travels by the force of gravity, while in HPLC, the solvent travels by the force of a high pressure pump, which reduces separation time [8]. Compared to LC, HPLC uses smaller particles inside the column, thereby providing superior resolving power. Ultrahigh-performance liquid chromatography (UHPLC) is a type of LC, which uses smaller column particles and provides higher resolving power than HPLC [9].

In LC-MS, a sample is first passed through an LC column and the resulting components and subsequently analyzed by MS. The mass spectrometer ionizes and fragments the components and causes separation based on mass-to-charge ratio [10]. Analysis of sphingolipids with MS allows the detection of multiple analytes in a single sample, which is particularly advantageous for sphingolipids due to their extensive ability for metabolic interconversion [11].

LC-MS is the primary method of analysis for sphingolipids due to the technique's high sensitivity, structural specificity, and quantitative precision. The specific use of HPLC-MS and UHPLC-MS for sphingolipids provides the additional benefit of minimizing ionization suppression from other species and reducing interference from isomers, isobars, and isotopes [3].

Lipids can be extracted from aqueous humor by multiple methods including MTBE. The MTBE method of lipid extraction has been instituted among the major classes of lipid extraction including Folch and Bligh and Dyer. MTBE lipid extraction is particularly advantageous when using minute biological samples [12, 13]. The MTBE method also reduces the risk of contamination from aqueous components in an extraction. For example, the Folch method results in an aqueous upper layer and organic lower layer. In this method, there is a potential risk of contamination from the aqueous upper layer during extraction of the organic lower layer. The MTBE method has a reduced risk of this contamination because it creates an organic upper layer and an aqueous lower layer [14].

In this chapter, we describe a method of analysis for sphingosine and sphinganine from the aqueous humor. We provide a modified MTBE extraction protocol specific for the aqueous humor and subsequently describe the analysis of sphingosine and sphinganine with UHPLC-MS.

2 Materials

2.1 Methyl Tertiary Butyl Ether (MTBE) Lipid Extraction

1. Methanol.
2. Butylhydroxytoluene (BHT).
3. 0.15 M Ammonium Acetate.
4. Eppendorf tubes.
5. MTBE.
6. Rocker.
7. Centrifuge.
8. Speed vacuum.

2.2 Ultrahigh-Performance Liquid Chromatography-Mass Spectrometry (UHPLC-MS) Analysis

1. 1:1 chloroform/methanol solution.
2. Sonicator.
3. HPLC vials w/ clear insert.
4. Internal standards: sphingosine and sphinganine.
5. 50:50 water/acetonitrile.
6. 5 mM ammonium formate.
7. 0.1% formic acid.
8. 88:10:2 isopropanol/acetonitrile/water.
9. Thermo Scientific Accucore™ Vanquish™ C18+ Reversed Phase HPLC Column, 1.5 μm, 2.1 mm × 100 mm.
10. Thermo Scientific Vanquish Horizon UHPLC System.
11. Thermo Scientific Q Exactive Orbitrap Mass Spectrometer.
12. Thermo Scientific Xcalibur™ Software.

3 Methods

3.1 Methyl Tertiary Butyl Ether (MTBE) Lipid Extraction

3.1.1 Preparation of Reagents

1. Create a methanol + BHT solution with a final concentration of 0.516 mg/mL.
   (a) Add 51.6 mg of BHT.
   (b) Add methanol to reach a final volume of 100 mL of solution.
2. Create a 0.15 M ammonium acetate solution.
3. Chill methanol + BHT solution, ammonium acetate, and MTBE to 4 °C.

3.1.2 Extraction of Lipids

1. Aliquot 50 μL of each aqueous humor sample into an Eppendorf tube.
2. Add methanol + BHT solution to the sample.
3. Transfer the sample to a 12 mL glass tube.
4. Add 4 mL of MTBE to the sample, maintain a ratio of 10:3 MTBE/methanol + BHT.
5. Incubate the sample overnight at 4 °C in the dark with rotation (*see* **Note 1**).

3.1.3 Isolation of Lipids

1. Pre-cool centrifuge to 4 °C.
2. Add 1.25 mL of 0.15 M ammonium acetate to the sample, maintaining a ratio of 20:5:6 MTBE/methanol + BHT/ammonium acetate.
3. Centrifuge for 10 minutes at 2000× *g* at 4 °C. Centrifugation will result in an upper (organic) layer and a lower (aqueous) layer.
4. Collect the upper (organic) layer in a 12 mL glass tube.

3.1.4 Desiccation of Lipids

1. Add 2 mL of organic phase into a glass vial.
2. Place the vial in a speed vacuum until the sample is completely dry.
3. Repeat steps 1 and 2 (of Subheading 3.1.4) until all of the organic phase is desiccated.

3.2 Ultrahigh-Performance Light Chromatography-Mass Spectrometry (UHPLC-MS) Analysis

3.2.1 Sample Preparation

1. Resuspend the desiccated organic phase in 50 μL of 1:1 chloroform/methanol solution.
2. Sonicate the solution to resuspend the pellet.
3. Transfer 30 μL of the solution to an HPLC vial with an insert.
4. Store the remaining 20 μL of solution at −80 °C.
5. Add 50 μL of internal standard to each lipid sample.

3.2.2 Ultrahigh-Performance Liquid Chromatography Preparation

1. Prepare the mobile-phase solvents (*see* **Note 2**).
   (a) Mobile Phase A: 50:50 water/acetonitrile, 5 mM ammonium formate, 0.1% formic acid.
   (b) Mobile Phase B: 88:10:2 isopropanol/acetonitrile/water, 5 mM ammonium formate, 0.1% formic acid.
2. Set the column temperature to 55 °C.
3. Set the pre-heater temperature to 55 °C.
4. Set the flow rate to 260 μL/min.
5. Set the flow gradient to increase Solvent B from 10% to 100% over the course of 30 min.
6. Set injection volume to 8 μL during sequence set up with Xcalibur.

3.2.3 High-Efficiency Source (HES) I-II Ion Source Setting Preparation

1. Spray voltage
   (a) Positive acquisition mode: 4 kV
   (b) Negative acquisition mode: 1.5 kV
2. Capillary temperature
   (a) Positive acquisition mode: 275 °C
   (b) Negative acquisition mode: 250 °C
3. Sheath Gas Flow Rate
   (a) Positive acquisition mode: 5
   (b) Negative acquisition mode: 0
4. Auxiliary Gas Flow Rate: 0 for both modes.
5. S-lens radio frequency (RF) level: 70 for both modes.
6. Probe heater temperature: 275 °C

3.2.4 Q-Exactive Mass Spectrometry Setting Preparation

1. Full MS
   (a) Scan Range: 250–1200 m/z
   (b) Microscans: 1
   (c) Resolution: 70,000
   (d) Automatic Gain Control (AGC) Target: 1e6
   (e) Maximum Ion Injection Time (IT): 240 ms
2. dd-MS^2
   (a) Microscans: 1
   (b) Resolution: 17,500
   (c) Automatic Gain Control (AGC) Target: 1e5
   (d) Maximum Ion Injection Time (IT): 50 ms
   (e) Loop Count/TopN: 10

(f) Isolation Window: 1/0 m/z

(g) Dynamic Exclusion: 8 s

(h) Normalized Collision Energy (NCE): 24

4 Notes

1. The tubes should be laid down parallel to the rocker's surface, to allow the liquid to move back and forth inside the tube.
2. Start by dissolving the ammonium formate in water and subsequently dilute with the solvent.

References

1. Hannun YA, Obeid LM (2018) Sphingolipids and their metabolism in physiology and disease. Nat Rev Mol Cell Biol 19(3):175–191. https://doi.org/10.1038/nrm.2017.107
2. Han X (2016) Lipidomics for studying metabolism. Nat Rev Endocrinol 12(11):668–679. https://doi.org/10.1038/nrendo.2016.98
3. Sullards MC, Liu Y, Chen Y et al (2011) Analysis of mammalian sphingolipids by liquid chromatography tandem mass spectrometry (LC-MS/MS) and tissue imaging mass spectrometry (TIMS). Biochim Biophys Acta Mol Cell Biol Lipids 1811(11):838–853. https://doi.org/10.1016/j.bbalip.2011.06.027
4. Edwards G, Aribindi K, Guerra Y et al (2014) Sphingolipids and ceramides of mouse aqueous humor: comparative profiles from normotensive and hypertensive DBA/2J mice. Biochimie 105:99–109. https://doi.org/10.1016/j.biochi.2014.06.019
5. Pulliero A, Seydel A, Camoirano A et al (2014) Oxidative damage and autophagy in the human trabecular meshwork as related with ageing. PLoS One 9(6):e98106. https://doi.org/10.1371/journal.pone.0098106
6. He Y, Ge J, Tombran-Tink J (2008) Mitochondrial defects and dysfunction in calcium regulation in glaucomatous trabecular meshwork cells. Invest Ophthalmol Vis Sci 49(11):4912–4922. https://doi.org/10.1167/iovs.08-2192
7. Simon MV, Basu SK, Qaladize B et al (2021) Sphingolipids as critical players in retinal physiology and pathology. J Lipid Res 62:100037. https://doi.org/10.1194/jlr.TR120000972
8. Marks DJB, Smith PJ (2014) High performance liquid chromatography. Br J Hosp Med 75(2):C18–C21
9. Nishi H, Nagamatsu K (2014) New trend in the LC separation analysis of pharmaceuticals—high performance separation by ultra high-performance liquid chromatography (UHPLC) with Core-shell particle C18 columns. Anal Sci 30(2):205–211. https://doi.org/10.2116/analsci.30.205
10. Zhou B, Xiao JF, Tuli L et al (2012) LC-MS-based metabolomics. Mol BioSyst 8(2):470–481
11. Haynes CA, Allegood JC, Park H et al (2009) Sphingolipidomics: methods for the comprehensive analysis of sphingolipids. J Chromatogr B Analyt Technol Biomed Life Sci 877(26):2696–2708. https://doi.org/10.1016/j.jchromb.2008.12.057
12. Kuyukina MS, Ivshina IB, Philp JC et al (2001) Recovery of Rhodococcus biosurfactants using methyl tertiary-butyl ether extraction. J Microbiol Methods 46(2):149–156. https://doi.org/10.1016/S0167-7012(01)00259-7
13. Folch J, Lees M, Sloane Stanley GH (1957) A simple method for the isolation and purification of total lipides from animal tissues. J Biol Chem 226(1):497–509. https://doi.org/10.1016/S0021-9258(18)64849-5
14. Eggers LF, Schwudke D (2016) Encyclopedia of lipidomics. Springer, Dordrecht

Chapter 5

Nanoparticles and Their Applications in Lipid Signaling

Zeinab Fotouhi Ashin, Ahmed S. Yacoub, Marwa Saady, Sophia E. Thompson, Kamal Awad, Pranesh Aswath, Marco Brotto, and Venu Varanasi

Abstract

This chapter provides an overview of the diverse range of applications associated with nanoparticles. The application of nanoparticles in the medical field has garnered considerable attention due to their unique properties and versatile compositions. They have shown promise in the treatment of cancer, fungal and viral infections, and pain management. These systems provide numerous benefits, such as increased drug stability, improved bioavailability, and targeted delivery to specific tissues or cells. The objective of this chapter is to provide a brief analysis of the differences between nanoparticles and lipid particles, focusing particularly on the importance of nanoparticle size and composition in their interactions with lipids. Additionally, the applications of nanoparticles in lipid signaling will be discussed, considering the vital roles lipids play in cellular signaling pathways. Nanoparticles have shown immense potential in the regulation and control of medical pathways. In this case, we will focus on the manufacture of liposomes, a type of nanoparticle composed of lipids. The reason behind the extensive investigation into liposomes as drug delivery vehicles is their remarkable biocompatibility and adaptability. This section will provide insights into the methods and techniques employed for liposome formulation.

Key words Lipid nanoparticle, Lipid signaling, Drug delivery, Solid lipid nanoparticles, Liposomes, Nanostructured lipid carriers

1 Introduction

1.1 Understanding the Science of Nanoparticles

Nanoparticles (NPs) are extremely small particles having diameters in nanometer scale; smaller than 100 nm. These materials can be 0D, 1D, 2D, or 3D depending on the overall shape. This classification is heavily reliant on electron mobility in the nanomaterials (NMs) along the dimensions. Electrons in 0D NMs, for example, are entrapped in a dimensionless environment, but electrons in 1D NMs can flow along the x-axis, which is less than 100 nm. Similarly, electrons move along the x-y-axis and the x, y, and z-axes in 2D and 3D NMs, respectively [1].

Marco Brotto and Sanjoy K. Bhattacharya (eds.), *Lipid Signalling: Methods and Protocols*, Methods in Molecular Biology, vol. 2816,
https://doi.org/10.1007/978-1-0716-3902-3_5,

Nanoparticles are divided into several categories based on their size, shape, and material characteristics. Some classifications divide nanoparticles into organic categories such as dendrimers, liposomes, and polymeric nanoparticles; or inorganic categories including fullerenes, quantum dots, and gold nanoparticles. Other classifications divide nanoparticles into carbon-based, ceramic, semiconducting, and polymeric nanoparticles. Furthermore, nanoparticles can be hard (such as titania [titanium dioxide], silica [silicon dioxide] particles, and fullerenes) or soft (such as liposomes, vesicles, and nanodroplets) [2, 3].

The significance of these materials was apparent when researchers discovered that their microscopic size can influence a substance's physiochemical qualities and capabilities [3]. NPs consist of three layers: the surface layer, which can be functionalized with a range of small molecules, metal ions, surfactants, and polymers. The shell layer, which is chemically distinct from the core in all aspects, and the core, which is the NP's central section and commonly refers to the NP itself [4].

1.2 Nanoparticles vs. Lipid Particles

Unlike nanoparticles, lipid particles consist of a core of hydrophobic lipids, mainly triglycerides and cholesterol esters. This core is surrounded by a shell of proteins, phospholipids, and free cholesterol. The proteins that make up the shell are known as apolipoproteins, which boost lipoprotein synthesis and function, and they are responsible for structural stability as well as interacting with cellular receptors and enzymes [5]. Because lipids are insoluble in water, they must be stored in a soluble form for optimal transit in the bloodstream. Lipid particles function as carriers for carrying lipids throughout the body in combination with proteins (lipoproteins). These lipoproteins are essential for dietary fat absorption and transport.

Typically, the structure consists of a glycerol backbone, two hydrophobic fatty acid tails, and a hydrophilic phosphate group. Phospholipids are hence amphipathic. The bilayer arrangement of phospholipids in the cell membrane protects the cell and acts as a barrier to some chemicals. The hydrophobic portion is facing in, while the hydrophilic portion is facing out. This configuration aids in controlling which molecules can enter and leave the cell. Small polar and nonpolar molecules, like oxygen and water, can easily diffuse into and out of the cell, for instance. Large, polar molecules like glucose cannot move about freely and so need the assistance of transport proteins [6].

Based on their density, which is determined by the quantities of lipids and proteins they contain, lipid particles are categorized. Higher lipid content results in decreased density, which is inversely correlated with density. Chylomicrons have the lowest density; they are large triglyceride-rich particles that are formed in the small intestine, followed by increasing densities of very-low-density lipoproteins (VLDLs), low-density lipoproteins (LDLs), and high-density lipoproteins (HDLs). The liver produces VLDLs, which are triglyceride-rich particles, while LDL is rich in cholesterol and formed from the combination of VLDL and intermediate-density lipoproteins (IDLs). Finally, HDL particles are rich in cholesterol and phospholipid [5, 6].

1.3 Interaction Between Nanoparticles and Lipids

Lipid particles can interact with nanoparticles through different contexts. For instance, adsorption on surface and surface interactions allow nanoparticles to interact with lipid particles. Charge, lipophilicity, and functional groups on nanoparticle surfaces can all influence how they interact with lipids. Hydrophobic bonds are possible between lipid particles, notably lipoproteins, and hydrophobic nanoparticles. Electrostatic forces can also occur between charged nanoparticles and the charged areas of lipids [7]. The second context would involve mimicking the structure of the cell membrane in which nanoparticles will be coated with lipids (phospholipids) forming a lipid bilayer around the nanoparticle, thus increasing their stability, biocompatibility, and effectiveness. Lipid coatings can provide a protective layer around the nanoparticle core, avoiding aggregation and increasing the time it spends in circulation in the body [8].

Moreover, nanoparticles can be modified using ligands or antibodies that bind to receptors on lipid particles specifically, LDL lipid particles are the most notably used for this interaction. This permits the nanoparticles to be taken up by cells that express these receptors, potentially improving therapeutic or imaging agent delivery that is best used in targeting drug delivery [9–11]. It is possible for lipid transfer to occur between lipid particles and nanoparticles as another type of lipid particle and nanoparticle interaction. Lipids from lipoproteins, for example, can be transferred to lipid-based nanoparticles, causing changes in the composition and properties of the nanoparticles. Enzymes, transporters, or other systems may mediate this transfer [12]. Interaction between lipid particles and nanoparticles can result in the formulation of lipid-based nanoparticles (LNPs). Liposomes, micelles, solid lipid nanoparticles (SLNs), and nanostructured lipid carriers (NLCs) are all examples of nanoparticle formulations that contain lipids. These formulations can ensure the stability and controlled release of the delivered drugs. LNPs can interact with lipid particles, particularly lipoproteins [13, 14].

1.4 Lipid Nanocarrier (LNPs) for Drug Delivery

LNPs are well positioned in the advancement of genomic therapies because they can encapsulate a wide range of pharmaceutical loads, including small molecules, nucleic acids, monoclonal antibodies, imaging agents, and peptides [15]. LNP-based medicines have also been approved by the FDA for treating additional diseases, including COVID-19 vaccination (Spikevax, Comirnaty) and

Amyloidosis (Onpattro), by delivering mRNA and siRNA, respectively [16, 17]. Additionally, different types of LNPs have been designed to deliver different types of drugs, for example, immunosuppressant cancer drugs, HAV viral vaccine, anesthetic drugs, and antiviral HIV-1 drugs [18]. LNPs have various advantages, including protecting drugs from in vivo degradation, increasing solubility and efficacy, allowing targeted drug delivery, modulating drug release, and changing drug biodistribution. They offer sheltering of the API from damaging enzymes and allowing transit across biological membranes. Thus, nanocarriers have the potential to overcome fundamental constraints of existing medicinal agents, such as ineffectiveness, enzymatic degradation, limited bioavailability, and off-target adverse effects.

The size of the nanoparticles influences their stability, biodistribution, cellular uptake, and overall therapeutic efficacy. Smaller LNPs (size 100 nm) have higher cellular uptake and a longer circulation period, whereas bigger LNPs (size >100 nm) have a higher drug-loading capacity but lower cellular uptake [19]. Furthermore, LNP surface charge is a vital metric indicating LNP stability in suspension. A significant surface charge would result in electrostatic repulsion between particles and increase their stability in the dispersion, while a surface charge value close to zero indicates a low surface charge and a stronger tendency for particle aggregation or coalescence, which could contribute to the destabilization of the LNP suspension [20]. Surface morphology analysis of LNPs can provide useful insights into the physical structure and properties of the LNP's outer surface. A well-defined and smooth surface morphology is extremely beneficial for LNPs in biological contexts because it leads to increased stability and reduces the risk of opsonization, which occurs when serum proteins attach to the surface and stimulate immune system clearance. This flat surface reduces undesired interactions with biological components, allowing optimal LNP biodistribution and therapeutic efficacy [21]. LNPs with rough or uneven surface morphology may interact with biological entities more frequently, thereby affecting LNP biodistribution patterns and impacting therapeutic effectiveness [21].

1.5 Examples of Nanoparticles

The first example of LNPs are known as liposomes; these spherical vesicles are made up of lipid bilayers. Their capacity to contain hydrophilic and hydrophobic medical entities within their water core or fatty lipid membrane makes them one of the most extensively utilized drug delivery methods. Drug delivery can be more precise, stable, and solubilized with the help of liposomes [22]. SLNs as another example of nanoparticles formed from waxes or triglycerides, which are solid lipids. They are used to encapsulate pharmaceuticals and for sustained drug delivery. SLNs have benefits such increased medication absorption, improved drug stability, and defense against degradation [23]. While NLCs combine liquid and solid lipids, compared with SLNs, this special structure offers a higher drug-loading capacity and better drug release kinetics. NLCs are utilized to deliver medical drugs, including targeted therapies and pharmaceuticals with poor solubility [24, 25].

Hybrid of lipids and polymer nanoparticles are nanomaterials, which mix lipids and polymers to take advantage of the benefits of both components. The lipid part allows for biological compatibility and drug encapsulation, while the polymer part provides stability and sustained release. They are used in medical treatment delivery, imaging agents, and gene therapy [26]. Another reported example is ethosomes that are lipid-based nanoparticles with significant ethanol concentration. Ethosomes are advantageous for transdermal medication delivery applications because the incorporation of ethanol improves the penetration of pharmaceuticals that are encapsulated through the skin. They play a key role in topical preparations, which enhance medication absorption [27]. The latter example is transferosomes that are lipid vesicles with high flexibility and have the capacity to pass through pores in the skin and other small openings. They help medications get through biological barriers better, such the skin or mucosal surfaces. Gene therapy, vaccine, and transdermal delivery are all possible uses for transferosomes [28].

1.6 Applications of Nanoparticles in Lipid Signaling

Nanoparticle technology is recognized as a promising tool for various biomedical applications in biology. Many innovative techniques are being developed for the detection of analytes, pathogens, and cells using nanoparticles. Furthermore, the utilization of nanoparticles has proven to be effective in the fields of biomedical imaging, biosensing, and as carriers for drug delivery [29].

For instance, SLNs have the potential to disrupt inflammatory pathways. An instance of cationic lipid-based nanoparticles was found to activate the toll-like receptor 4 (TLR4) pathway. Moreover, it was observed that in RAW264.7 macrophages, various SLNs, predominantly composed of glyceroltrimyristate and cetylpyridinium chloride, elicited cytotoxic effects and resulted in reduced levels of interleukin 6 and TNF-α expression [30].

Toll-like receptors (TLRs) like TLR4 are found in immune cells like macrophages. These receptors help the innate immune system produce cytokines and chemokines in response to infection and tissue damage. TLRs are also found in adipocytes, hepatocytes, and skeletal muscle cells [31–33]. These non-immune cells can boost immunity by secreting chemokines and cytokines [34]. Recent studies have examined human, mice, and skeletal muscle cell cultures. Many methods, including gain and loss of function, were used in these studies. These studies found that TLR4 signaling reduces fatty acid oxidation. Thus, this muscle activity may raise blood triglycerides [35]. Fasting and TLR4 loss increased skeletal muscle oxidative capacity and decreased triglycerides and non-esterified free fatty acids [35].

Lipid sensors are used to detect changes in lipid composition and concentration, which are important in lipid signaling pathways: for example, the detection of cardiovascular disease (CVD) biomarkers. The cardiac troponin complex comprises three regulatory proteins, namely troponin I (cTnI) and troponin T (cTnT). These proteins are released into the blood flow as a result of cardiomyocyte injury. The ongoing research of nanoparticle-based diagnostics is focused on the detection of troponins and other biomarkers associated with CVD [36]. For example, a research investigation utilized electrochemical impedance spectroscopy in conjunction with a peptide sequence (CFYSHSFHENWPS) that was deliberately engineered to have a strong binding affinity to cardiac troponin I (cTnI). This peptide sequence was immobilized onto gold nanoparticles, which were in turn coated onto a glassy carbon electrode [37]. The charge transfer resistance that was discovered had a clear positive relationship with the quantity of cTnI, hence facilitating the detection and quantification of the analyte in actual blood samples without the need for labeling. Potentiometric sensors, which utilize the differences in electrical potential between a working electrode and a reference electrode, have also been employed in the diagnosis of (CVD) to quantify the quantity of analytes [38].

Gold nanoparticles are frequently employed in photoacoustic imaging within the field of bioimaging to enhance the visualization of atherosclerotic plaque [39]. This methodology utilizes gold nanoparticles as contrast agents, specifically engineered to selectively bind to vascular cell adhesion molecule 1. This targeting mechanism organizes the accumulation of these nanoparticles in portions of the vasculature that are susceptible to illness. The nanoparticles absorb pulsed laser beams, resulting in thermal expansion and the production of sonic sounds that are detected by an ultrasound transducer [40].

2 Materials

1. Phosphate-buffer saline (10 mM, pH 7.4, and 25 °C).
2. Methanol (highly flammable liquid and vapor. Toxic if swallowed, in contact with skin, or if inhaled. Causes damage to organs (eyes, central nervous system). Purchased from Merck.
3. Chloroform (This volatile liquid exhibits volatility and is under suspicion for its potential carcinogenic and reproductive toxicity. Thus, safety precautions must be implemented carefully.). Purchased from Merck.
4. Ethyl acetate.

5. Nitrogen gas (Contains gas under pressure; may explode if heated. Thus, safety precautions must be implemented carefully.)
6. Dipalmitoyl phosphatidylcholine (DPPC) is provided by Avanti Polar Lipids (Alabaster, Alabama, USA).
7. 1-Palmitoyl-2-oleoyl-sn-glycero-3-phosphocholine (POPC) is supplied by Avanti Polar Lipids, a company based in Alabaster, Alabama, USA.
8. Cholesterol.
9. PEG-2000 di-stearoyl phosphatidylethanolamine (PEG-2000-DSPE).
10. Maleimide PEG-2000 di-stearoyl phosphatidylethanolamine (mal-PEG-2000-DSPE).
11. Liquid nitrogen tank.
12. Rotary vacuum system.
13. Extruder.

3 Methods

The production of nanoparticles can be achieved through various methods, wherein the specific procedure employed is contingent upon the type of nanoparticles and the chosen synthesis technology [41]. The following is a comprehensive overview of the sequential procedures entailed in the synthesis of liposome.

3.1 Liposome Preparation

1. Combine dipalmitoyl phosphatidylcholine (DPPC), cholesterol, PEG-2000 di-stearoyl phosphatidylethanolamine (PEG-2000-DSPE), maleimide PEG-2000 di-stearoyl phosphatidylethanolamine (mal-PEG-2000-DSPE) in molar ratio (1:1:0.2:0.1) within a round-bottom flask. Dissolve them in chloroform/methanol (1:1, v/v) (*See* **Note 1**).
2. Eliminate the chloroform/ methanol from the suspension through rotary evaporation under the inert nitrogen gas at a temperature of 35 °C over a duration of 2 h to form a dry lipid film. (*See* **Note 2**).
3. Rehydrate the lipid film using a 1 ml solution of acetate buffer (pH 7).
4. Stir the liposome solution vigorously and subject it to 10 freeze and thaw cycles using liquid nitrogen. This process resulted in the formation of unilamellar liposomes with a size in the nano range.
5. In order to achieve a reduction in the size of liposomes and homogenization, extrude for 31 times through two stacked polycarbonate membranes with 100 nm pores.

3.2 The Encapsulation of Drugs into Liposome

In order to encapsulate drugs into liposomes, it is necessary to stick to a series of procedural processes. In this study, curcumin and 2,4,6-trichlorphenyl oxalate (TCPO) were chosen as representative medicines.

1. Prepare the stock solution of curcumin and TCPO, which is poorly soluble in water, in HPLC-grade methanol and ethyl acetate, respectively.
2. Generate sample by dissolving POPC, curcumin, and TCPO in a weight ratio of 10:0.5:0.5 (w/w) in a solution of CHCl3/ MeOH in a 1:1 (v/v) ratio.
3. Use a rotary evaporator to remove the chloroform and form a thin lipid film (using rotary vacuum pump for at least 2 h).
4. Once the lipid film is formed, the lipid suspensions should be incubated for about 10–30 min at environment temperature in order to remove any residual solvent.
5. Resuspend the lipid film with phosphate-buffered saline (PBS) containing NaCl 140 mM and EDTA 0.1 mM, and pH 7.4 at 25 °C.
6. Perform vigorous vortexing on the liposome suspension along with crystal beads.

3.3 Extruder Assembly and Extrusion

All of the necessary parts of the extruder are highlighted in Fig. 1. To assemble the extruder apparatus, follow these steps:

1. Start by placing two internal membrane supports upright on a flat surface with the O-rings facing up.
2. Moisten two filter supports with phosphate buffer (pH 7) and place them on the inner-diameter Teflon orifice of the O-rings.
3. Next, assemble the filter and internal membrane supports and insert this assembly into the extruder outer casing with the O-ring facing upward.
4. Cover the filter support and O-ring with a polycarbonate membrane, placing it in the extruder outer casing.
5. On top of the remaining internal membrane support, place a second pair of filter supports pre-wetted with phosphate buffer (pH 7).
6. Carefully insert the second internal membrane support into the casing, ensuring that the O-ring is oriented in a downward position.
7. Install the Teflon bearing into the retainer nut.
8. Attach the retainer nut to the threaded end of the extruder outer casing and carefully tighten it. (*See* **Note 3**).
9. Place the fully constructed extruder apparatus into the platform, ensuring that the hex nut is placed securely in place and the swing-arm clips keep the syringes (Fig. 2).

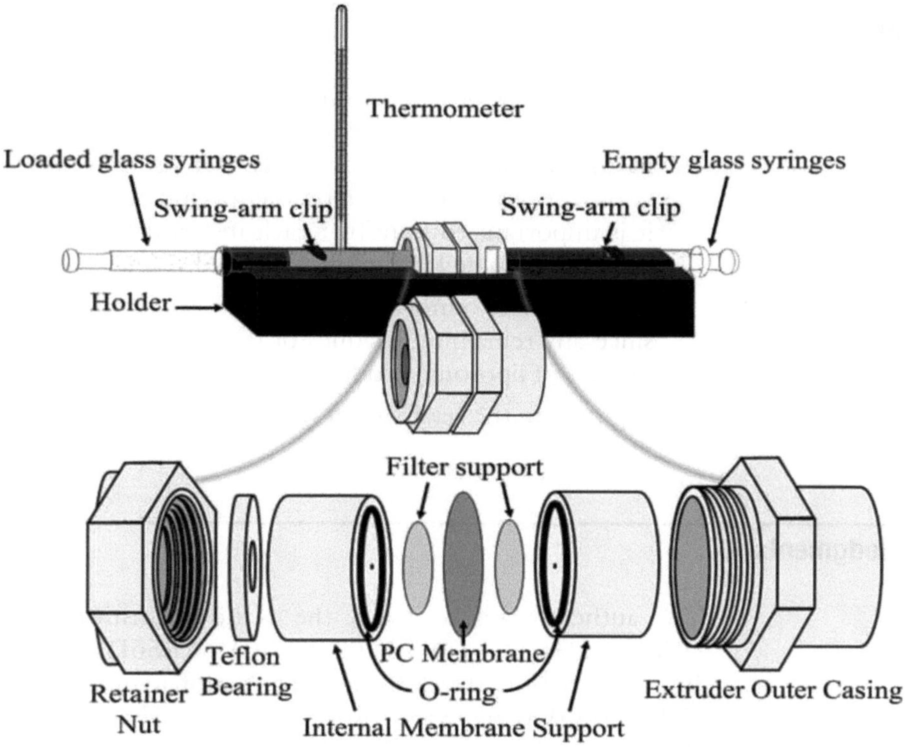

Fig. 1 Schematic of mini-extruder assembly components

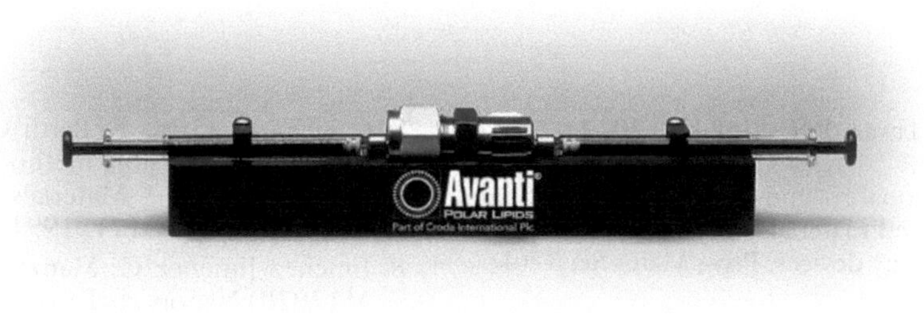

Fig. 2 The finalized configuration of the extruder machine

10. Transfer the processed sample into a gas-tight syringe.
11. Put the lipid-containing syringe into one of the ends of the extruder and the blank syringe into the opposite side.
12. Push down on the plunger of the lipid-containing syringe to transfer the mixture to the unfilled syringe, continuing until the lipid suspension turns clear (around 31 times).
13. Make sure that the final injection of the lipid solution fills the second syringe, in order to prevent contamination with larger fragments.
14. Remove the lipid-filled syringe from the extruder and transfer the lipid solution into a clean glass container.

4 Notes

1. The quantity of chloroform required is contingent upon the lipid solubility. The solubility limit for a certain lipid is provided in the datasheet by Avanti Polar Lipids. The lipid ratio of liposomes is contingent upon the specific experimental design, it is important to note that each individual liposome is composed of a single kind of phospholipid.
2. The complete removal of the solvent is of utmost importance, since any remaining residues of the solvent might disrupt the process of liposome production.
3. Tightening the retainer nut requires no tools and may be done manually.

Acknowledgments

The authors want to thank the National Institutes of Health (R01DE031872, 1R03DE023872-01, 1R56DE027964-01A1-01, NIH S10OD025230) and the National Institutes of Aging (NIA 7-PO1AG039355; 5R01AG056504, 5R01AG060341) for their generous support for the authors during this study. The authors also want to thank the Osteo Science Foundation (Peter Geistlich Research Award) for their generous support.

References

1. Tiwari JN, Tiwari RN, Kim KS (2012) Zero-dimensional, one-dimensional, two-dimensional and three-dimensional nanostructured materials for advanced electrochemical energy devices. Prog Mater Sci 57(4): 724–803
2. Jeevanandam J et al (2018) Review on nanoparticles and nanostructured materials: history, sources, toxicity and regulations. Beilstein J Nanotechnol 9:1050–1074
3. Khan I, Saeed K, Khan I (2019) Nanoparticles: properties, applications and toxicities. Arabian J Chem 12(7):908–931
4. Nomoev AV et al (2015) Structure and mechanism of the formation of core–shell nanoparticles obtained through a one-step gas-phase synthesis by electron beam evaporation. Beilstein J Nanotechnol 6:874–880
5. Feingold KR (2021) Introduction to lipids and lipoproteins. BTI – Endotext
6. Ahmed S, Shah P, Ahmed O (2023) Biochemistry, lipids. BTI – StatPearls
7. Cagliani R, Gatto F, Bardi G (2019) Protein adsorption: a feasible method for nanoparticle functionalization? Materials 12. https://doi.org/10.3390/ma12121991
8. Jiménez-Jiménez C, Manzano M, Vallet-Regí M (2020) Nanoparticles coated with cell membranes for biomedical applications. Biology 9. https://doi.org/10.3390/biology9110406
9. Kularatne RN, Crist RM, Stern ST (2022) The future of tissue-targeted lipid nanoparticle-mediated nucleic acid delivery. Pharmaceuticals 15. https://doi.org/10.3390/ph15070897
10. Magalhães J et al (2019) Lipid nanoparticles biocompatibility and cellular uptake in a 3D human lung model. Nanomedicine 15(3): 259–271
11. Marques AC et al (2020) Functionalizing nanoparticles with cancer-targeting antibodies: a comparison of strategies. J Control Release 320:180–200
12. Fernandes F et al (2021) Critical review of lipid-based nanoparticles as carriers of

neuroprotective drugs and extracts. Nanomaterials 11. https://doi.org/10.3390/nano11030563

13. Hald Albertsen C et al (2022) The role of lipid components in lipid nanoparticles for vaccines and gene therapy. Adv Drug Deliv Rev 188: 114416
14. Zhao Y, Huang L (2014) Chapter Two - lipid nanoparticles for gene delivery. In: Huang L, Liu D, Wagner E (eds) Advances in genetics. Academic, pp 13–36
15. Samimi S et al (2019) Chapter 3 – Lipid-based nanoparticles for drug delivery systems. In: Mohapatra SS et al (eds) Characterization and biology of nanomaterials for drug delivery. Elsevier, pp 47–76
16. Guo S et al (2021) Membrane-destabilizing ionizable lipid empowered imaging-guided siRNA delivery and cancer treatment. Exploration 1(1):35–49
17. Wang Z et al (2022) Nanotechnology-facilitated vaccine development during the coronavirus disease 2019 (COVID-19) pandemic. Exploration 2(5):20210082
18. Wang Y, Grainger DW (2022) Regulatory considerations specific to liposome drug development as complex drug products. Front Drug Deliv 2:901281
19. Puri A et al (2009) Lipid-based nanoparticles as pharmaceutical drug carriers: from concepts to clinic. Crit Rev Ther Drug Carrier Syst 26(6): 523–580
20. Rasmussen MK, Pedersen JN, Marie R (2020) Size and surface charge characterization of nanoparticles with a salt gradient. Nat Commun 11(1):2337
21. Zhigaltsev IV, Cullis PR (2023) Morphological behavior of liposomes and lipid nanoparticles. Langmuir 39(9):3185–3193
22. Nakhaei P et al (2021) Liposomes: structure, biomedical applications, and stability parameters with emphasis on cholesterol. Front Bioeng Biotechnol 9:705886
23. Tenchov R et al (2021) Lipid nanoparticles—from liposomes to mRNA vaccine delivery, a landscape of research diversity and advancement. ACS Nano 15(11):16982–17015
24. Naseri N, Valizadeh H, Zakeri-Milani P (2015) Solid lipid nanoparticles and nanostructured lipid carriers: structure, preparation and application. Adv Pharm Bull 5(3):305–313
25. Yacoub AS et al (2022) Artificial intelligence-assisted development of in situ forming nanoparticles for arthritis therapy via intra-articular delivery. Drug Deliv 29(1): 1423–1436
26. Mukherjee A et al (2019) Lipid–polymer hybrid nanoparticles as a next-generation drug delivery platform: state of the art, emerging technologies, and perspectives. International J Nanomed 14:1937–1952
27. Abdulbaqi IM et al (2016) Ethosomal nanocarriers: the impact of constituents and formulation techniques on ethosomal properties, in vivo studies, and clinical trials. 2016(1178–2013 (Electronic))
28. Fernández-García R et al (2020) Transferosomes as nanocarriers for drugs across the skin: quality by design from lab to industrial scale. Int J Pharm 573:118817
29. Wang EC, Wang AZ (2014) Nanoparticles and their applications in cell and molecular biology. Integr Biol 6(1):9–26
30. Olbrich C, Schöler N, Tabatt K, Kayser O, Müller RH (2004) Cytotoxicity studies of Dynasan 114 solid lipid nanoparticles (SLN) on RAW 264.7 macrophages-impact of phagocytosis on viability and cytokine production. J Pharm Pharmacol 56(7):883–891
31. Lang CH et al (2003) Endotoxin stimulates in vivo expression of inflammatory cytokines tumor necrosis factor alpha, interleukin-1β, -6, and high-mobility-group protein-1 in skeletal muscle. Shock 19(6):538
32. Matsumura T et al (2000) Endotoxin and cytokine regulation of toll-like receptor (TLR) 2 and TLR4 gene expression in murine liver and hepatocytes. J Interferon & Cytokine Res 20(10):915–921
33. Shi H et al (2006) TLR4 links innate immunity and fatty acid–induced insulin resistance. J of Clin Investig 116(11):3015–3025
34. Olefsky JM, Glass CK (2010) Macrophages, inflammation, and insulin resistance. Ann Rev Physiol 72(1):219–246
35. Frisard MI et al (2010) Toll-like receptor 4 modulates skeletal muscle substrate metabolism. Am J Physiol Endocrinol Metab 298(5): E988–E998
36. Sharma S, Jackson PG, Makan J (2004) Cardiac troponins. J Clin Pathol 57(10):1025
37. Wang B et al (2016) Label-free electrochemical impedance peptide-based biosensor for the detection of cardiac troponin I incorporating gold nanoparticles modified carbon electrode. J Electroanalytical Chem 781:212–217

38. Moreira FTC et al (2011) Artificial antibodies for troponin T by its imprinting on the surface of multiwalled carbon nanotubes: its use as sensory surfaces. Biosens Bioelectron 28(1): 243–250
39. Singh P et al (2018) Gold nanoparticles in diagnostics and therapeutics for human cancer. Int J Mol Sci 19:1979. https://doi.org/10.3390/ijms19071979
40. Rouleau L et al (2013) VCAM-1-targeting gold nanoshell probe for photoacoustic imaging of atherosclerotic plaque in mice. Contrast Media Mol Imaging 8(1):27–39
41. Deb A, Vimala R (2018) Camptothecin loaded graphene oxide nanoparticle functionalized with polyethylene glycol and folic acid for anticancer drug delivery. J Drug Deliv Sci Technol 43:333–342

Chapter 6

Market Needs and Methodologies Associated with Patient Lipidomic Diagnoses and Analyses

Ahmed S. Yacoub, Zeinab Fotouhi Ashin, Kamal Awad, Sindhu Guntur, Michael Wilson, Merina Daniel, Pranesh Aswath, Marco Brotto, and Venu Varanasi

Abstract

This chapter conducts an in-depth exploration of the impact of musculoskeletal (MSK) disorders and injuries, with a specific emphasis on their consequences within the older population demographic. It underscores the escalating demand for innovative interventions in MSK tissue engineering. The chapter also highlights the fundamental role played by lipid signaling mediators (LSMs) in tissue regeneration, with relevance to bone and muscle recovery. Remarkably, Prostaglandin E_2 (PGE_2) emerges as a central orchestrator in these regenerative processes. Furthermore, the chapter investigates the complex interplay between bone and muscle tissues, explaining the important influence exerted by LSMs on their growth and differentiation. The targeted modulation of LSM pathways holds substantial promise as a beneficial way for addressing muscle disorders. In addition to these conceptual understandings, the chapter provides a comprehensive overview of methodologies employed in the identification of LSMs, with a specific focus on the Liquid Chromatography-Mass Spectrometry (LC-MS). Furthermore, it introduces a detailed LC MS/MS-based protocol tailored for the detection of PGE_2, serving as an invaluable resource for researchers immersed in this dynamic field of study.

Key words Bone, Lipid mediators, Lipid signaling, Muscle, Bone-muscle cross talk, Lipidomics, Prostaglandin E_2 quantification

1 Introduction

1.1 Epidemiology and Market Considerations for Musculoskeletal (MSK) Tissue Engineering

MSK disorders and injuries affect 1.7 billion people globally and afflict 75% of the US population aged 65 and above. Figure 1 shows the percentages of this age population out of the total US population from 2000 to 2020 as reported by the United Nations Population Division in 2022 revision [1]. The percentage of this population has increased with time, showing a steady increase from 2000 to 2006 and surged from 2007 onward, contrasting the early and late sections of the decade. The United Nations also

Marco Brotto and Sanjoy K. Bhattacharya (eds.), *Lipid Signalling: Methods and Protocols*, Methods in Molecular Biology, vol. 2816, https://doi.org/10.1007/978-1-0716-3902-3_6,

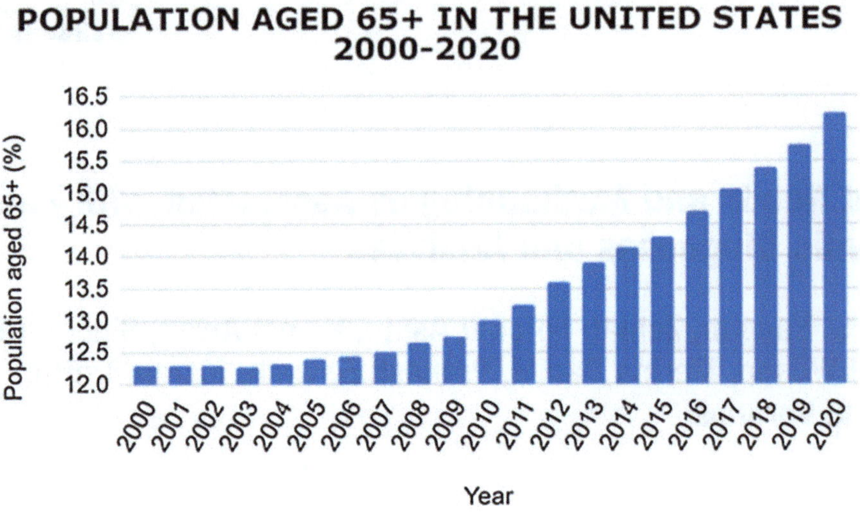

Fig. 1 Population Ages 65 and above. (Source: World Bank https:/data.worldbank.org/indicator/SP.POP.65UP.TO.ZS?end=2020&locations=US&start=2000&view=chart)

predicted a 56% increase in the population aged 60, to 1.4 billion by 2023. These data indicate the growing need to address these disorders and injuries [1].

For example, Fig. 2 shows mortality in Spain in relation to muscular injuries. These injuries are linked to muscular dystrophy, cardiac disorders, cachexia, and orthopedic problems. These injuries and conditions are further exacerbated by aged individuals as seen in Fig. 3. Thus, soft tissue implants are needed to prolong life in men and women with catastrophic muscle loss or even to help repair and regenerate the injured muscle after traumatic injuries. For example, muscle tissue implants can help improve aesthetics and proportions [2]. Harvard created a muscle-contracting implant in April 2023. These implants improved muscle use and patient quality of life. Further, these implants dominate the muscle tissue market, leaving no soft tissue implants for shear force injuries [3].

In the case of bone tissue healing, common diseases are osteoporosis and osteopenia. Osteopenia is a condition that gradually decreases bone mass due to a lack of calcium [4]. It is usually a precursor to osteoporosis, a severe loss of bone mineral density due to drastic changes in bone physiology [5]. Further loss of bone mass is exacerbated by a lack of vitamin D and inadequate protein intake. A study conducted by the Centers for Disease Control and Prevention (CDC) from 2017 to 2018 concluded that 43.1% of people aged 50 and over had osteopenia, and 12.6% had developed osteoporosis. As shown in Fig. 4, most of the population in the study had either disease. These results are concerning, as the

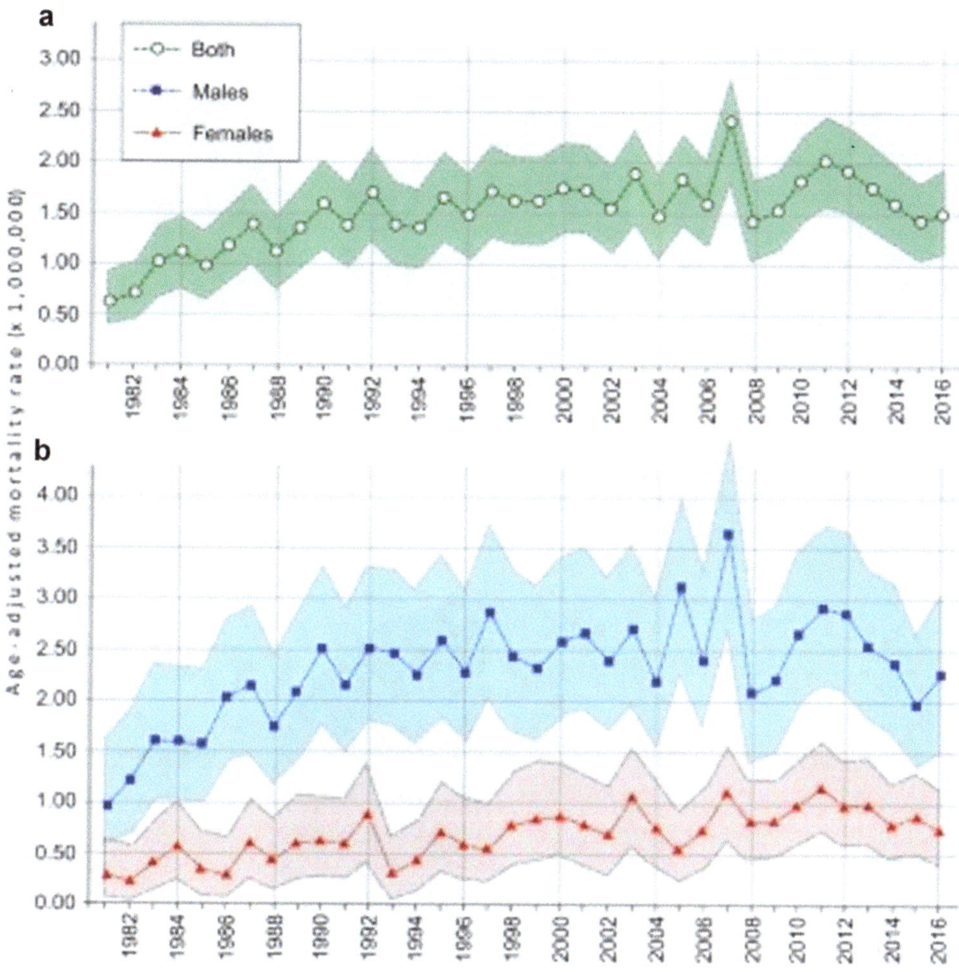

Fig. 2 Muscle injury incidents in Spain from 1982–2016. (**a**) Overall incidents of muscle injuries in both males and females. (**b**) Breakdown of muscle injury incidents between men and women. Citation: https:/www.nature.com/articles/s41598-022-07814-z

demand to care for such a large group of people may exceed the supply of autologous grafts that can stimulate healing.

Reflecting on Fig. 1, the population of individuals over 65 will increase greatly. Given that this information is a conservative estimate, the number of people with bone diseases will exponentially increase as the population ages. The above discussion exemplifies the market need. Yet, the autologous tissue to heal bone and muscle injuries will be in short supply relative to the prevalence of these incidents. Figure 5 shows the global market value for orthopedic implants, including reconstructive joint replacements, spinal implants, dental implants, trauma, and OrthoBiologics [6]. We can predict a proportional trend in the market's growth as a function of time. Reflecting on the discussion above about the exponential increase in the aging population, it is projected that the healthcare

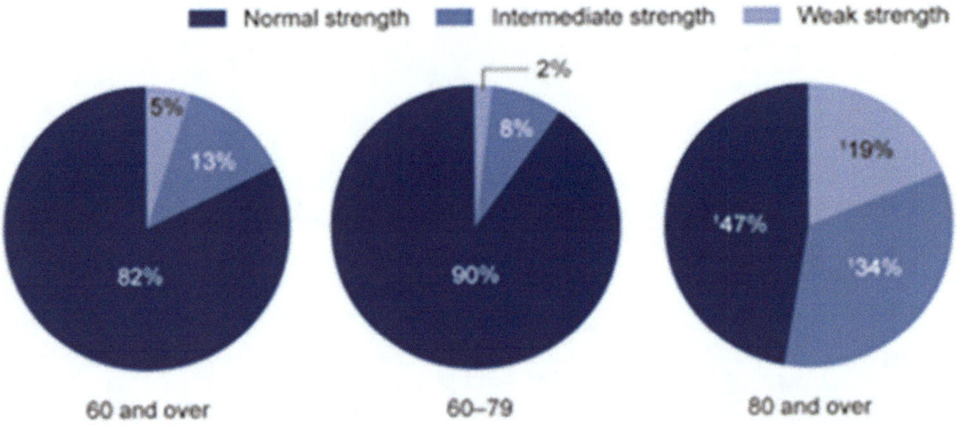

Fig. 3 Correlation of muscle strength to age. From left to right: ages 60 and over, 60 - 79, 80 and over. Muscle strength is reported as normal, intermediate, and weak. (Source: CDC/NCHS, National Health and Nutrition Examination Survey, 2011–2012)

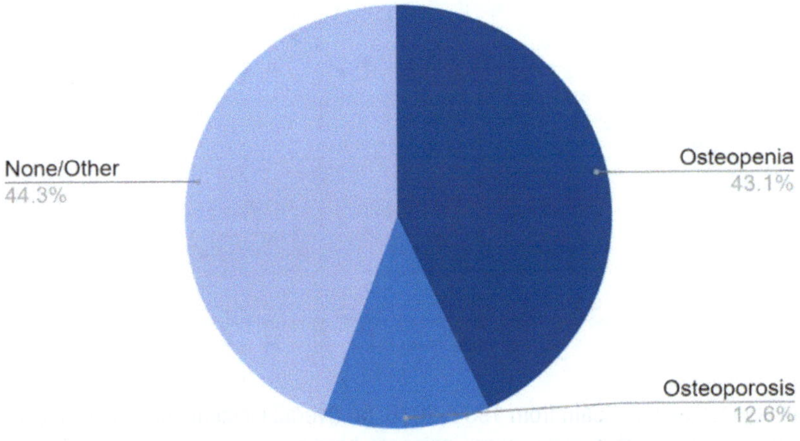

Fig. 4 Percentage of people aged 50+ with bone diseases

demand will markedly outpace the supply of autologous treatments. Based on our estimates on both soft and hard tissue markets, it is likely that current treatment modalities will not meet the future needs of patients. Thus, there is a growing demand to increase the supply of alternative treatments that can target healing mechanisms of MSK tissue regeneration.

1.2 Lipid Signaling Mediators and MSK Tissue Regeneration

Lipids have been well recognized as important signaling molecules that can elicit physiological responses. LSMs, mostly derived from essential polyunsaturated fatty acids (PUFAs), are important mediators that can regulate many physiological responses. These bioactive metabolites are involved in various processes, including inflammation, tissue development, and regeneration. Modifications of the mediators are produced by a crucial group of enzymes that

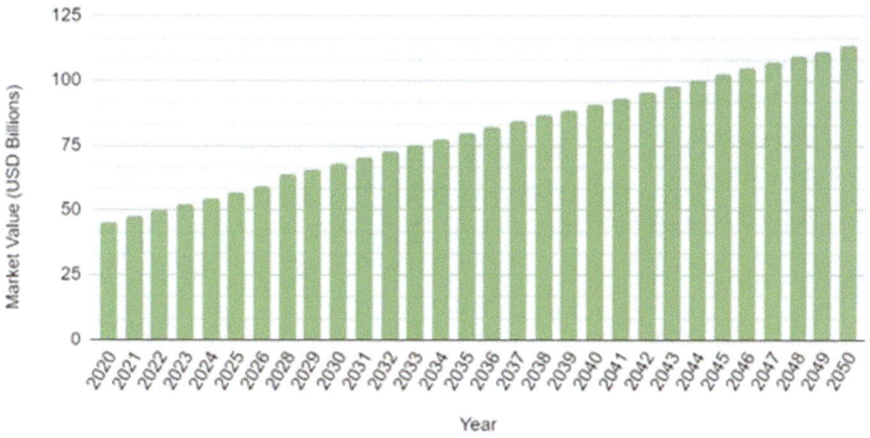

Fig. 5 Orthopedic implants projected market 2020–2050

include phospholipases, prostaglandin synthase, 5-lipoxygenase, phosphoinositide 3-kinase, sphingosine kinase, and sphingomyelinase [7]. Several important LSMs have been well documented for their impact on bone and muscle regeneration. For instance, Prostaglandin E2 (PGE_2) plays a leading role in bone and skeletal muscle repair and regeneration. PGE_2 participates in skeletal muscle repair during all phases of healing including inflammation, regeneration, and fibrosis Fig. 6 [8]. The Cyclooxygenase2/Prostaglandin E2 (COX-2/PGE_2) pathway, which is mostly induced in pathological situations, is important in promoting skeletal muscle regeneration and reducing fibrosis formation [9]. PGE_2 can not only accelerate the fracture healing process via EP2/4 but also promote the repair of muscles and tendons following injury by EP4 (type of receptor for prostaglandin E2) [10]. In the process of muscle repair, PGE_2 can improve muscle function, regulate muscle-derived stem cells, as well as facilitate muscle formation [11, 12]. Additionally, PGE_2 can facilitate the proliferation and differentiation of muscle-derived stem cells (MDSCs) [13].

The above observations reported in muscle have a coupled effect with bone. The Wnt/β-Catenin pathway and PGE_2 key LSMs secreted by osteocytes have been found to enhance in vitro myogenesis and ex vivo muscle contractility. Traumatic bone injuries, such as open bone fractures or those resulting from surgical trauma, have a detrimental impact on the adjacent muscles, necessitating the need for tissue repair in both compartments. The skeletal muscle acts as a flap or supplementary periosteum to vascularize bone defects and boost satellite cell osteogenic potential [14].

Modulating the release of bioactive LSMs could be a potential strategy for promoting muscle regeneration and repair. Targeting

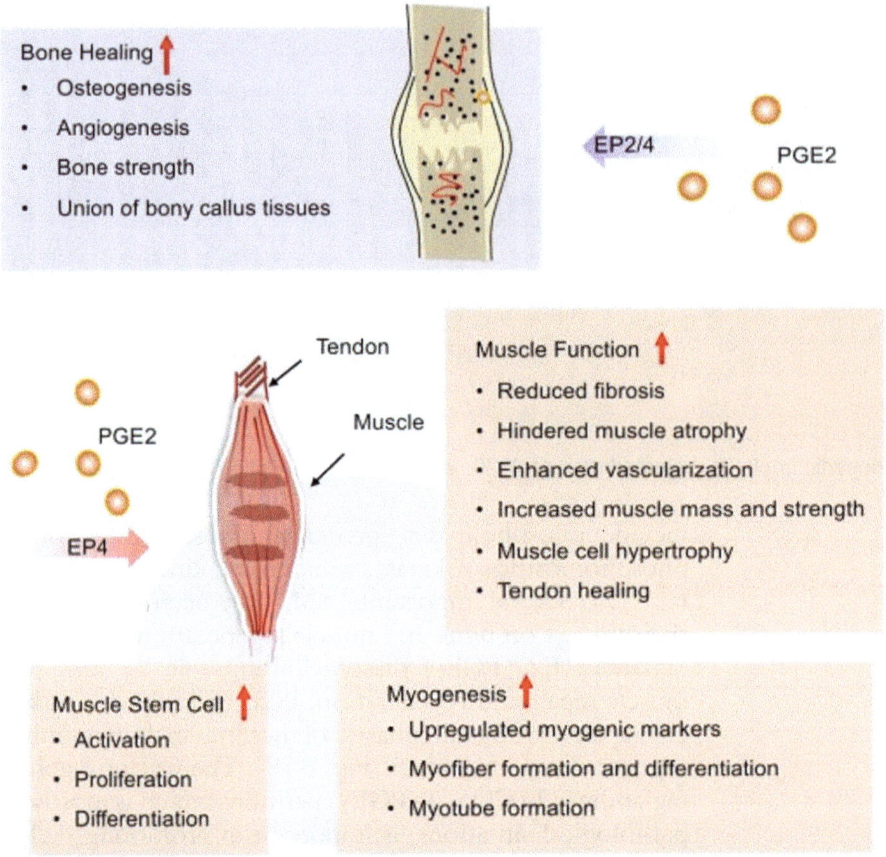

Fig. 6 Prostaglandin E2 (PGE2) plays a role in the process of skeletal muscle repair throughout all stages of healing, encompassing inflammation, regeneration, and fibrosis

COX-1/2 pathways could be a promising approach for treating muscle disorders characterized by impaired myogenesis, such as muscular dystrophy [15]. Production of specific LSM plays a key role in enhancing or suppressing certain aspects of myogenesis, such as the activation of satellite cells, proliferation of myoblasts, and myogenic differentiation. Therefore, COX-1 and COX-2 are the key enzymes to synthesize prostaglandins (PGs) and thromboxane A2 (TXA2) from arachidonic acid (AA) [15].

The above examples of lipid signaling mediators illustrate their crucial role in regulating the regeneration of new bone and muscle during tissue repair. It will be necessary for new healing strategies to target these lipids signaling mediators to help promote improved outcomes for patients suffering from a lack of rapid or proper bone and muscle healing.

2 Musculoskeletal Lipid Signaling Mediators

In this section, we cover lipid signal mediators involved in the regulation of bone and muscle homeostasis and regeneration.

2.1 Bone Lipid Signaling Mediators

Bone LSMs are a class of bioactive molecules that play significant roles in the regulation of bone metabolism and signaling processes. These molecules play a regulatory role in multiple cellular processes within bone tissue, encompassing bone formation, bone resorption, as well as the functions of osteoblasts and osteoclasts [16]. Several bone LSMs have been identified, including prostaglandins, leukotrienes, sphingosine-1-phosphate (S1P), lysophosphatidic acid (LPA), and ceramides.

2.1.1 Prostaglandins (PGs)

Prostaglandins are a group of lipid compounds called eicosanoids. These lipids are derived from arachidonic acid [17]. PGs are produced by the action of cyclooxygenase enzyme (COX) in different cells, such as immune cells, smooth muscle cells, osteocytes, and platelets [18]. The effects of PGs are mediated through G-protein-coupled receptors known as EP1, EP2, EP3, and EP4 receptors. These receptors are expressed in bone cells like osteoblasts and osteoclasts [19]. Prostaglandin E2 (PGE_2) is particularly noteworthy as it promotes differentiation, collagen synthesis, mineralization and also enhances activity and bone resorption [20].

Several cytokines, growth factors, and hormones have been linked to increased PGE_2 production in osteoblasts by influencing cPLA2a, COX 2, or mPGES 1 enzymes [21]. Among these molecules, some like IL 1 and IL 6 have been found to stimulate bone resorption while others like IL 4 and IL 13 inhibit bone resorption by suppressing COX 2 PGE_2 production in mouse osteoblasts [22, 23]. These studies consistently emphasize the role of cPLA2a/COX 2/mPGES 1 mediated PGE_2 production by osteoblasts, in bone resorption.

Despite its involvement in bone resorption through the differentiation of osteoclasts mediated by osteoblasts, PGE_2 also promotes bone formation when tested on cultures consisting solely of osteoblasts. For instance, PGE_2 stimulates the development of bone nodules, which are structures created in vitro to resemble calcification that occurs naturally in living organisms. Additionally, it enhances the activity of alkaline phosphatase (ALP), which serves as a marker for differentiation, in rat calvarial osteoblasts and mouse osteoblastic MC3T3 E1 cells [24].

2.1.2 Leukotrienes

Leukotrienes are synthesized from arachidonic acid via the enzymatic action of lipoxygenases. They are classified into several types, including leukotriene B4 (LTB4), leukotriene C4 (LTC4), leukotriene D4 (LTD4), and leukotriene E4 (LTE4) [25]. Leukotrienes act on specific receptors, such as leukotriene B4 receptors (BLT1 and BLT2) receptors for LTB4. In bone, leukotrienes are involved in regulating osteoclast differentiation and bone resorption. LTB4 promotes osteoclast formation and activity [26]. Leukotriene inhibitors, such as montelukast, are used to treat asthma and allergic rhinitis but may also impact bone metabolism [27].

2.1.3 Sphingosine-1-Phosphate (S1P)

S1P is a bioactive sphingolipid that acts as a potent signaling mediator. It is generated through the phosphorylation of sphingosine-by-sphingosine kinases. S1P binds to a family of G-protein-coupled receptors known as S1P receptors (S1P1-S1P5). In bone, S1P regulates osteoblast and osteoclast functions [28]. It promotes osteoblast proliferation, differentiation, and mineralization through S1P1 and S1P2 receptors. S1P also influences osteoclastogenesis and bone resorption via S1P1 and S1P3 receptors. Moreover, S1P signaling contributes to skeletal development and bone homeostasis [29].

2.1.4 Lysophosphatidic Acid (LPA)

LPA is generated through the hydrolysis of lysophospholipids, such as lysophosphatidylcholine or lysophosphatidylethanolamine. LPA acts on some receptors, including LPA1-LPA6 receptors. In bone, LPA affects osteoblast and osteoclast activities [30]. It promotes osteoblast proliferation, differentiation, and mineralization through LPA1 and LPA2 receptors. LPA also stimulates osteoclast differentiation and bone resorption through LPA2 and LPA3 receptors. Dysregulated LPA signaling has been implicated in bone-related disorders, including osteoporosis and rheumatoid arthritis [31].

2.2 Muscle Lipid Signaling Mediators

Lipoxins, Resolvins, Protectins, and Maresins are involved in inflammation resolution after skeletal muscle ischemia-reperfusion injury. Lipid mediators coordinate tissue homeostasis to mend tissues and reduce inflammation. Arachidonic acid produces anti-inflammatory lipoxins. They regulate macrophage polarization and protect tissues from oxidative stress. Omega-3 fatty acid-derived resolvins promote solving. Resolvins before ischemia have been shown to inhibit polymorphonuclear leukocyte (PMN) infiltration. In addition, omega-3 fatty acid-derived protectins and maresins reduce inflammation and aid in tissue regeneration. These lipid mediators help repair skeletal muscle tissues and reduce inflammation. In general, lipid mediators exert a considerable influence on the promotion of tissue healing and the mitigation of inflammation in skeletal muscle tissue [32].

2.3 Lipid Signaling Mediators Involved in Bone Muscle Cross Talk

Bone–muscle cross talk is a complex and dynamic process. Besides the mechanical interaction between bone and muscle, there is an endocrine cross talk involving various signaling molecules and lipid mediators such as PGs, sphingolipids, endocannabinoids, and phospholipids. For example, signaling lipid mediators released from 5 lipoxygenase (5LO) pathways influence both bone and muscle cells, interfering in their proliferation and differentiation capacities. Various research investigations have centered on the importance of LSMs, particularly PGs, in the cross talk between bone and muscle cells. Studies indicated that cyclooxygenases (COX), specifically COX-1 and COX-2, orchestrate LSM synthesis from AA. COXs also interact with lipoxygenases (LOX) and Cytochromes P450 (CYPs) to alter lipid mediator signaling [14]. Targeted siRNAs knocked down COX-1 and COX-2 in mouse primary myoblasts and C2C12 cells. They examined how these enzymes affected myogenic development and LSM signaling in skeletal muscle cells. LC-MS/MS and targeted lipidomics were used to determine lipid profile alterations after COX-1 and COX-2 inhibition during myogenesis [15].

3 Materials and Methods

3.1 Materials

Cayman Chemical (MI, USA) provided the reference standard for PGE_2 and its deuterated analogs, PGE_2-d4. Gibco, Life Technologies (NY, USA) supplied Hepes balanced salt solution and Dulbecco's modified Eagle's medium (DMEM). Atlanta Biologicals (GA, USA) furnished fetal bovine serum (FBS) and all other necessary cell culture reagents. Fisher Scientific (NJ, USA) offered LC-MS/MS grade acetonitrile and methanol. Sigma-Aldrich (MO, USA) provided formic acid. Deionized water was generated in-house using a Milli-Q Gradient water purification system (MA, USA).

3.2 Analytical Techniques for Identifying Lipid Signaling Mediators

Several techniques have been employed for the detection and quantification of LSMs. Several frequently employed techniques in the field include Liquid Chromatography-Mass Spectrometry (LC-MS), Gas Chromatography-Mass Spectrometry (GC-MS) [33], Nuclear Magnetic Resonance (NMR) Spectroscopy, or Enzyme-Linked Immunosorbent Assay (ELISA). In this section, we will discuss the LCMS as a superior technique for detecting these LSMs.

LC–MS/MS-Based Method for the Detection of the Lipid Signaling Mediators

This protocol for the detection of Prostaglandin E2 (PGE_2) is an example for the LSMs utilizing LC-MS/MS method as mentioned in a previous study by Gandhi et al. [34]. Figure 7 shows the basic conceptualization of how these markers can be identified from several types of biological experiments.

I. *Method Development: LC-MS/MS Optimization*

The LC-MS/MS system employed in this technique consisted of a triple quadrupole mass spectrometer operating in multiple reaction monitoring mode (TSQ Quantum, Thermo Scientific, CA, USA). An elevated ESI operated in negative mode. The spray voltage was adjusted to 2000 kV, and the vaporizer temperature was maintained at 450 °C. The cycle time was set at 0.3 s, while the capillary temperature was established at 325 °C. Sheath gas and

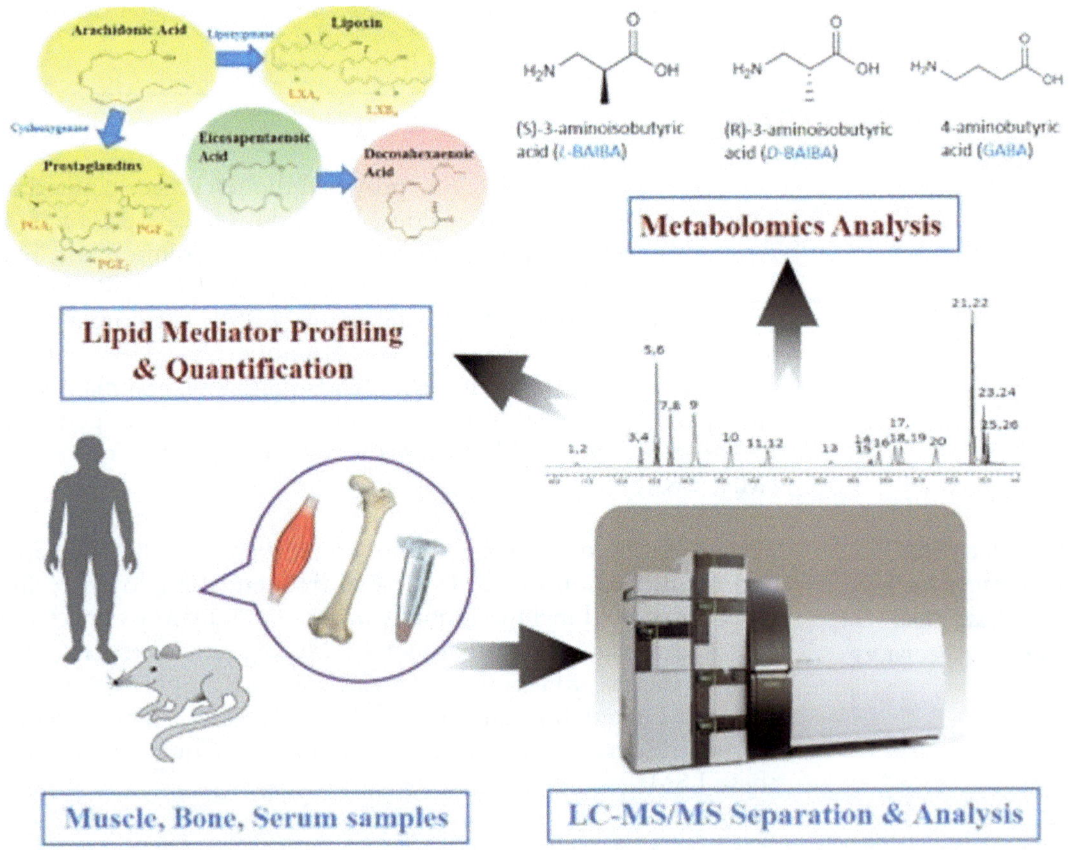

Fig. 7 Schematic representation illustrates the stepwise process of lipidomic analysis using Liquid Chromatography-Mass Spectrometry (LGMS). The method encompasses sample preparation, chromatographic separation, mass spectrometry, and data analysis, enabling the precise identification and quantification of LSMs

AUX gas pressures were configured at 40 and 20 psi, respectively. The analyte underwent full-scan mode to detect its molecular ion peaks, followed by collision-activated ionization on the molecular ion peak using MS/MS mode.

For each analyte, the tube lens and collision energies were fine-tuned individually. To prevent unwanted detector saturation, a divert valve was employed to collect data from 1 to 4.6 min. Details of the molecular ion and fragment ion peaks, along with tube lens settings and collision energies, are obtained. Data processing utilized Xcalibur software version 3.0.63.

Chromatographic separation was executed on a Waters Acquity UPLC system (MA, USA). Analyte was separated on a Kinetex C18 column (2.1 × 50 mm) with an internal particle size of 2.6 μm (Phenomenex, CA, USA). To accelerate run times, a linear gradient comprising 0.1% formic acid in Milli Q Gradient water and 0.1% formic acid in acetonitrile was employed. Efficient resolution was achieved by sustaining an LC flow of 0.45 mL (about 0.02 oz)/min, following this gradient: 0–3 min at 30%, 3–3.65 min transitioning from 30% to 90%, 3.65–4.8 min at 90%, and 4.8–5 min shifting from 90% to 30%. The total run time was 5 min, and a 20 μL sample injection volume was used. The column temperature was stabilized at 20 °C.

II. *Standard Preparation*

1. Standards were procured in the form of solutions at concentrations of either 100 μg/mL or 1 mg/mL.
2. Working stock solutions at a concentration of 1 μg/mL were prepared using acetonitrile as the solvent and then stored at a temperature of −80 °C.
3. In the case of in vitro samples, calibration curves were established within Dulbecco's modified Eagle's medium (DMEM) supplemented with 2% fetal bovine serum (FBS), ranging from 0 to 10 ng/mL.
4. These samples were also spiked with a 10 ng/mL concentration of a deuterated analog, which functioned as an internal standard (IS).

III. *Carry-Over Assessment*

Potential residual sample carry-over from the injection needle was evaluated by injecting blank samples after the injection of the high calibrator standard (10 ng/mL). A comparison was made between the peak area of the blank sample and the peak area of the samples at the limit of quantification (LOQ) for each analyte. To prevent any inadvertent sample carry-over, it was necessary for the peak area of the blank sample to remain below 20% of the peak area observed in the LOQ samples.

IV. *LC-MS/MS Sample Preparation*

In vitro samples (Cells suspension or collected culture media):

1. Ninety microliters of cell culture supernatants were combined with 10 μL of an internal standard (IS) mixture, which contained deuterated PGE_2 at a concentration of 100 ng/mL and was prepared using acetonitrile as the solvent.
2. The samples underwent protein precipitation by adding 300 μL each of acetonitrile and ethyl acetate. Following a brief vortexing step of approximately 30 s, the samples were subjected to centrifugation at 3300 rpm for 20 min at 4 °C.
3. The resulting supernatant was carefully collected into a clean tube. Subsequently, the supernatant was dried under N2 gas at 40 °C and then reconstituted using a 30% acetonitrile solution containing 0.01% formic acid in water, yielding a volume of 125 μL.
4. A 20-μL portion of this reconstituted solution was injected onto the LC column for analysis.

Tissue samples (Soft tissue specimens) [35]:

1. The solid tissues were homogenized using the bead beating method.
2. The homogenate was agitated on ice and kept away from light for 1–2 h.
3. Then, centrifuged at 6000× *g* at 4 °C for 10 min.
4. The supernatant then transferred to a 1.5 mL (about 0.05 oz) centrifuge tube, and all solvents were removed using a centrifugal evaporator.
5. Dried extracts stored at −80 °C instantly for LC– MS/MS analysis.

V. *Data analysis*

To ascertain the concentration of unknown samples, the proportions of analyte peak area to IS peak area (y-axis) were graphed against the concentration (x-axis). By utilizing least square regression analysis with 1/X^2 as the weight factor, calibration curves were generated for each prostaglandin. Statistical significance was assessed through Student's t-test with a significance threshold set at $p < 0.05$, employing GraphPad Prism (6.0). The results are presented as mean ± standard deviation (SD), with a total of 4–5 samples per group for in vitro assays.

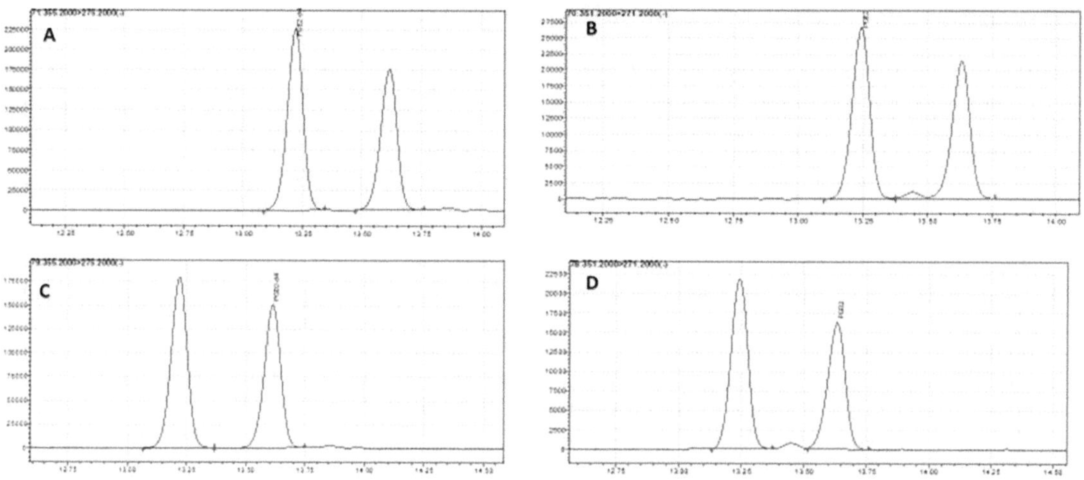

Fig. 8 Chromatograms of some of the detected PGs and their deuterated standards (**a**) PGE2-d4, (**b**) PGE2, (**c**) PGD2-d4 and (**d**) PGD2

Table 1
Identified PGs with their retention time and MS parameters

				MS Parameters		
PGs LMs	**Retention Time (mm)**	**MRM**	**m/z**	**Q1(V)**	**CE(Q)**	**Q3(V)**
PGE2-d4	*13.25*	*MRM(−)*	*355.20>275.20*	*21.0*	*17.0*	*17.0*
PGE2	*13.25*	*MRM(−)*	*351.20>271.20*	*21.0*	*16.0*	*18.0*
PGD2-d4	*13.6*	*MRM(−)*	*355.20>275.20*	*19.0*	*17.0*	*12.0*
PGD2	*13.6*	*MRM(−)*	*351.20>271.20*	*22.0*	*18.0*	*18.0*

VI. *Examples of the data from the LC-Ms analysis*

Figure 8 shows the LC-MS analysis of Prostaglandins E2, D2 and their deuterated analogs PGE_2 -d4 and PGD2-d4. Table 1 shows examples of the parameters used after method optimization.

Acknowledgments

The authors want to thank the National Institutes of Health (R01DE031872, 1R03DE023872-01, 1R56DE027964-01A1-01, NIH S10OD025230) and the National Institutes of Aging (NIA 7-PO1AG039355; 5R01AG056504, 5R01AG060341) for their generous support for the authors during this study. The authors also want to thank the Osteo Science Foundation (Peter Geistlich Research Award) for their generous support.

References

1. Division UNP Population ages 65 and above (% of total population). 2022. Available from: https://data.worldbank.org/indicator/SP.POP.65UP.TO.ZS?view=chart
2. Montemurro P et al (2023) Anatomical variations of the pectoralis muscle and its importance for breast implant surgery. Plast Reconstr Surg Glob Open 11(1):e4761
3. Tinken TM et al (2010) Shear stress mediates endothelial adaptations to exercise training in humans. Hypertension 55(2):312–318
4. Varacallo M, Seaman TJ, Jandu JS, et al. Osteopenia. [Updated 2023 Aug 4]. In: StatPearls [Internet]. Treasure Island (FL): StatPearls Publishing; 2023 Jan-. Available from: https://www.ncbi.nlm.nih.gov/books/NBK499878/
5. Föger-Samwald U et al (2020) Osteoporosis: pathophysiology and therapeutic options. EXCLI J 19:1017–1037
6. Precedence Research Orthopedic Implants Market Size T, Growth, Report 2030. https://www.precedenceresearch.com/orthopedic-implant-market
7. Wymann MP, Schneiter R (2008) Lipid signalling in disease. Nat Rev Mol Cell Biol 9(2): 162–176
8. Cheng H et al (2021) Role of prostaglandin E2 in tissue repair and regeneration. Theranostics 11(18):8836–8854
9. Shen W et al (2008) Interaction between macrophages, TGF-beta1, and the COX-2 pathway during the inflammatory phase of skeletal muscle healing after injury. J Cell Physiol 214(2):405–412
10. Xie C et al (2009) Rescue of impaired fracture healing in COX-2−/− mice via activation of prostaglandin E2 receptor subtype 4. Am J Pathol 175(2):772–785
11. Li M et al (2003) A novel, non-prostanoid EP2 receptor-selective prostaglandin E2 agonist stimulates local bone formation and enhances fracture healing. J Bone Miner Res 18(11): 2033–2042
12. Pajarinen J et al (2019) Mesenchymal stem cell-macrophage crosstalk and bone healing. Biomaterials 196:80–89
13. Dong Y et al (2020) Effects of SW033291 on the myogenesis of muscle-derived stem cells and muscle regeneration. Stem Cell Res Ther 11(1):76
14. Biguetti CC, Couto MCR, Silva ACR, JVTC S, Rosa VM, Shinohara AL, Andreo JC, MAH D, Wang Z, Brotto M, Matsumoto MA (2020) New surgical model for bone–muscle injury reveals age and gender-related healing patterns in the 5 lipoxygenase (5LO) knockout mouse. Front Endocrinol 11:484. https://doi.org/10.3389/fendo.2020.00484
15. Mo C et al (2019) Multi-staged regulation of lipid signaling mediators during myogenesis by COX-1/2 pathways. Int J Mol Sci 20(18)
16. Kita Y, Tokuoka SM, Shimizu T (2017) Mediator lipidomics by liquid chromatography-tandem mass spectrometry. Biochim Biophys Acta Mol Cell Biol Lipids 1862(8):777–781
17. Park JY, Pillinger MH, Abramson SB (2006) Prostaglandin E2 synthesis and secretion: the role of PGE2 synthases. Clin Immunol 119(3): 229–240
18. Morteau O (2001) Prostaglandins and inflammation: the cyclooxygenase controversy. In: Górski A, Krotkiewski H, Zimecki M (eds) Inflammation. Springer, Dordrecht, pp 67–81
19. An Y, Yao J, Niu X (2021) The signaling pathway of PGE_2 and its regulatory role in T cell differentiation. Mediat Inflamm 2021: 9087816
20. Kawaguchi H et al (1995) The role of prostaglandins in the regulation of bone metabolism. Clin Orthop Relat Res 313:36
21. Hikiji H et al (2008) The roles of prostanoids, leukotrienes, and platelet-activating factor in bone metabolism and disease. Prog Lipid Res 47(2):107–126
22. Prystaz K et al (2018) Distinct effects of IL-6 classic and trans-signaling in bone fracture healing. Am J Pathol 188(2):474–490
23. McCormick SM, Heller NM (2015) Commentary: IL-4 and IL-13 receptors and signaling. Cytokine 75(1):38–50
24. Nagata T et al (1994) Effect of prostaglandin E2 on mineralization of bone nodules formed by fetal rat calvarial cells. Calcif Tissue Int 55(6):451–457
25. Cuzzo B., Lappin SL, Leukotrienes. [Updated 2023 Aug 14]. In: StatPearls [Internet]. Treasure Island (FL): StatPearls Publishing; 2023 Jan-. Available from:. https://www.ncbi.nlm.nih.gov/books/NBK526114/
26. Tian W et al (2020) Leukotrienes in tumor-associated inflammation. Front Pharmacol 11
27. Scow DT, Luttermoser GK, Dickerson KS (2007) Leukotriene inhibitors in the treatment of allergy and asthma. Am Fam Physician 75(1):65–70
28. Cartier A, Hla T (2019) Sphingosine 1-phosphate: lipid signaling in pathology and therapy. Science 366(6463):eaar5551

29. Meshcheryakova A, Mechtcheriakova D, Pietschmann P (2017) Sphingosine 1-phosphate signaling in bone remodeling: multifaceted roles and therapeutic potential. Expert Opin Ther Targets 21(7):725–737
30. Hao Y et al (2020) Lysophospholipids and their G-coupled protein signaling in Alzheimer's disease: from physiological performance to pathological impairment. Front Mol Neurosci 13:58
31. Sims SM et al (2013) Lysophosphatidic acid: a potential mediator of osteoblast-osteoclast signaling in bone. Biochim Biophys Acta 1831(1): 109–116
32. Barnig C et al (2022) Resolution of inflammation after skeletal muscle ischemia-reperfusion injury: a focus on the lipid mediators lipoxins, resolvins, protectins and maresins. Antioxidants (Basel) 11(6):1213
33. Panthi S et al (2019) Detection of lipid mediators of inflammation in the human tear film. Eye Contact Lens 45(3):171–181
34. Gandhi AS et al (2017) Quantitative analysis of lipids: a higher-throughput LC-MS/MS-based method and its comparison to ELISA. Future Sci OA 3(1):Fso157
35. Wang Z et al (2023) Mass spectrometry approaches for detection and determination of prostaglandins from biological samples. In: Bhattacharya SK (ed) Lipidomics: methods and protocols. Springer, New York, pp 299–311

Chapter 7

Measurement of Intracellular Ca^{2+} for Studying GPCR-Mediated Lipid Signaling

Yan Chang, James Y. Yi, Marco Brotto, and Zui Pan

Abstract

Intracellular Ca^{2+} can be conveniently monitored by sensitive Ca^{2+} fluorescent dyes in live cells. The Gαq involved lipid signaling pathways and, thus, can be studied by intracellular Ca^{2+} imaging. Here we describe the protocols to measure intracellular Ca^{2+} for studying PEG2-EP1 activity in esophageal smooth muscle cells. The ratiometric Fura-2 imaging provides quantitative data, and the Fluo-4 confocal microscopic imaging has high-spatial resolution.

Key words PGE-2, G-protein-coupled receptor, Second messenger, Lipid signaling mediator, Ca^{2+} signaling, Esophageal smooth muscle, Ratiometric dye, Fura-2 AM, Fluo-4 AM

1 Introduction

There are a variety of lipid signaling mediators that act upon G-protein-coupled receptors (GPCRs) such as prostaglandins and lysophospholipids [1]. These GPCRs play important roles in many physiological and pathophysiological processes in different tissues. Therefore, their ligands have been searched for and evaluated as potential therapeutic drugs for the pertinent diseases. Activated GPCRs trigger various signal transduction pathways by releasing different second messengers depending on the type of G protein α subunit (Gα) underneath the plasma membrane, namely Gαs, Gαi, Gαq, and $G_1\alpha_{2/13}$ [2]. Among them, Gαq mediates the generation of two second messengers, diacylglycerol (DAG) and inositol 1,4,5-trisphosphate (IP_3). The latter subsequently results in the release of intracellular calcium (Ca^{2+}) from the endoplasmic reticulum (ER) to the cytosol. Intracellular Ca^{2+} can be conveniently monitored by sensitive Ca^{2+} fluorescent dyes in live cells, so the measurement of intracellular Ca^{2+} has become a standard method for studying Gαq involved ligand-GPCR activity.

Marco Brotto and Sanjoy K. Bhattacharya (eds.), *Lipid Signalling: Methods and Protocols*, Methods in Molecular Biology, vol. 2816, https://doi.org/10.1007/978-1-0716-3902-3_7,

As a major metabolite of arachidonic acid, prostaglandin E2 (PGE2) exists in various tissues and participates in numerous physiological and pathophysiological processes [3, 4]. These include tumor occurrence, invasion and metastasis, angiogenesis, chronic inflammation, tumor immunity, and cell apoptosis. PGE2 carries out its lipid signaling function through four GPCRs: EP1, EP2, EP3, and EP4 [5]. EP1 is coupled with the Gαq subunit, and its activation results in elevation of intracellular Ca^{2+}. EP2–EP4 are coupled with either Gαs or Gαi for the cAMP pathway [6]. Depending on the expression of the specific type of EPs, PGE2 may even induce opposite effects in any given cell. For example, PGE2 can regulate either contraction or relaxation in smooth muscles [7]. In this chapter, the protocols for the measurement of intracellular Ca^{2+} by using two common Ca^{2+} probes, i.e., Fura-2 AM (a ratiometric dye for quantitative measurement) and Fluo-4 AM (a fast-responding dye, suitable for confocal microscopy) are provided. Live cell imaging revealed the changes in intracellular Ca^{2+} from responding to PGE2 treatment in cultured human esophageal smooth muscle cells.

2 Materials

All solutions are made using ultrapure water, which are purified by Barnstead™ MicroPure™ Water Purification System (Thermo Scientific, USA). All reagents are prepared at room temperature and stored either at 4 °C or − 20 °C as indicated. The waste is disposed following Institutional Chemical Environmental Management regulations.

2.1 Balanced Salt Solution (BSS)

BSS solution is prepared as follows: 140 mM NaCl, 2.8 mM KCl, 2 mM $MgCl_2$, 10 mM HEPES, 0.5 mM EGTA; pH 7.2 adjusted by 5 M NaOH. It is stored at 4 °C.

BSS-2Ca^{2+} solution: 2 mM $CaCl_2$ is added into BSS.

The solutions are warmed up to room temperature prior to use (*see* **Note 1**).

2.2 Ca^{2+} Fluorescent Dyes

Fura-2 AM and Fluo-4 AM esters are obtained from Biotium, Inc. (CA, USA). Both dyes are dissolved in DMSO at the stock concentration of 5 mM and stored at −20 °C (*see* **Note 2**).

2.3 PGE2 Stock

PGE2 is dissolved in DMSO at the concentration of 50 mM and stored at −20 °C. The PGE2 stock solution can be diluted into BSS-2Ca^{2+} solution to reach its working concentrations during experiments.

3 Methods

3.1 Cell Culture

The esophageal smooth muscle (ESM) cells were a generous gift from Dr. Rhonda Souza, which were obtained during Heller myotomy from organ donor controls at Baylor Scott and White Research Institute [8, 9]. The primary culture was established as previously described [10]. ESM cells are cultured in a 5% CO_2 atmosphere at 37 °C with mixed H-DMEM medium (Corning, USA) containing 10% fetal bovine serum (FBS, VWR, USA), 25 ng/mL FGF (Pepro Tech, USA), 10 ng/ml EGF (sigma, USA), 5 μg/mL insulin (sigma, USA), 0.4 μg/mL hydrocortisone and 1% penicillin/streptomycin (Corning, USA) [9].

3.2 Fura-2 AM Dye Loading

1. Seed the cells in glass-bottom dishes with #1.5 (MatTek, US) and culture the cells for at least 24 h prior to Ca^{2+} measurement.
2. Remove the cell culture medium and rinse with BSS-2Ca^{2+} solution.
3. Add 1 mL BSS-2Ca^{2+} solution containing 5 μM Fura-2 AM ester into the Petri dish (*see* **Note 3**) [11].
4. Incubate the cells for 40 min at 37 °C and then put the Petri dish at room temperature for another 15 min to allow ester hydrolysis (*see* **Note 4**).
5. Rinse with BSS-2Ca^{2+} twice and maintain the cells in BSS-2Ca^{2+} solution.

3.3 Ratiometric Ca^{2+} Measurement by the Spectrofluorometer

1. Mount the dish on the stage of Nikon TE200 inverted microscope, which is connected to a dual-wavelength spectrofluorometer (Horiba Scientific, NJ) with excitation wavelengths at 352 and 378 nm and emission at 510 nm (*see* **Note 5**) [12].
2. Set up a gravity perfusion system with different solutions: BSS-2Ca^{2+}, BSS-2Ca^{2+}/5 μM PGE2 (*see* **Note 6**). Computer-controlled automated perfusion systems can also be used.
3. Use visualization mode to select the cells of interest, which have uniform green fluorescence in cytosol.
4. Position the opening tip of the perfusion system with 40x objective (Nikon TE200 Super Fluo, NA 1.3) until the tip is right above the region of interest at a 45-degree angle (*see* **Note 7**).
5. Simultaneously record the fluorescence signals of F_{352nm} and F_{378nm}. Ratio (F_{352nm}/F_{378nm}) is displayed in a separate window. Record the resting stage (cells in BSS-2Ca^{2+} solution) for 3 min and then switch the perfusion solution to BSS-2Ca^{2+}/ 5 μM PGE2.

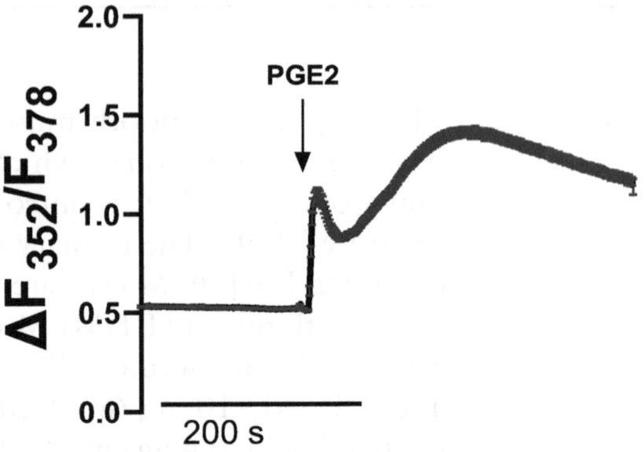

Fig. 1 Ratiometric Ca^{2+} measurement in ESM cells responding to PGE2. Intracellular Ca^{2+} concentration was represented by the fluorescent ratio of 352 nm/378 nm of Fura-2

6. Intracellular Ca^{2+} concentration is proportional to the ratio (F_{352nm}/F_{378nm}) as shown in Fig. 1. A Ca^{2+} transient is displayed as the first peak responding to addition of PGE2. The second peak is caused by Ca^{2+} influx through store-operated Ca^{2+} entry.

A Ca^{2+} transient (peak) is induced by PGE2 treatment. The arrow indicates when PGE2 is added to bath solution.

3.4 Fluo-4 AM Dye Loading

1. Seed the cells in glass-bottom dishes with #1.5 (MatTek, US) and culture the cells for at least 24 h prior to Ca^{2+} measurement.
2. Remove the cell culture medium and rinse with BSS-2Ca^{2+} solution.
3. Add 1 mL BSS-2Ca^{2+} solution containing 5 μM Fluo-4 AM ester into the Petri dish (*see* **Note 3**) [11].
4. Incubate the cells for 30 min at room temperature (*see* **Note 4**).
5. Rinse with BSS-2Ca^{2+} twice and maintain the cells in phenol red–free cell culture medium (*see* **Note 8**).

3.5 Intracellular Ca^{2+} Imaging by Confocal Microscopy

1. Mount the Petri dish on the stage of Nikon Ti2 inverted microscope, which is connected to Nikon A1R HD laser scanning confocal system (Nikon, Japan). An environmental chamber is set up at 37 °C in the presence of 5% CO_2 (Okolab, Italy) (*see* **Note 9**).
2. Use 40× objective (Plan Fluor, NA 1.3) with FITC filter set (Ex 488 nm, Em 500–550 nm) [13]. Capture an image of a cell of interest to ensure the proper pinhole size and scan speed with good fluorescent signal/noise ratio. Figure 2 shows ESM cells with green fluorescence in cytoplasm with low background.

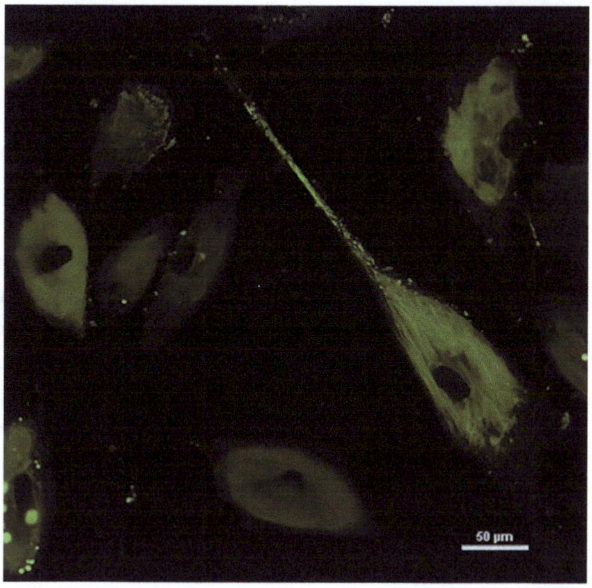

Fig. 2 Representative live cell image of ESM cells loaded with 5 µM Fluo-4 AM. The recording is performed in culture condition under a 40X objective (NA 1.3) with Nikon A1R HD confocal microscope

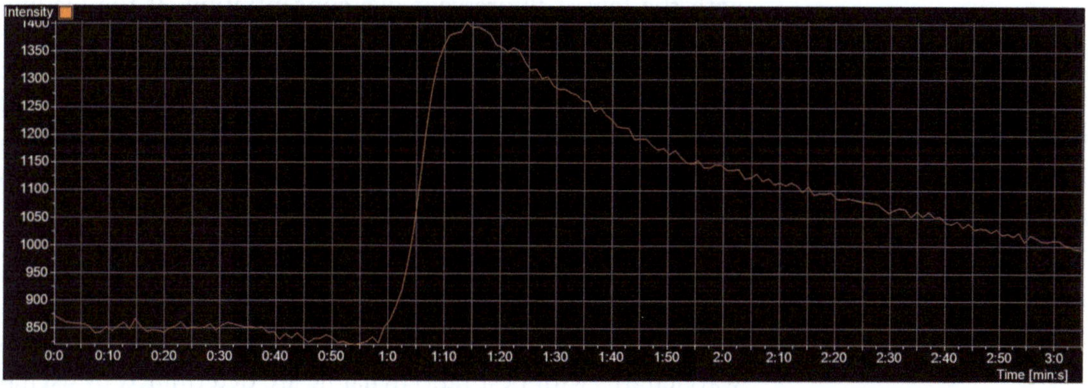

Fig. 3 The fluorescent changes over time reflect the changes in intracellular Ca^{2+} in an ESM cell responding to PGE2 treatment

3. Acquire time-lapse live cell images using high-speed resonant scanning at "Denoise.ai" mode with pinhole size 24.27 µm, 512 × 512 pixel, 0.96 s/frame, duration 10 min. The baseline of intracellular Ca^{2+} signal is recorded for 1 min, then PGE2 solution is added into the Petri dish.
4. The fluorescent intensity is analyzed by NIS-Elements software (Nikon, Japan). Record the fluorescent signal at baseline for 1 min before adding PEG2. As shown in Fig. 3, increased fluorescent intensity upon PGE2 treatment indicates elevation of intracellular Ca^{2+} in the ESM cell.

4 Notes

1. Prepare the BSS-2Ca^{2+} solution freshly and filtrate the solution if there is precipitate. Adjust the solution pH as accurate as possible.
2. All fluorescent dyes must be protected from moisture and light. Putting the dye in a dark box or wrapping the container in foil paper can protect the dye from light. The fluorescent dyes are usually received in powder form and stored in a freezer. Before preparing the work solution, it is important to let the vial containing the dye powder to sit at room temperature for a few minutes to avoid moisture from getting into the powder.
3. During the dye loading, the cells must be protected from light. This can be achieved by simply wrapping the Petri dish with foil paper.
4. The fluorescent dye loading concentrations, incubation time, and temperature may vary depending on the cell type. The exact concentration, time, and temperature should be predetermined. The rule of thumb is to achieve a reasonable signal with the lowest fluorescent dye concentration. If the loading time is too long, most of the dye will be trapped into subcellular compartments. Therefore, it is important to control the loading time and ensure that the dye has an even fluorescent pattern in the cells. The loading temperature also matters. Loading at 37 °C helps the dye cross the plasma membrane, and the following room temperature treatment allows the cytosolic esterase to cleave ester and convert non-fluorescent AM form to its fluorescent form.
5. The exact excitation wavelengths to be used can slightly vary depending on the optics of each individual system. The two wavelengths with best ratio dynamic are determined by excitation spectrum scanning of Fura-2 salt in solutions containing Ca^{2+} ranging from 0 to 39.8 μM (or higher concentration at which Fura-2 fluorescence is saturated). The detailed selection method was described previously [12].
6. Place the PGE2 solution on ice before usage to protect from degradation. Freshly prepare the BSS-2Ca^{2+}/PGE2 solution just before the experiment.
7. The perfusion system should be tested before the experiments and ensure that the flow can steadily sweep the area.
8. Prewarm the phenol red–free culture medium in a 37 °C water bath before experiment.

9. Turn on the 37 °C incubator, which is conjugated on the confocal microscope at least 30 min prior to imaging to allow the temperature to stabilize.

Acknowledgments

This work was partially supported by National Institutes of Health (NIH) Grants S10 OD025230 and R15 HL168628 (to ZP). J.Y.Y was supported by NIH R25 HL163861 through UTA SURPINT Research Program Fellowship.

References

1. Krishna Deepak RNV, Verma RK, Hartono YD, Yew WS, Fan H (2021) Recent advances in structure, function, and pharmacology of class A lipid GPCRs: opportunities and challenges for Drug Discovery. Pharmaceuticals (Basel) 15:12
2. Dorsam RT, Gutkind JS (2007) G-protein-coupled receptors and cancer. Nat Rev Cancer 7:79–94
3. Mo C, Romero-Suarez S, Bonewald L, Johnson M, Brotto M (2012) Prostaglandin E2: from clinical applications to its potential role in bone- muscle crosstalk and myogenic differentiation. Recent Pat Biotechnol 6:223–229
4. Mo C, Zhao R, Vallejo J, Igwe O, Bonewald L, Wetmore L, Brotto M (2015) Prostaglandin E2 promotes proliferation of skeletal muscle myoblasts via EP4 receptor activation. Cell Cycle 14:1507–1516
5. Sugimoto Y, Narumiya S (2007) Prostaglandin E receptors. J Biol Chem 282:11613–11617
6. Markovic T, Jakopin Z, Dolenc MS, Mlinaric-Rascan I (2017) Structural features of subtype-selective EP receptor modulators. Drug Discov Today 22:57–71
7. Bai X, Ihara E, Otsuka Y, Tsuruta S, Hirano K, Tanaka Y, Ogino H, Hirano M, Chinen T, Akiho H, Nakamura K, Oda Y, Ogawa Y (2019) Involvement of different receptor subtypes in prostaglandin E2-induced contraction and relaxation in the lower esophageal sphincter and esophageal body. Eur J Pharmacol 857: 172405
8. Nelson M, Zhang X, Genta RM, Turner K, Podgaetz E, Paris S, Cardenas J, Gu J, Leeds S, Ward M, Nguyen A, Konda V, Furuta GT, Pan Z, Souza RF, Spechler SJ (2021) Lower esophageal sphincter muscle of patients with achalasia exhibits profound mast cell degranulation. Neurogastroenterol Motil 33: e14055
9. Nelson M, Zhang X, Pan Z, Spechler SJ, Souza RF (2021) Mast cell effects on esophageal smooth muscle and their potential role in eosinophilic esophagitis and achalasia. Am J Physiol Gastrointest Liver physiol 320:G319–G327
10. Odiase E, Zhang X, Chang Y, Nelson M, Balaji U, Gu J, Zhang Q, Pan Z, Spechler SJ, Souza RF (2021) In esophageal squamous cells from eosinophilic esophagitis patients, Th2 cytokines increase eotaxin-3 secretion through effects on intracellular calcium and a non-gastric proton pump. Gastroenterology 160(2072–2088):e2076
11. Zhu H, Zhang H, Jin F, Fang M, Huang M, Yang CS, Chen T, Fu L, Pan Z (2014) Elevated Orai1 expression mediates tumor-promoting intracellular Ca2+ oscillations in human esophageal squamous cell carcinoma. Oncotarget 5: 3455–3471
12. Pan Z, Zhao X, Brotto M (2012) Fluorescence-based measurement of store-operated calcium entry in live cells: from cultured cancer cell to skeletal muscle fiber. J Vis Exp
13. Cui C, Chang Y, Zhang X, Choi S, Tran H, Penmetsa KV, Viswanadha S, Fu L, Pan Z (2018) Targeting Orai1-mediated store-operated calcium entry by RP4010 for anti-tumor activity in esophagus squamous cell carcinoma. Cancer Lett 432:169–179

Chapter 8

Isolation of Mitochondria from Murine Skeletal Muscle

Li Dong, Xuejun Li, Ang Li, Jianxun Yi, and Jingsong Zhou

Abstract

Skeletal muscle is one of the largest tissues in human body. Besides enabling voluntary movements and maintaining body's metabolic homeostasis, skeletal muscle is also a target of many pathological conditions. Mitochondria occupy 10–15% volume of a muscle myofiber and regulate many cellular processes, which often determine the fate of the cell. Isolation of mitochondria from skeletal muscle provides opportunities for various multi-omics studies with a focus on mitochondria in biomedical research field. Here we describe a protocol to efficiently isolate mitochondria with high quality and purity from skeletal muscle of mice using Nycodenz density gradient ultracentrifugation.

Key words Skeletal muscle, Mitochondria isolation, Nycodenz density gradient ultracentrifugation

1 Introduction

Skeletal muscle is one of the largest tissues in human body that attaches the bone of the skeleton, enabling voluntary movements of the body. As it comprises around 40% of whole-body lean mass, skeletal muscle is also essential for maintaining the homeostasis of the whole-body metabolism [1, 2]. Besides performing normal physiological functions, skeletal muscle is also a primary or secondary target of many pathological conditions, such as brain and spinal cord trauma, muscle injury, neuromuscular diseases, muscular dystrophies, aging (sarcopenia), and cachexia (cancer- or chronic illness-induced muscle atrophy).

Performing muscle functions costs energy in the form of adenosine triphosphate (ATP). Mitochondria are a major site to produce ATP at the expense of consuming other nutrient substrates and oxygen [3, 4]. The high energy demand of muscle contraction is met by a large endowment of mitochondria that occupy 10–15% volume of a muscle myofiber [5]. Mitochondria appear in almost all eukaryotic cells. They are double-membraned organelles with phosphatidylinositol enriched in the outer membrane and cardiolipin enriched in the inner membrane [6, 7]. The outer membrane

Marco Brotto and Sanjoy K. Bhattacharya (eds.), *Lipid Signalling: Methods and Protocols*, Methods in Molecular Biology, vol. 2816, https://doi.org/10.1007/978-1-0716-3902-3_8,

exhibits higher permeability to small molecules and ions. The inner membrane folds into many invaginations (cristae) where the permeability is tightly controlled that allows the electron transport chain and ATP synthase to function appropriately to produce ATP [8].

Mitochondria participate in regulating many cellular processes and often determine the survival or death of a cell. Particularly in skeletal muscle, mitochondria are the major site of regulating the metabolic status [9]. With the ability of Ca^{2+} uptake, mitochondria can also shape the cytosolic Ca^{2+} level during excitation-contraction coupling [10, 11] and are the major site of integrating the Ca^{2+} signaling and ROS (reactive oxygen species, a by-product of mitochondrial oxidative phosphorylation) production, which determines the physiological or pathological status of a muscle myofiber [12, 13]. This gained skeletal muscle mitochondria as a therapeutic and preventive target for various disorders [9, 12, 14, 15]. Thus, isolation of mitochondria from skeletal muscle with high quality and purity became a highly demanding practice in various biomedical studies in muscle research field.

Animal models have been widely used in basic and preclinical research studies, among which various mouse models gained their popularity due to their short gestation time, enormous importance in carrying human genetic traits, similarities of disease translation, etc. [16, 17]. In this chapter, we present a protocol to isolate mitochondria from skeletal muscle of mice. This protocol was adapted from a previously published method [18] with modifications that consumes shorter time and uses a lower speed of centrifugation. The quality and purity of isolated mitochondria samples using the current protocol are demonstrated by the immunoblotting assessments (Fig. 4). The isolated mitochondria samples can be further used for mitochondrial functional assessments [19], immunoblotting [20], lipidomics and proteomics studies [21], etc.

2 Materials

2.1 Animal Model

Four-month-old mice with genetic background of B6SJL (Jackson Laboratory, Cat# 100012) were used for the mitochondrial isolation from skeletal muscle in this study. Protocols on usage of mice were approved by the Institutional Animal Care and Use Committee (IACUC) of the University of Texas at Arlington.

2.2 Chemicals/ Materials

General chemicals used in the current experiment were purchased from Themo Fisher Scientific. The key items are listed as follows:

Protease/phosphatase Inhibitors (100X, ThermoFisher Scientific, Cat# 78446)

Nycodenz (Axis-Shield Poc AS, Cat# 1002424)

Anti-TOM20 antibody (Proteintech, Cat# 11802-1-AP)

Anti-tubulin antibody (Cell Signaling Technology, Cat# 2144)

2.3 Preparation of Solutions

1. Homogenization Buffer Stock (500 mL): 100 mM Tris-HCl, pH 7.4, 100 mM sucrose, 10 mM EDTA, 46 mM KCl. Stored at 4 °C.
2. 50% (w/v) Nycodenz Stock (50 mL): 5 mM Tris pH 7.4, 3 mM KCl, 0.3 mM EDTA, 50% Nycodenz. Stored at 4 °C.
3. Dilution Buffer (100 mL): 128 mM NaCl, 5 mM Tris pH 7.4, 3 mM KCl, 0.3 mM EDTA. Stored at 4 °C.
4. Working Solutions: use the dilution buffer to dilute 50% Nycodenz to make 30%, 25% and 23% Working Solutions. Stored at 4 °C.
5. Muscle Dissection Stock Solution (1 L): 145 mM NaCl, 5 mM KCl, 2 mM $MgCl_2$, 10 mM HEPES, pH 7.2. Stored at 4 °C.

2.4 Equipment and Accessories (Figs. 1 and 2)

1. Grinder (Heidolph RZR1) with a PTFE pestle
2. Sorvall Legend X1R centrifuge (ThermoFisher Scientific, Cat# 75004263)
3. SW60 Ti rotor with 44.5 buckets (Beckman coulter, Cat# 335650)
4. Sorvall WX + Ultra series centrifuge (ThermoFisher Scientific, Cat#75000100)
5. Eppendorf Centrifuge 5417R (Eppendorf)
6. 10 mL Wheaton homogenizer glass tubes (Wheaton, Cat# 357979)
7. 50 mL centrifuge tubes (ThermoFisher Scientific, Cat# 3118-0050)
8. 11 × 60 mm Ultra-Clear tubes (Beckman coulter, Cat# 344062)
9. Student fine scissors (Fine Science Tool, Cat# 91460-11)
10. 35 mm tissue culture dishes (ThermoFisher Scientific, Cat# 130180)
11. 1.5 mL Eppendorf tubes (FisherScientific, Cat# 05-408-129)

3 Methods

Most of the procedures are carried out at room temperature unless otherwise specified.

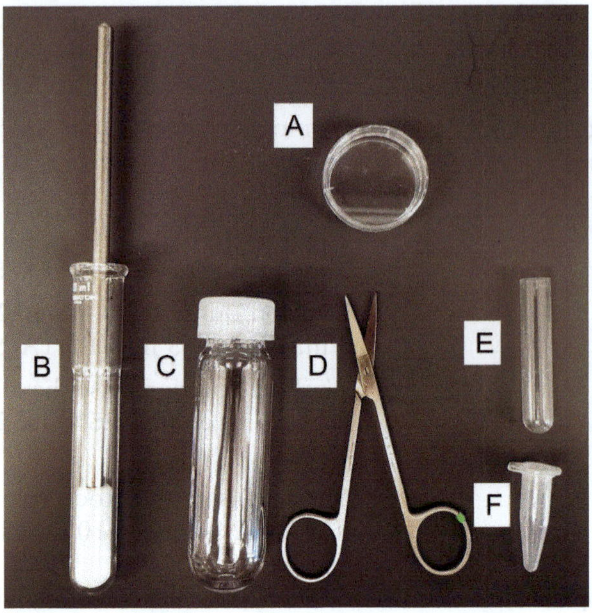

Fig. 1 Accessories used in this experiment. (**a**) 35 mm tissue culture dish. (**b**) 10 mL Wheaton homogenizer glass tube with PTFE pestle. (**c**) 50 mL centrifuge tube. (**d**) Student fine scissors. (**e**) 11 × 60 mm Ultra-clear tube. (**f**) 1.5 mL Eppendorf tube

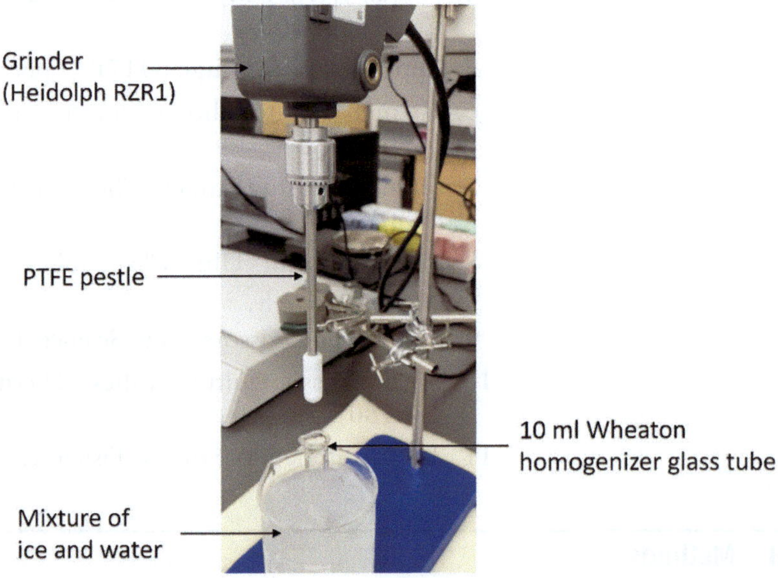

Fig. 2 Muscle sample preparation. Chopped muscle sample was added into a 10 mL Wheaton homogenizer glass tube. The speed of Grinder (Heidolph RZR1) was set to 8. The tube was kept in an ice bucket (500 mL beaker containing mixture of ice and water) during the grinding process

3.1 Preparation of Solutions Before the Experiment

1. Prepare the Homogenization Buffer: Add 5 mM BSA and protease/phosphatase inhibitors (1X) to Homogenization Buffer Stock (i.e., total 150 mL for 10 muscle samples). Add 3 mL of Homogenization Buffer to each 35 mm culture dish and leave the dishes on ice. One dish per muscle sample.
2. Prepare Washing Buffer: Add protease/phosphatase inhibitors (1X) to Homogenization Buffer Stock (i.e., total 50 mL for 10 muscle samples).
3. Prepare the muscle dissection solution: Add 10 mM glucose to the Muscle Dissection Stock Solution, i.e., add 90 mg glucose in 50 mL Muscle Dissection Stock Solution. Leave the solution on ice.

3.2 Muscle Sample Preparation

1. Skeletal muscle dissection: Euthanize the mouse by quick cervical dislocation. Dissect the skeletal muscle from the hindlimbs in a 35 mm culture dish containing ice cold Muscle Dissection Solution. Different types of muscle can be dissected separately for mitochondrial isolation (*see* **Note 1**). For this particular experiment, the muscle sample used was a mix of tibialis anterior, extensor digitorum longus, soleus, and gastrocnemius. Place each dissected muscle sample in a 35 mm culture dish containing 3 mL Homogenization Buffer on ice.
2. Transfer each muscle sample to one 1.5 mL Eppendorf tube and weigh on a balance (up to 0.8 g muscle per tube) (*see* **Note 2**). Add 1–2 vol of Homogenization Buffer (i.e., 1 mL for 0.5 g muscle) and cut the muscle into fine pieces with the student fine scissors. Add Homogenization Buffer up to 1.2 mL for an easy transfer of the muscle sample in the next step.
3. Transfer the chopped muscle sample to a 10 mL Wheaton homogenizer glass tube. Add Homogenization Buffer up to a total of 9–10 vol of the muscle sample weight (i.e., 5 mL for 0.5 g muscle sample).
4. Set the speed of the grinder at 8. Grind the muscle sample with a PTFE pestle for 10 times by moving the tube up and down. The tube should be kept inside an ice bucket (a 500 mL beaker containing mixture of ice and water) during the procedure. (Fig. 2).

3.3 Isolation of Mitochondria

1. Transfer the sample to a 50 mL centrifuge tube. Balance tubes and spin the samples at 800 g, 4 °C for 10 min (X1R centrifuge, 2623 rpm). Carefully transfer the supernatant (first supernatant, S1) to a new 50 mL centrifuge tube.
2. Centrifuge S1 at 10,000 g, 4 °C for 10 min (X1R centrifuge, 9274 rpm). Remove the supernatant (S2) and save it later for cytosolic protein collection. The pellet with brownish color contains mitochondria.

3. Resuspend the pellet in 1.5 mL 25% Nycodenz.
4. For each 11 × 60 mm Beckman Ultra-Clear tube, first add 1.25 mL 30% Nycodenz solution at the bottom, then carefully overlay the 1.5 mL 25% Nycodenz solution containing the resuspended pellet without disturbing the lower layer, and finally, overlay 1.25 mL 23% Nycodenz solution carefully on the top of the 25% Nycodenz solution without disturbing the lower layer.
5. Place the Ultra-Clear tubes onto Beckman SW60 Ti rotor with 44.5 buckets. Centrifuge at 19,800 rpm, 4 °C for 90 min (Ultra series centrifuge). The mitochondria form a light brown band at the interface between 30% and 25% Nycodenz (Fig. 3). Use 1 mL pipette to remove the top layer until ~10 mm left above the mitochondrial band. Use 100 μL pipette to gently transfer the mitochondrial band to a new 1.5 mL Eppendorf tube.
6. Add 1 vol of washing buffer, and spin at 10,000 rpm 4 °C for 10 min (Eppendorf Centrifuge 5417R). Discard the supernatant, add 300 μL of washing buffer, resuspend the pellet, and spin at 10,000 rpm 4 °C for 5 min. Repeat the wash once more and store the pellet at −80 °C for future biochemical studies.
7. Take 1 mL of supernatant S2 for centrifugation at 16,400 rpm, 4 °C for 20 min (Eppendorf Centrifuge 5417R) to remove debris, and then collect the supernatant (S3) as the cytosolic fraction. Store S3 at −80 °C for future biochemical studies.
8. The quality and purity of the isolated mitochondria were checked by immunoblotting assessments using antibodies specific to mitochondrial (anti-TOM20) or cytosolic (anti-tubulin) proteins. As demonstrated in Fig. 4a, TOM20, specifically expressed on mitochondria membrane, was only detected in the purified mitochondria fraction but not in the cytosolic fractions, while tubulin, a component of the cytosolic microtubules, was only detected in the cytosolic fraction. Figure 4b shows the Coomassie blue stain of the same blots used in Fig. 4a, which reflects the different pattern of the protein bands between the cytosolic and mitochondrial proteins (~100 μg protein was loaded per lane). Those results indicate a successful isolation of mitochondria from the cytosolic fraction.

4 Notes

1. Based on the experiment design, different types of muscle (i.e., extensor digitorum longus, soleus, or gastrocnemius, etc.) can be dissected separately for the mitochondrial isolation.

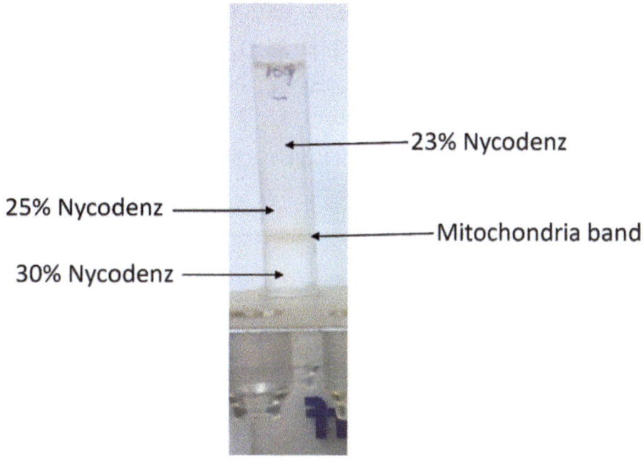

Fig. 3 The brownish band at the interface between 30% and 25% Nycodenz fractions contains the purified mitochondria after the Nycodenz gradient ultracentrifugation

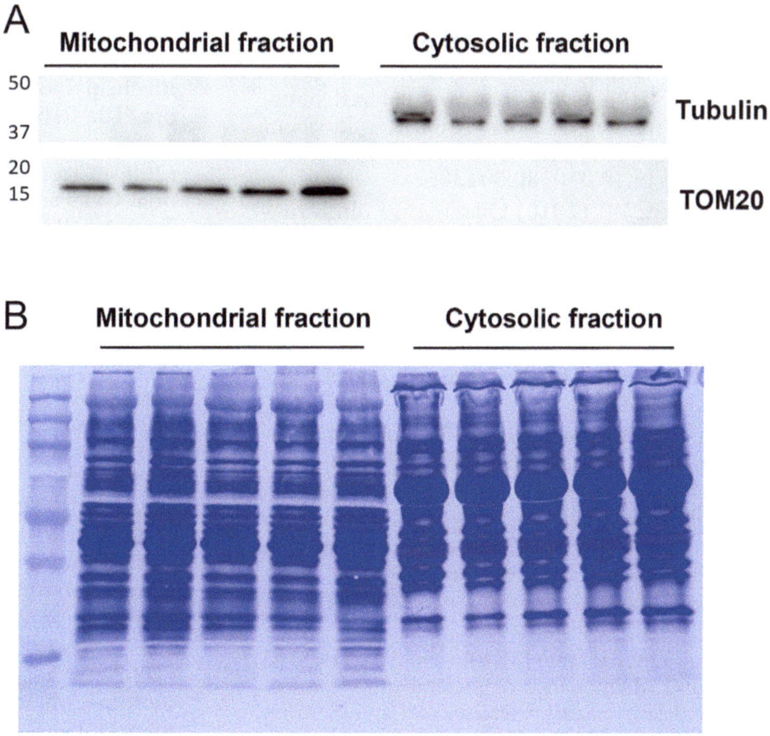

Fig. 4 (**a**) Western blot demonstrates a successful purification of mitochondria from the skeletal muscle samples. TOM20, a protein specifically expressed on mitochondria membrane, was only detected in the mitochondrial fractions, while tubulin, a cytosolic protein, was only detected in the cytosolic fractions. (**b**) Coomassie blue stain of the same blots as in (**a**), showing different patterns of protein bands between the mitochondria and cytosolic fractions. ~ 100 μg protein per lane. 1:10000 for both anti-TOM20 and anti-Tubulin antibodies

2. Don't use more than 0.8 g muscle mass/per sample. Overloaded sample may result in too much crude mitochondria in the 25% Nycodenz solution, which can disturb interface between 23% and 25% Nycodenz solutions during the overlaying step.

Acknowledgments

This work was supported by grants from NIH R01NS105621 and R01NS129219 to JZ.

References

1. Frontera WR, Ochala J (2015) Skeletal muscle: a brief review of structure and function. Calcif Tissue Int 96:183–195
2. Stump CS, Henriksen EJ, Wei Y, Sowers JR (2006) The metabolic syndrome: role of skeletal muscle metabolism. Ann Med 38:389–402
3. Hatefi Y (1993) ATP synthesis in mitochondria. Eur J Biochem 218:759–767
4. Knowles JR (1980) Enzyme-catalyzed phosphoryl transfer reactions. Annu Rev Biochem 49:877–919. https://doi.org/10.1146/annurev.bi.49.070180.004305
5. Eisenberg BR (2010) Quantitative ultrastructure of mammalian skeletal muscle. Compr Physiol:73–112
6. Ahmadpour ST, Maheo K, Servais S, Brisson L, Dumas JF (2020) Cardiolipin, the mitochondrial signature lipid: implication in cancer. Int J Mol Sci 21. https://doi.org/10.3390/ijms21218031
7. Horvath SE, Daum G (2013) Lipids of mitochondria. Prog Lipid Res 52:590–614. https://doi.org/10.1016/j.plipres.2013.07.002
8. Kuhlbrandt W (2015) Structure and function of mitochondrial membrane protein complexes. BMC Biol 13:89. https://doi.org/10.1186/s12915-015-0201-x
9. Hood DA, Memme JM, Oliveira AN, Triolo M (2019) Maintenance of skeletal muscle mitochondria in health, exercise, and aging. Annu Rev Physiol 81:19–41
10. Yi J, Ma C, Li Y, Weisleder N, Rios E, Ma J, Zhou J (2011) Mitochondrial calcium uptake regulates rapid calcium transients in skeletal muscle during excitation-contraction (E-C) coupling. J Biol Chem 286:32436–32443. https://doi.org/10.1074/jbc.M110.217711
11. Rudolf R, Mongillo M, Magalhaes PJ, Pozzan T (2004) In vivo monitoring of ca(2+) uptake into mitochondria of mouse skeletal muscle during contraction. J Cell Biol 166:527–536. https://doi.org/10.1083/jcb.200403102
12. Zhou J, Li A, Li X, Yi J (2019) Dysregulated mitochondrial ca(2+) and ROS signaling in skeletal muscle of ALS mouse model. Arch Biochem Biophys 663:249–258. https://doi.org/10.1016/j.abb.2019.01.024
13. Feissner RF, Skalska J, Gaum WE, Sheu SS (2009) Crosstalk signaling between mitochondrial Ca2+ and ROS. Front Biosci (Landmark Ed) 14:1197–1218
14. Hesselink MK, Schrauwen-Hinderling V, Schrauwen P (2016) Skeletal muscle mitochondria as a target to prevent or treat type 2 diabetes mellitus. Nat Rev Endocrinol 12: 633–645
15. Russell AP, Foletta VC, Snow RJ, Wadley GD (2014) Skeletal muscle mitochondria: a major player in exercise, health and disease. Biochim Biophys Acta 1840:1276–1284. https://doi.org/10.1016/j.bbagen.2013.11.016
16. Mukherjee P, Roy S, Ghosh D, Nandi S (2022) Role of animal models in biomedical research: a review. Lab Ani Res 38:18
17. Bockamp E, Maringer M, Spangenberg C, Fees S, Fraser S, Eshkind L, Oesch F, Zabel B (2002) Of mice and models: improved animal models for biomedical research. Physiol Genomics 11:115–132
18. Wong M, Gertz B, Chestnut BA, Martin LJ (2013) Mitochondrial DNMT3A and DNA methylation in skeletal muscle and CNS of transgenic mouse models of ALS. Front Cell Neurosci 7:279. https://doi.org/10.3389/fncel.2013.00279

19. Iuso A, Repp B, Biagosch C, Terrile C, Prokisch H (2017) Assessing mitochondrial bioenergetics in isolated mitochondria from various mouse tissues using seahorse XF96 analyzer. In: Mitochondria: Practical Protocols, pp 217–230
20. Xiao Y, Karam C, Yi J, Zhang L, Li X, Yoon D, Wang H, Dhakal K, Ramlow P, Yu T et al (2018) ROS-related mitochondrial dysfunction in skeletal muscle of an ALS mouse model during the disease progression. Pharmacol Res 138:25–36. https://doi.org/10.1016/j.phrs.2018.09.008
21. Liu X, Xu G (2018) Recent advances in using mass spectrometry for mitochondrial metabolomics and lipidomics-a review. Anal Chim Acta 1037:3–12

Chapter 9

Effects of a Laparotomy on Targeted Lipidomics Profiles in a Mouse Model of Surgical Stress During Aging

Jian Huang, Kamal Awad, Lauren Harlow, Carlos A. Toro, Marco Brotto, and Christopher P. Cardozo

Abstract

Laparotomy (EL) is one of the most common procedures performed among surgical specialties. Previous research demonstrates that surgery is associated with an increased inflammatory response. Low psoas muscle mass and quality markers are associated with increased mortality rates after emergency laparotomy. Analysis of lipid mediators in serum and muscle by using liquid chromatography–mass spectrometry (LC-MS)-based lipidomics has proven to be a sensitive and precise technique. In this chapter, we describe an LC-MS/MS protocol for the profiling and quantification of signaling lipids formed from Eicosapentaenoic Acid (EPA) and Eicosatetranoic acid (ETA) by 5, 12, or 15 lipoxynases. This protocol has been developed for and validated in serum and muscle samples in a mouse model of surgical stress caused by laparotomy.

Key words Laparotomy, Surgical stress, Lipidomics, Omics, Muscle, Serum

1 Introduction

Trauma is the fourth leading cause of death in the overall population and is the primary cause of death during the first half of the human life span [1]. Approximately 9–14.9% of all trauma cases involve the abdomen [2]. Abdominal trauma is one of the preventable causes of death in polytrauma patients [3], with laparotomy traditionally regarded as the standard treatment [4]. Surgery in older adults is associated with increased morbidity, mortality, and healthcare costs compared to younger individuals [5]. In the United States, more than 50% of all surgeries are performed in adults greater than 65 years of age. According to one estimate, 7% of patients who underwent noncardiac surgery experienced postoperative complications that led to a 78% increase in hospital cost and a 114% increase in hospital length of stay [6]. Postoperative

Marco Brotto and Sanjoy K. Bhattacharya (eds.), *Lipid Signalling: Methods and Protocols*, Methods in Molecular Biology, vol. 2816, https://doi.org/10.1007/978-1-0716-3902-3_9,

complications were more common with increasing age, with 20% of those greater than 80 years old experiencing one or more complications [7, 8].

An emergency laparotomy (EL) is one of the most common procedures performed among surgical specialties. EL is associated with considerable 30-day morbidity and mortality despite many efforts to improve the short- and long-term outcomes following EL [9–14]. Physiologic changes and the increased burden of chronic diseases among the elderly result in them being at greater risk of perioperative complications, functional deterioration and subsequent institutionalization, and increased morbidity and mortality of EL in elderly patients, all resulting in a worsened quality of life [15–18]. Each year, millions of patients undergo abdominal surgery worldwide [19]. Despite significant clinical efforts, up to a half of those patients still report inadequate recovery after surgery, indicating the need for novel strategies to improve postoperative outcomes including functional recovery [20]. Previous research demonstrates that surgery is associated with an increased inflammatory response [21, 22]. Low psoas muscle mass and quality markers are associated with increased 30-day mortality rates after emergency laparotomy. Therefore, muscle markers can be used as a new feasible tool to identify most at-risk patients requiring further interventions [23].

Surgery, including abdominal procedures, precipitates muscle weakness, weight loss, and inactivity in the postoperative convalescent period. In older adults, surgery induces muscle catabolism leading to muscle wasting [24], decreases limb muscle endurance and strength [25], delays recovery of muscle function [26], and causes persistent fatigue [27] and weight loss [28]. These adverse postoperative responses are associated with poor outcomes such as delayed return of physical activity, impaired recovery, and increased hospital readmission and mortality [29–32]. Surgery induces complex and acute inflammatory and oxidative stress responses that are essential in mediating immune defense and tissue repair [33–36]. And inflammation and oxidative stress in muscles contribute to skeletal muscle weakness and sarcopenia in older individuals [37, 38]. LMs have different major functions in living systems, such as being major constituents of biological membranes, efficient energy sources, modifiers of proteins, and signaling molecules [39], and play important roles through their signaling pathways on the regulation of pathophysiological states such as inflammation, metabolic syndrome, and cancer [39–42]. Lipid mediators (LMs) can regulate skeletal muscle mass and function and potentially lead to muscle wasting in response to various physiological and pathological conditions [43–46]. To better understand the biological mechanism after abdominal surgery, Ko et al. established a mouse model of surgical trauma introduced via laparotomy C57BL/6 J mice and showed that this procedure induces coordinated, postoperative blood cell expression of unique inflammatory and oxidative or metabolic stress genes [21]. Cata et al. created a rat model of postoperative stress, and they found changes in lipid LMs that modulate inflammation in intestinal mucosa after laparotomy [22]. Recently, Asche-Godin et al. characterized the effects of laparotomy on postoperative physical function, skeletal muscle gene expression, and LM levels in C57BL/6 N mice across the life span (3, 20, or 24 months of age) to investigate the molecular underpinnings of the interaction of age and surgical stress on postoperative decline in physical function [47].

Their results demonstrated that laparotomy-induced an age-associated postoperative decline in forelimb grip strength that was the greatest in the oldest mice. Unexpectedly, aparotomy-induced age-associated differential gene expression and canonical pathway activation with the greatest in the youngest mice suggest that in old mice, pathways activated by surgery in young mice are either refractory to effects of surgical stress or are already fully activated. Analysis of lipid mediators in serum and gastrocnemius muscle by using liquid chromatography–mass spectrometry (LC-MS)-based lipidomics approach showed that serum levels of 12-HEPE, 12-HETE, 12-HTT, PGD2, PGF2a, and TXB2 were elevated in 24-month-old mice associated with increased muscle levels of 12-HEPE, 12-HETE, PDG2, PGE_2, and TXB2. These alterations of LMs in skeletal muscles (gastrocnemius) suggest an age-related increased role for LMs in the biology of aging. The presence of clear and robust changes in serum levels of biologically active LMs suggests that muscle-derived LMs may regulate other tissues such as fat, liver, bone, or brain.

In this chapter, we will focus on describing the detailed protocol we have used to profile LMs produced from the polyunsaturated fatty acids eicosapentanoic acid (EPA) and eicosatetranoic acid (ETA). EPA is a type of omega-3 fatty acid found in fish and algae. EPA, along with docosahexaenoic acid (DHA), is known for its potential health benefits. EPA and ETA can be converted in the body into various lipid mediators by 5, 12, or 15 lipoxygenases to generate bioactive compounds involved in various physiological processes. EPA has been the subject of research for its potential health benefits, particularly in relation to inflammation, cardiovascular health, and mental well-being.

2 Materials

1. FST, Item No. 11042-08, Octagon Forceps
2. FST, Item No. 14171-10, Kaye Scissors—Extra Large Loops
3. FST, Item No. 14106-09, Fine Scissors—Extra Large Loops
4. Integra Miltex, Disposable Safety Scalpel, Blade no. 11

5. Ethicon, Vicryl suture 5-0, V421H, FS-2 needle, 70 cm undyed
6. 3 M Vetbond Tissue Adhesive
7. Covetrus, NDC 11695-6777-2, Isoflurane, USP
8. Inhaled anesthetic vaporizer
9. Water-heated surgical station with nose cone
10. Analgesic (carprofen, subcutaneous injection)
11. Antibiotic (Baytril, subcutaneous injection)
12. Animal clipper
13. Water-heated warming blanket set to 37 °C
14. Timer or stopwatch
15. 2 mL screw cap cryovials
16. 1.6 mL Eppendorff tubes or their equivalent.

3 Methods

To systematically study the interactions between surgical stress and both functional and biochemical outcomes in older mice, we have developed a very tightly controlled mouse model of laparotomy.

3.1 Surgical Procedures

1. Weigh the animal and record its pre-surgical body weight.
2. Induce anesthesia by inhalation of 3% isoflurane in an induction chamber.
3. Transfer the animal to a water-heated surgical station equipped with a mouse nosecone and appropriate air scavenging system.
4. Remove the hair over the abdomen with clipper.
5. Clean the skin with three applications of betadine 10% solution (povidone-iodine) followed by three applications of 70% ethanol.
6. Mark the site of the incision with a 1 cm mark on the skin along the midline midway between the thorax and pelvis. Place this line in the same location for each experimental animal using the xyphoid process as landmark.
7. Using octagonal forceps, lift the skin away from the abdomen at the top of your 1 cm marking.
8. Using a sterile, disposable scalpel, make a 1 cm vertical midline incision traversing the skin and subcutaneous tissue. Take care not to rupture the abdominal wall, peritoneum cavity or puncture any organs.

9. Using the octagonal forceps and your Kaye scissors, bluntly dissect the skin away from the abdominal wall.
10. Abdominal organs will then be partially exposed by opening the abdominal cavity via a midline incision of the abdominal wall. The incision can be created by first using the octagonal forceps to pinch the abdominal wall, lifting it away from the organs, and using a disposable scalpel to make a 1 cm vertical midline incision traversing the abdominal wall.
11. Alternatively, you can also use a pair of fine, sharp, large-loop scissors to create this midline incision. As before, use the forceps to pinch the abdominal wall and to gently pull it up. You will then push the bottom point of the scissor through the wall, right below your forceps, and proceed to use small gentle snips to create the incision.

    Note: The abdominal wall incision using scissors, as opposed to a scalpel, may be useful for smaller animals and those animals where it is difficult to pull the abdominal wall away from the organs far enough to safely push the tip of scalpel through the abdominal wall without risking organ damage. We saw this phenomenon of the organs "sticking" to the abdominal wall, even after it was pinched by the forceps and pulled away, most often in animals (usually aged) that showed signs of dehydration.
12. The organs were then left partly exposed for 2 min.
13. The abdominal wall will be closed from the cranial-to-caudal end by placing four interrupted surgical sutures using absorbable sutures.
14. Skin and subcutaneous tissue overlaying the 1 cm vertical midline abdominal surgical incision will be closed using absorbable sutures. Two to four interrupted, surgical sutures, placed from the cranial-to-caudal end and then secured by placing a drop of tissue adhesive directly on the knot of the suture, allow it to dry completely before returning the mouse to the recovery cage.
15. The expected duration of the laparotomy procedure is 8 min, with the animal under anesthesia for a total of 15 min.
16. Time these steps to keep consistent the total time under anesthesia at 15 min, and the total time between initial incision and end of skin closure at 8 min.
17. Place the animal in a clean cage that has been pre-warmed using the water-heated warming pad after administering analgesic and antibiotic.
18. Monitor for recovery until fully mobile.

Recommended Procedures: Best Practices

- Administer 1 mL of Lactated Ringers solution or normal saline subcutaneously at the end of the procedure to replace any intravascular volume lost during the procedure.
- Perform the surgeries within a pre-specific timeframe each day to minimize effects of diurnal cycles on the lipidomics outcomes.
- Place the animals on a nutrient-controlled diet such as that from Research Diets (D12450J; 10% fat, 20% protein, 70% carbohydrate, 3.82 kcal/g energy density) to minimize effects of batch-to-batch variations in lipids content of chow.
- Record daily food intake and animal weight during the week prior to surgery and daily thereafter.

3.2 Gastrocnemius Muscle Dissection and Serum Collection

1. Weigh the animal and induce a deep surgical plane of anesthesia by inhalation of 3% isoflurane in an induction chamber.

   Note: for hindlimb skeletal muscle dissection, it is recommended to perform practice surgeries to learn to recognize muscle structures and hindlimb anatomy. Muscles are connected to bones through tendons. For accurate muscle weights, be sure to peel muscle from bone in the same way each time and to cut the junction between muscle and tendon in the same place each time.
2. Using forceps, lift the skin anteriorly where the upper hindlimb joins the abdomen. Make a 1 cm incision through the skin oriented from left to right.
3. Insert the tips of the scissors under the skin and separate skin from tissues below it using blunt dissection. To do this, lift the skin proximal to the incision, then insert tip of the scissors between the skin and the hindlimb and make an opening with the blunt side of the scissors touching the internal side of the skin to enlarge the incision toward the posterior side of the hind limb.
4. Cut the skin gently with scissors. Extend the subcutaneous pocket using blunt dissection. Continue working toward the toes until the Achilles tendon is exposed.
5. Gently hold the skin of the anterior part of the distal hind limb, then peel the skin from hip toward the toes of the mouse hindlimb until the muscles are completely visible from the hip to the toes.
6. Gently separate the Achilles tendon from other connective tissues. To do this, use the forceps to pick up the distal part of the Achilles tendon, and hold it firmly. Insert the tip of the scissors behind the Achilles tendon—between the tendon and the tibia/fibula. Separate the tendon from surrounding tissues

using blunt dissection. The scissors should be pressed against the bones, spreading the blades to create a space between the muscle and the bones. Keeping the scissors closed, slide them along the bones and toward the knee until you reach the back of the knee.

Note: If performed correctly the gastrocnemius, plantaris and soleus muscles will be easily and cleanly separated from the bones.

7. With the forceps, hold the proximal edge of the Achilles tendon (where the gastrocnemius ends and the tendon begins), then cut distal to it.
8. Use forceps to hold the Achilles on the distal part of the gastrocnemius in parallel to the bones, using the scissors on the other hand, cut the non-muscle tissue around the proximal part of the gastrocnemius, along the bones until the muscle is fully detached. Use the scissors to clear the gastrocnemius of any non-muscle tissues and separate from other muscles.
9. Weight the muscle.
10. Put the gastrocnemius inside a pre-labeled screw top tube and immediately freeze in liquid nitrogen.
11. Repeat the dissection of the gastrocnemius on the other hindlimb. A proper dissection will yield an intact muscle. These tissues can be frozen and stored at −80 °C until further analyses.
12. For blood collection, reopen the abdominal laparotomy incision using tweezers and a scalpel or fine scissors being careful not to nick the intestines or any other organ.
13. Using a surgical scissors, open the left thoracic cavity cutting upward along the midline from the abdomen.
14. Collect the cardiac blood with a 1 mL syringe via ventricular puncture using a 24 g needle.
15. Transfer the blood to a 1.6 mL collection tube and place on ice for 20 min to allow it to clot.
16. Separate the cellular elements and serum using a refrigerated centrifuge at 4 °C at a speed of 2000 × *g* for 10 min.
17. Remove the serum carefully using a 1 mL micropipette. Be careful not to disturb the sedimented cells and save the serum in a clean pre-labeled 2 mL screw top tube. Store the serum at −80 °C until analysis.

3.3 Lipidomics Analysis

In this section, we will report the detailed protocol for the profiling and detection of EPA LMs using a liquid chromatographic separation coupled with a tandem mass spectroscopy (LC-MS/MS) protocol. LC/ESI-MS/MS method using the Shimadzu LCMS-8060 system (Shimadzu Scientific Instruments, Inc., Columbia, MD,

Table 1
EPA lipid mediators

No.	Compound	No.	Compound
1	Δ17-6-keto prostaglandin F1α	14	(±)18-HEPE
2	Resolvin E1	15	15(S)-HEPE
3	8-iso prostaglandin F3α	16	11(S)-HEPE
4	Thromboxane B3	17	8(S)-HEPE
5	Prostaglandin F3α	18	9(S)-HEPE
6	11-dehydro thromboxane B3	19	12(S)-HEPE
7	Prostaglandin E3	20	5(S)-HEPE
8	Prostaglandin D3	21	15(S)-HpEPE
9	Lipoxin A5	22	12(S)-HpEPE
10	Leukotriene B5	23	5(S)-HpEPE
11	(±)17,18-DiHETE	24	(±)17(18)-EpETE
12	(±)14(15)-DiHETE	25	(±)14(15)-EpETE
13	(±)5(6)-DiHETE	26	Eicosapentaenoic acid (EPA)

USA) will be described. This method provides a simultaneous quantification for lipids and with on-column detection at femtomolar level without the necessity of chemical derivatization. In this protocol, we mainly focused on the quantification and profiling of EPA metabolites (see Table 1).

3.3.1 Solid-Phase Extraction (SPE)

In this method, we have used our developed and detailed SPE protocol published in this book in the chapter entitled "Detailed Protocol for Solid Phase Extraction for Lipidomic Analysis." This protocol is adapted for different sample types (such as cells, serum, plasma, body fluids, soft tissues, and bone marrow), please see the mentioned chapter for full details on the SPE.

3.3.2 Standards of EPA Lipid Mediators

Standards of the targeted LMs (Table 1) and their related deuterated Internal Standards (IS) such as eicosapentaenoic acid-d5 (EPA-d5) were used in this protocol. These compounds can be purchased from any company, but the purity must exceed 98%. Methanol, acetonitrile, water, and other solvents used for the LC mobile phase must be of HPLC-MS analytical grade. Reagents like formic acid used during the protocol could be a reagent grade (>95%). The following steps are taken to prepare the IS stock solution and the standards working solution:

Table 2
Preparation of IS solution

	Stock conc. (μg/μL)	Stock volume (μL)	Final conc. in mixture (μg/mL)	Volume (μL) for 1 mL IS stock
Internal standards (IS)				
Eicosapentaenoic acid-d5 (EPA-d5) 50 μg in 100 μL ethanol	0.5	100	10	20
Methanol				980
Final volume of the prepared IS stock				1000
	Stock Conc. (μg/μL)	**Stock volume (μL)**	**Final Conc. in mixture (μg/mL)**	**Volume (μL) for 1 mL stock**
Standards stock solution (Std)				
Eicosapentaenoic acid (EPA) 50 μg in 100 μL ethanol	1	100	2.5	2.5
(±)17,18-DiHETE, 50 μg in 100 μL ethanol	1	100	2.5	2.5
12(S)-HpEPE, 50 μg in 100 μL ethanol	1	100	2.5	2.5
Methanol				992.5

1. *Internal Standards Stock Solution*: The IS stock solution is prepared by mixing the specific volume of each isotope labeled internal standard with methanol for a final IS stock according to Table 2. After preparing the IS stock, it is stored as aliquots in small tubes and frozen at −80 °C for future use.
2. *Standards Stock Solution (Std)*: The standard compounds are mixed in methanol to prepare the stock solution of standards mixture. Here, we used Eicosapentaenoic Acid (EPA), (±) 17,18-DiHETE, and 12(S)-HpEPE as standard compounds with final concentration of 2.5 μg/mL of each compound in the final stock volume of 1 mL in methanol as described in Table 2. After preparing the standards, it is stored as aliquots in small tubes and frozen at −80 °C.
3. *Working Solution and Calibration Curve:* Using the prepared Std stock, the working solution can be prepared by serial dilutions of eight concentrations ranging from 1 μg/mL to 1 ng/mL in methanol. At least 50 μL of the highest concentration (1 μg/mL) is prepared to be used for the lower concentrations and the calibration curve as well. To prepare the final mixture for the calibration curve, 5 μL of each working solution +5 μL

of the IS solution +90 μL methanol is mixed in 300 μL of the target polyspring insert, placed in clear autosampler glass vials, then closed using the PTEF/silicon screw cap and vortexed for 10 seconds to ensure proper mixing.

3.3.3 LC-MS/MS Analysis

1. *Instrument:* Shimadzu LCMS-8060 system with the UF-Qarray ion guide that increases LC/MS/MS sensitivity by enhancing ion signal intensity and reducing noise. By improving the ion sampling device, the ion guide, and the vacuum efficiency, the LCMS-8060 delivers a new vision of sensitivity and makes a real difference to working better and faster. Multiple Reaction Monitoring (MRM) is successfully employed to provide sensitive and specific analysis of lipidomics, metabolomics, and proteomics.
2. *Mobile phase:* Pump A: 0.1% formic acid in water, and Pump B: 100% Acetonitrile.
   (a) The LC system is equipped with system controller: CBM-20A, three LC-30AD pumps, an SIL-30AC autosampler, and a CTO-30A column oven. The LC separation is conducted on a C8 column (Ultra C8, 150 × 2.1 mm, 3 μm, RESTEK, Manchaca, TX, USA). The oven temperature for the analytical column is set to 40 °C.
   (b) The triple-quadrupole mass spectrometer (LCMS-8060 system) is operated and optimized under electrospray ionization in negative mode. MS parameters are optimized as follows: interface: DUIS (ESI) and (APCI), voltage, 4.0 kV; interface temperature, 280 °C; desolvation line (DL) temperature, 275 °C; heating block temperature, 400 °C; drying gas (N2), 10 L/min; nebulizing gas (N2), 3 L/min; heating gas (air), 10 L/min; and collision-induced dissociation (CID) gas (Ar), 230 kPa.
   (c) Considering the MRM technique, the acquisition is divided into multiple segments. Start with MRM optimization method for lipidomics to obtain the m/z transitions and their MS parameters (Q1, CE, and Q3 voltages) based on the best MRM responses. These parameters are presented in Table 3 for the EPA LMs.
   (d) Once all parameters have been optimized, the optimized method is run with full range to scan and monitor all IS and Std LMs in a concentrated sample of IS + Std mixture.
   (e) Once all LMs are identified and detected, the exact retention time in the method is recorded and updated if needed.
   (f) The optimized method is saved, and the calibration curve samples are running; chromatograms are obtained from the calibration curve and shown in Fig. 1.

Table 3
Mobile phase and LC program parameters

Time (min)	Mobile phase A	Mobile phase B	Flow rate (mL/min)
0.00–5.00	75	25	0.4
5.00–10.00	65	35	0.4
10.00–20.00	25	75	0.4
20.00–20.10	5	95	0.4
20.10–25.00	5	95	0.4
25.00–25.10	90	10	0.4
25.10–30.00	90	10	0.4

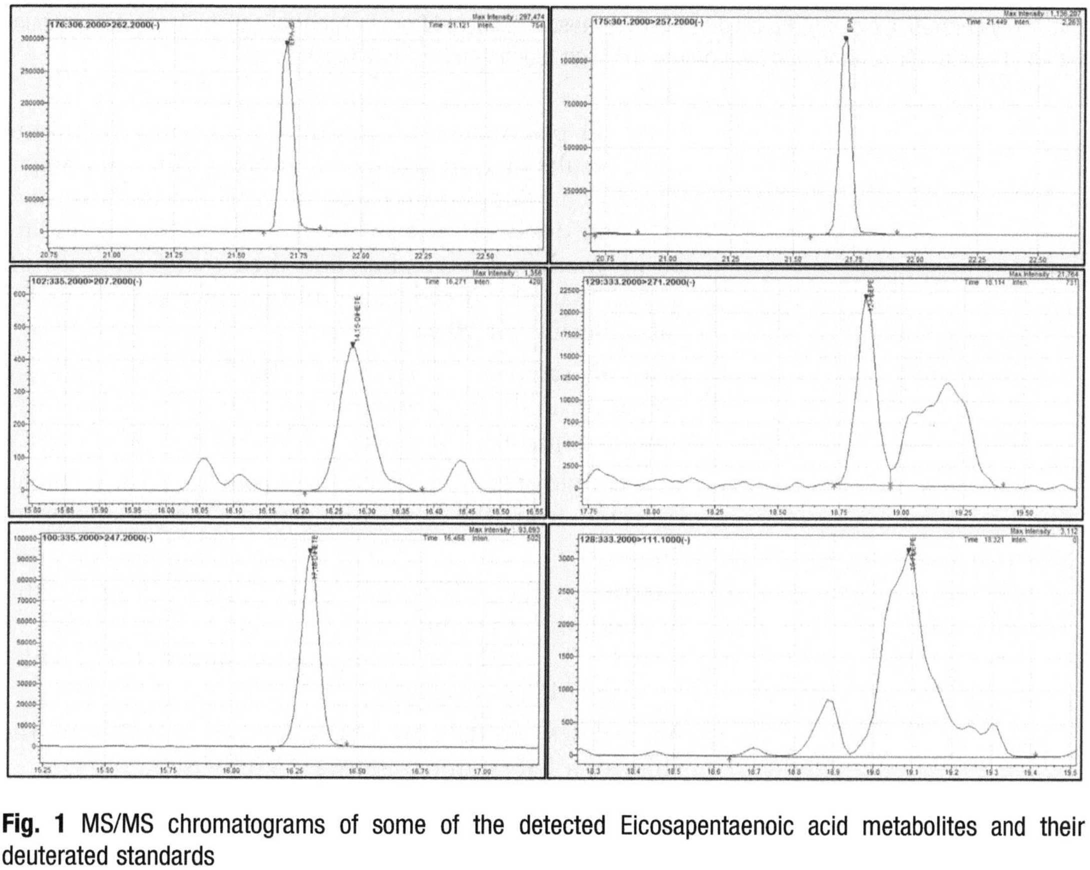

Fig. 1 MS/MS chromatograms of some of the detected Eicosapentaenoic acid metabolites and their deuterated standards

(g) The browser tab in the LabSolutions software is used to analyze and check the calibration curve accuracy with at least $R^2 > 99.99\%$.

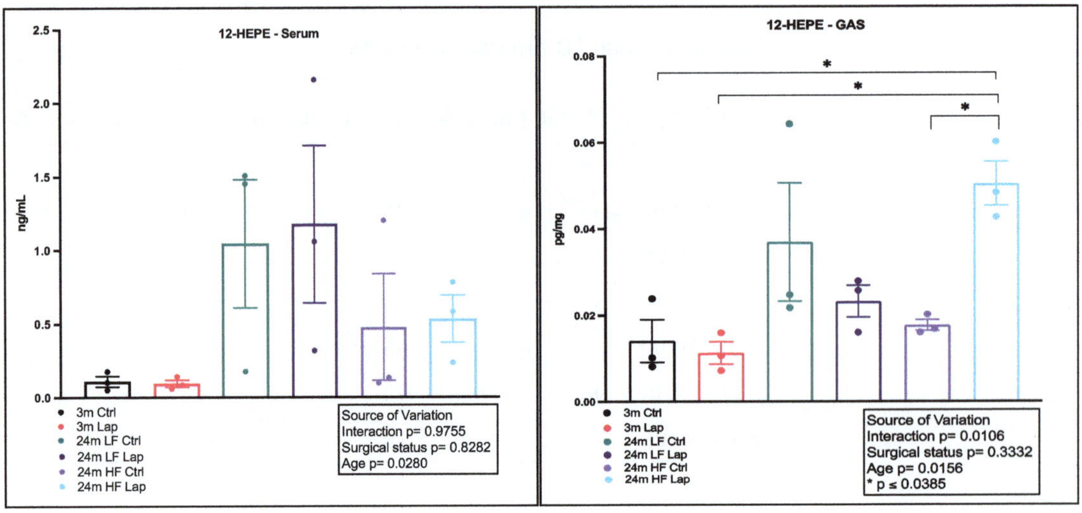

Fig. 2 Concentration of 12-HEPE is serum versus muscle collected from the control and the laparotomy mouse model. Groups: *Ctr* control, *Lap* laparotomy, *HF* high functioning, *LF* low functioning

(h) Once all parameters are optimized and the calibration curve results met the standards, samples will begin preparation. The samples prepared by SPE are reconstituted with the dried extracts in the 1.5 mL tubes by adding 50 μL methanol, vertexing, and then transferring 20 μL of each sample directly into autosampler glass vials for auto injection of 10 μL.

(i) After all samples are ran successfully, the raw data are uploaded into the browser with the calibration curve results to quantify the detected LMs in the tested samples.

(j) An example of the quantified LMs from the serum samples collected from the laparotomy model[47] is presented in Fig. 2.

(k) External calibration curves are performed daily for this method, with the limits of quantitation (LOQ) of approximately 0.1 nM for each analyte under the present LC-MS/MS conditions.

(l) All analyses and data processing are performed using the Shimadzu LabSolutions software V5.99 SP2.

4 Notes

1. For MS/MS analysis, solid tissue samples such as muscles should typically weigh between 50 and 100 mg. Extracts from more than 100 mg samples can potentially cause noticeable ion-suppression effects and contaminate the LC-MS system. Therefore, for tissue samples larger than 100 mg, cutting

and using an appropriately sized section before extraction is necessary.

2. Wash volumes used in SPE are determined by the mass of the sorbent bed. Typically, 1 mL of solvent is needed per 100 mg of bed mass. Larger wash volumes can be used to ensure that the extract is as pure as possible before analysis.

Acknowledgemt and Fundings

This work was supported by the National Institute on Aging (NIA 5K08AG050808-05 and 1AG060341-01), VA Rehabilitation and Development Service (Grant #B2020-C), and the James J. Peters VA Medical Center. National Institutes of Health Grants: NIA 2-PO1AG039355, NIA-R01AG056504; National Institute of Diabetes and Digestive and Kidney Diseases R01DK119066 and National Institutes of Neurological Disorders and Stroke (NINDS) 2-R01NS105621. The authors are thankful for the generous support from the George W. and Hazel M. Jay and Evanston Research Endowments, and the UTA Shimadzu Institute for Research Technologies (https://www.uta.edu/sirt/).

References

1. Rhee P et al (2014) Increasing trauma deaths in the United States. Ann Surg 260(1):13–21
2. Smith J et al (2005) Abdominal trauma: a disease in evolution. ANZ J Surg 75(9):790–794
3. Parajuli P et al (2018) Role of laparoscopy in patients with abdominal trauma at level-I trauma center. Surg Laparosc Endosc Percutan Tech 28(1):20–25
4. Ozkan OV et al (2016) Laparoscopy in abdominal trauma. Curr Trauma Rep 2(4):238–246
5. Partridge JS, Harari D, Dhesi JK (2012) Frailty in the older surgical patient: a review. Age Ageing 41(2):142–147
6. Khan NA et al (2006) Association of postoperative complications with hospital costs and length of stay in a tertiary care center. J Gen Intern Med 21(2):177–180
7. Hamel MB et al (2005) Surgical outcomes for patients aged 80 and older: morbidity and mortality from major noncardiac surgery. J Am Geriatr Soc 53(3):424–429
8. Pofahl WE, Pories WJ (2003) Current status and future directions of geriatric general surgery. J Am Geriatr Soc 51(7 Suppl):S351–S354
9. Hajibandeh S et al (2019) Meta-analysis of the effect of sarcopenia in predicting postoperative mortality in emergency and elective abdominal surgery. Surgeon 17(6):370–380
10. Aggarwal G et al (2019) Evaluation of the collaborative use of an evidence-based care bundle in emergency laparotomy. JAMA Surg 154(5):e190145
11. Saunders DI et al (2012) Variations in mortality after emergency laparotomy: the first report of the UK Emergency Laparotomy Network. Br J Anaesth 109(3):368–375
12. Broughton KJ et al (2017) The Perth emergency laparotomy audit. ANZ J Surg 87(11): 893–897
13. Mak M, Hakeem AR, Chitre V (2016) Pre-NELA vs NELA – has anything changed, or is it just an audit exercise? Ann R Coll Surg Engl 98(8):554–559
14. Jeppesen MM et al (2019) A nationwide cohort study of short- and long-term outcomes following emergency laparotomy. Dan Med J 66(1):A5523
15. Bentrem DJ et al (2009) Identification of specific quality improvement opportunities for the elderly undergoing gastrointestinal surgery. Arch Surg 144(11):1013–1020

16. Finlayson E et al (2012) Functional status after colon cancer surgery in elderly nursing home residents. J Am Geriatr Soc 60(5):967–973
17. Lawrence VA et al (2004) Functional independence after major abdominal surgery in the elderly. J Am Coll Surg 199(5):762–772
18. Chua MSH, Chan DKH (2020) Increased morbidity and mortality of emergency laparotomy in elderly patients. World J Surg 44(3): 711–720
19. Rose J et al (2015) Estimated need for surgery worldwide based on prevalence of diseases: a modelling strategy for the WHO Global Health estimate. Lancet Glob Health 3(Suppl 2):S13–S20
20. Kleif J, Gögenur I (2018) Severity classification of the quality of recovery-15 score – an observational study. J Surg Res 225:101–107
21. Ko F, Isoda F, Mobbs C (2015) Laparotomy in mice induces blood cell expression of inflammatory and stress genes. J Interf Cytokine Res 35(4):302–312
22. Cata JP et al (2021) A rat model to investigate quality of recovery after abdominal surgery. Pain Rep 6(2):e943
23. Brzeszczyński FF, Brzeszczyńska JI (2023) Markers of sarcopenia increase 30-day mortality following emergency laparotomy: a systematic review. Scand J Surg 112(1):58–65
24. Desborough JP (2000) The stress response to trauma and surgery. Br J Anaesth 85(1): 109–117
25. Edwards H, Rose EA, King TC (1982) Postoperative deterioration in muscular function. Arch Surg 117(7):899–901
26. Watters JM et al (1993) Impaired recovery of strength in older patients after major abdominal surgery. Ann Surg 218(3):380–390. discussion 390-3
27. Christensen T, Kehlet H (1993) Postoperative fatigue. World J Surg 17(2):220–225
28. Christensen T, Kehlet H (1984) Postoperative fatigue and changes in nutritional status. Br J Surg 71(6):473–476
29. Siu AL et al (2006) Early ambulation after hip fracture: effects on function and mortality. Arch Intern Med 166(7):766–771
30. Creditor MC (1993) Hazards of hospitalization of the elderly. Ann Intern Med 118(3): 219–223
31. Zargar-Shoshtari K, Hill AG (2009) Postoperative fatigue: a review. World J Surg 33(4): 738–745
32. Morley JE (1997) Anorexia of aging: physiologic and pathologic. Am J Clin Nutr 66(4): 760–773
33. Ni Choileain N, Redmond HP (2006) Cell response to surgery. Arch Surg 141(11): 1132–1140
34. Menger MD, Vollmar B (2004) Surgical trauma: hyperinflammation versus immunosuppression? Langenbeck's Arch Surg 389(6): 475–484
35. Delogu G et al (2001) Mitochondrial perturbations and oxidant stress in lymphocytes from patients undergoing surgery and general anesthesia. Arch Surg 136(10):1190–1196
36. Liu PT et al (1994) Studies in surgical trauma: oxidative stress in ischaemia-reperfusion of rat liver. Clin Sci (Lond) 86(4):453–460
37. Peake J, Della Gatta P, Cameron-Smith D (2010) Aging and its effects on inflammation in skeletal muscle at rest and following exercise-induced muscle injury. Am J Physiol Regul Integr Comp Physiol 298(6):R1485–R1495
38. Gomes MJ et al (2017) Skeletal muscle aging: influence of oxidative stress and physical exercise. Oncotarget 8(12):20428–20440
39. Murakami M (2011) Lipid mediators in life science. Exp Anim 60(1):7–20
40. Mo C et al (2012) Prostaglandin E2: from clinical applications to its potential role in bone- muscle crosstalk and myogenic differentiation. Recent Pat Biotechnol 6(3):223–229
41. Markworth JF, Maddipati KR, Cameron-Smith D (2016) Emerging roles of pro-resolving lipid mediators in immunological and adaptive responses to exercise-induced muscle injury. Exerc Immunol Rev 22:110–134
42. Leuti A et al (2020) Bioactive lipids, inflammation and chronic diseases. Adv Drug Deliv Rev 159:133–169
43. Mo C et al (2015) Prostaglandin E2 promotes proliferation of skeletal muscle myoblasts via EP4 receptor activation. Cell Cycle 14(10): 1507–1516
44. Marrelli MT et al (2020) The skeletal muscles of mice infected with Plasmodium berghei and Plasmodium chabaudi reveal a crosstalk between lipid mediators and gene expression. Malar J 19(1):254
45. Al Saedi A et al (2022) Lipid metabolism in sarcopenia. Bone 164:116539
46. Shochat C et al (2021) Deletion of SREBF1, a functional bone-muscle pleiotropic gene, alters bone density and lipid signaling in zebrafish. Endocrinology 162(1):bqaa189
47. Asche-Godin SL et al (2022) RNA-sequencing reveals a gene expression signature in skeletal muscle of a mouse model of age-associated postoperative functional decline. J Gerontol A Biol Sci Med Sci 77(10):1939–1950

Chapter 10

Determining Isoprenoid-Facilitated Monomeric GTPase Turnover in Primary Human Trabecular Meshwork Cultures

Evan B. Stubbs Jr.

Abstract

Members of the Rho family of small monomeric GTPases regulate a plethora of critical cellular functions including gene expression, cell cycle progression, and the dynamic modeling of the actin cytoskeleton. Diversity among Rho family members is derived, in part, from variations in their subcellular distribution. Localization of newly synthesized (naïve) Rho proteins to target subcellular compartments is largely governed by lipid modifications, including posttranslational prenylation. Here, using well-established and widely available contemporary methodologies, detailed protocols by which to semiquantitatively evaluate the functional consequence of posttranslational prenylation in human trabecular meshwork cells are described. We propose the novel concept that posttranslational prenylation *itself* is a key regulator of mammalian Rho GTPase protein expression and turnover.

Key words Farnesyl, Geranylgeranyl, Human, Trabecular meshwork, Monomeric GTPase

1 Introduction

Newly synthesized small monomeric GTPases, including those of the Rho family, contain a unique carboxy terminal -CAAX motif that, when prenylated posttranslationally, serves to facilitate target membrane association [1–3]. Catalyzed by farnesyl transferase or geranylgeranyl transferase-I, the posttranslational prenylation of Rho GTPases involves the addition of either a sesquiterpene (farnesyl) or diterpene (geranylgeranyl) isoprenoid, respectively, to the terminal cysteine residue of these small monomeric proteins. Whereas a small fraction of newly synthesized GTPases may associate with subcellular compartments, the majority of naïve monomeric GTPases remain cytosolic awaiting lipid modifications necessary for target membrane association and subsequent activation [1, 4, 5].

Produced in mammalian cells de novo as metabolic intermediates of the cholesterol (mevalonate) biosynthetic pathway [6],

Marco Brotto and Sanjoy K. Bhattacharya (eds.), *Lipid Signalling: Methods and Protocols*, Methods in Molecular Biology, vol. 2816, https://doi.org/10.1007/978-1-0716-3902-3_10,

isoprenoids and their derivatives are not unique to eukaryotes and belong to a family of naturally occurring biologically active terpenoids [7, 8]. Synthesized as key constituents of membranes, isoprenoids also serve as metabolic building blocks of vitamins, pheromones, and reproductive hormones [9]. Involved in oxidative phosphorylation and photosynthesis, and as integral components of signal transduction pathways, isoprenoids are endogenous to a myriad of organisms as diverse as bacteria, fungi, insects, plants, and mammals [10]. In mammalian cells, farnesyl and geranylgeranyl isoprenoids function as posttranslational modifiers of not only small monomeric Ras and Ras-related GTPases but also a variety of intracellular proteins including nuclear lamins A and B, rhodopsin kinase, and γ subunits of heterotrimeric GTP-binding proteins [2].

While the biochemical consequences of posttranslational protein prenylation remain to be fully elucidated, one key biophysical attribute is to enhance protein hydrophobicity, thereby facilitating intracellular membrane localization and subsequent activation [3, 11, 12]. An alternative role for isoprenoids within mammalian cells may involve regulation of gene expression or protein stability [13–16]. By limiting endogenous prenylation, monomeric Rho GTPases were found to accumulate in the cytosol [1]. The mechanism by which unprenylated Rho GTPases localize to the cytosol may involve not only biophysical limitations but also appear to be a consequence of altered gene expression and protein stability [5]. These and other studies have generated considerable interest in evaluating the role of protein prenylation as a potential therapeutic target for the management of seemingly unrelated disorders involving dysregulation of monomeric GTPase signaling. Here, laboratory protocols using well-established and widely available contemporary methodologies to semiquantitatively evaluate the functional consequences of posttranslational protein prenylation in human trabecular meshwork cells are detailed.

2 Materials

Aqueous solutions should be prepared with ultrapure water exhibiting a resistivity of 18.2 MΩ.cm at 25 °C. Here, water is ultrapurified in a stepwise manner using a Milli-Q Advantage A10 system by first passing feed water through a carbon filter and subsequently subjecting the filtered water to reverse osmosis. Water is then passed through an Elix Advantage Progard T2 cartridge to a resistivity of 15 MΩ.cm and stored in a 100 L reservoir unit at 23 °C until needed. Ultra-purification is achieved on-demand by passing purified water through a Q-Gard T1 purification cartridge, a Quantum TIX polishing cartridge, and finally, a BioPak polishing cartridge. Unless indicated otherwise, reagents used should be of electrophoresis grade purity. Materials used for cell culture should be either purchased sterile or filtered, as indicated, through a single-use Corning polystyrene filter unit with a 0.22 μm cellulose acetate low-protein binding membrane. Cells should be cultured using pyrogen-free cell culture–treated sterile polystyrene flasks and maintained at 37 °C under an humidified atmosphere of 5% CO_2 and 95% air.

2.1 Primary Human TM Cell Culture

1. Fresh cadaver corneoscleral rim.
2. Three 35 × 10 mm sterile surface modified polystyrene *Primaria* tissue culture dishes (Corning 353801).
3. Complete MEM cell culture medium (ThermoFisher Scientific): Low glucose (1.0 g/L) Minimal Essential Medium (11095-072) supplemented with 2 mM L-glutamine, 5% adult bovine serum, 20% fetal bovine serum, 1× concentration of MEM non-essential (11140-050) and 1× concentration of MEM essential amino acids (11130-051), 100 U/mL penicillin/100 μg/mL streptomycin (15140-122), and 2.5 μg/mL amphotericin B (15290-026).
4. Enzymatic Digestion Cocktail: Type I Collagenase (4 mg/mL; Worthington Biochemical, LS004196) and Dispase II (4 mg/mL; Sigma-Aldrich, D4693) dissolved in Hanks Balanced Salt Solution (ThermoFisher Scientific, 14170-161) supplemented with bovine serum albumin (4 mg/mL).

2.2 Transformed Human TM Cell Culture

1. Transformed human trabecular meshwork cell line (GTM3).
2. Pyrogen-free cell culture–treated 25 cm^2 (Corning 430639) or 75 cm^2 (Corning 430641U) sterile polystyrene flasks.
3. Complete DMEM: High-glucose (4.5 g/L) Dulbecco's Modified Eagle's Medium (DMEM) containing 4 mM GlutaMAX-I (ThermoFisher Scientific, 10566-024) and supplemented with 10% fetal bovine serum, 100 U/mL penicillin, and 100 μg/mL streptomycin.
4. Trypsin-EDTA (0.05%) with phenol red (ThermoFisher Scientific, 2530-062).

2.3 Test Reagents

All test reagents used here are commercially available.

1. Lovastatin stock solution (10 mM stock concentration, 1000×): This inactive lactone pro-drug that must first be converted by alkaline hydrolysis to the active dihydroxy acid prior to use in cell culture. Dissolve 4 mg of lovastatin in 0.1 mL absolute ethanol, add 0.15 mL of 0.1 N NaOH, and incubate for 2 h at 50 °C. Neutralize the resultant dihydroxy acid with 0.1 N HCl and bring to a final volume of 1 mL with sterile ultrapure water, aliquot, and store at −20 °C until use.

2. S-Mevalonate stock solution (10×): Prepare 50 mM stock solution of S-mevalonate (Sigma-Aldrich 44714) in MEM, aliquot, and store at −20 °C until use.
3. Farnesyl pyrophosphate (FPP) and geranylgeranyl pyrophosphate (GGPP) stock solutions: These isoprenoids are purchased from Sigma-Aldrich (FPP, F6892; GGPP, G6025) and are supplied as 1 mg/mL (2.2 mM) concentrates in a methanol: ammonium hydroxide solution.
4. Geranylgeranyl transferase inhibitor-298 (GGTI-298) stock solution (1000×): Prepare 10 mM stock solutions of GGTI-298 (Sigma-Aldrich, G5169) in DMSO, aliquot, and store at −20 °C until use.
5. Cycloheximide and Epoxomicin, and stock solutions (1000×): Prepare stock solutions of cycloheximide (5 mM; Sigma-Aldrich, C1988) in culture media, aliquot, and store at −20 °C until use. Prepare stock solutions of epoxomicin (10 mM; Sigma-Aldrich, E3652) in DMSO, aliquot, and store at −20 °C until use.

2.4 Electrophoresis and Electrotransfer

1. Precast Mini-PROTEAN tris-glycine 4–20% polyacrylamide gels (10 or 15 well, 8.6 × 6.8 cm).
2. Gel running buffer (pH 8.3): 10× concentrate containing 1.92 M L-glycine, 1% sodium dodecyl sulfate (SDS), and 0.25M Tris-base.
3. Laemmli sample buffer (2× concentrate) is commercially available.
4. 10× Cleland's Reagent (dithiothreitol, 0.5M)
5. Pre-stained protein molecular weight standards (3.5–260 kDa range).
6. Electrotransfer buffer (pH 8.3): 0.2 M L-glycine, 0.1% sodium dodecyl sulfate (SDS), 20% methanol, and 0.025 M Tris-base.
7. Nitrocellulose membranes (0.2 μm).
8. Mini trans-blot filter paper.

2.5 Immunoblotting

1. Ponceau S Stock Solution (10×): Ponceau S (2% w/v; Sigma-Aldrich, P3504) in a solution of trichloroacetic acid (30% w/v Sigma-Aldrich, T4885) and sulfosalicylic acid (30% w/v; Sigma-Aldrich, S2130).
2. Blocking buffer: *Carnation* instant nonfat powdered milk (5% w/v) suspended in 50 mM Tris-buffered isotonic saline (pH 7.4) supplemented with 0.05% Tween-20 (*see* **Note 1**).
3. Washing buffer: 50 mM Tris-buffered isotonic saline (pH 7.4) supplemented with 0.05% Tween-20.

4. Mouse anti-RhoA monoclonal (clone 26C4) & rabbit anti-RhoB polyclonal (Santa Cruz Biotechnology) and rabbit anti-GAPDH polyclonal (Trevigen) antibodies.
5. Goat anti-mouse IgG and goat anti-rabbit IgG horseradish peroxidase-conjugated secondary antibody (Jackson ImmunoResearch Laboratories).
6. SuperSignal West Pico Chemiluminescent Substrate.
7. High performance chemiluminescence film (Amersham Hyperfilm™ ECL).

3 Methods

3.1 Preparation and Culture of Human Primary TM Cells

Human primary TM cells are prepared fresh from surgically "discarded" donor corneoscleral rims as detailed further and previously described [17]. With the aid of a dissecting microscope, the TM is lifted from the rim as a continuous strip by gentle teasing using a pair of titanium straight fine toothed forceps (World Precision Instruments, 555048FT). To increase TM cell yield, the isolated TM strip is immediately immersed in a freshly prepared enzymatic digestion cocktail prior to cell culture. Authenticity of TM cell origin should be routinely performed for every primary cell preparation [18]. Due to observed phenotypic changes with advanced passage, primary TM cell cultures should be limited to less than seven passages [19]. Approval for the use of human cadaver material should be obtained from your local *Institutional Review Board* prior to tissue procurement. Advanced approval from your local *Institutional Biosafety Committee* as well as other required administrative committees is similarly recommended. The following tissue culture procedures should be conducted within a laminar flow hood using aseptic techniques.

1. In association with your Institutional Ophthalmology Department, establish a mechanism whereby whole or partial cadaver corneoscleral rims can be recovered as "discarded" material at time of corneal transplant. Once the corneal transplant button is excised, instruct the surgical team to return the unused "discarded" corneoscleral rim to the original corneal storage medium.
2. Using sterile technique, transfer the corneoscleral rim to a *Primaria* tissue culture dish containing 3 mL complete MEM. Be careful to place the cornea side down so as to expose the brown-colored remnant ciliary body and beige-colored TM [17]. Culture overnight at 37 °C in a humidified atmosphere of 95% air/5% CO_2.
3. The next day, prepare fresh enzymatic digestion cocktail and transfer 3 mL to a new *Primaria* tissue culture dish.

4. Under a well-illuminated dissecting microscope, identify the boundary between the clear/opaque cornea and the off-white sclera. The trabecular meshwork (TM) strip is nestled in a trough/valley adjacent to the cornea bordered by Schwalbe's line and the scleral spur.
5. Adjacent to the scleral spur, locate the remnants of the pigmented ciliary body band.
6. Use a pair of fine forceps to remove the remaining pigmented ciliary body band.
7. Once the remnant ciliary body band has been cleared, the beige-toned TM can be isolated by gently teasing one end out of the trough and lifting it away from the rim. With practice, one can remove the TM strip *en block* with one steady motion (*see* **Note 2**).
8. Transfer the isolated TM strip (explant) to the *Primaria* tissue culture dish containing the enzymatic digestion cocktail.
9. Incubate the TM strip for no less than 1.5 h at 37 °C in a humidified atmosphere of 95% air/5% CO_2.
10. Liberate the digested TM strip with gentle trituration and transfer the resultant cell suspension into two 1.5 cc sterile Eppendorf tubes.
11. Sediment the dispersed cells (1000 *g* × 15 min, 4 °C), discard ~80% of the supernatant, and resuspend the translucent cell "pellet" with 50 μL complete MEM. In many cases, the cell pellet will be difficult to visualize.
12. Transfer the cell suspension to the center of a new *Primaria* tissue culture dish and allow cells to contact and adhere to the dish by incubating ***undisturbed*** × 2 h at 37 °C in a humidified atmosphere of 95% air/5% CO_2.
13. Gently add in a circumferential manner 2 mL of complete MEM and culture overnight at 37 °C in a humidified atmosphere of 95% air/5% CO_2.
14. The next day, it is very important to gently rinse the cultured cells twice with 1 cc each of complete MEM per rinse and replace with 2 mL complete MEM.
15. Culture overnight at 37 °C in a humidified atmosphere of 95% air/5% CO_2.
16. Repeat **step 14**.
17. Replace media every 2–3 days.
18. After ~3 weeks, TM cell cultures will be semi-confluent (*see* **Note 3**). At this time, primary TM cells can be passaged. Aspirate the culture medium, wash once with 1 mL Hanks Balanced Salt solution to remove residual serum-containing media, and add 1 mL of 0.05% Trypsin-EDTA.

19. Gently rotate the flask to allow adherent cells to be coated with the trypsin-EDTA solution. Incubate for 3 min @ 37 °C and transfer dissociated cell suspension to a 15 mL conical centrifuge tube containing 6 mL complete MEM.
20. Centrifuge the cell suspension (751 *g* × 5 min, 4 °C) and gently resuspend the packed cell pellet with 0.5 mL complete MEM.
21. Transfer the entire 0.5 mL cell suspension to a T25 cm^2 tissue culture flask containing 4.5 mL complete MEM (considered the first passage and noted as p1). Gently swirl the flask to evenly distribute the cells. Be aware that TM cell cultures have been reported to change with increasing passage [19]. Thus, the use of TM cell cultures should be limited to no more than seven passages (p7) (*see* **Note 4**).

3.2 Culture of Human Transformed TM Cells

1. Aspirate and discard media from a confluent T75 cm^2 flask of primary human TM cells.
2. Add 5 ml of Hanks Balanced Salt Solution (HBSS) and gently rotate the flask to allow adherent cells to be "washed" free of serum. Aspirate and discard the HBSS.
3. Add 2 ml of 0.05% Trypsin-EDTA solution and allow to coat the washed cells.
4. Incubate for 3 min @ 37°C and transfer dissociated cell suspension to a 15 ml conical centrifuge tube containing 5 ml complete MEM.
5. Centrifuge the cell suspension (751 *g* × 5 min, 4°C) and gently resuspend the packed cell pellet with 5 ml of complete MEM containing 10% (v/v) DMSO.
6. Transfer cell suspension into storage vials (1 ml per vial), step freeze by placing vials in a cell freezer at −80°C × 48 h, store in a liquid nitrogen Dewar.

3.3 Culture of Human Transformed TM Cells

Simian virus 40 (SV40)-transformed human TM cells used here are from a male glaucomatous patient (GTM3) and were a generous gift from A. F. Clark [20]. The following tissue culture procedures should be conducted within a laminar flow hood using aseptic techniques.

1. Aspirate and discard the culture medium from a confluent T75 cm^2 flask of GTM3 cells.
2. Wash adherent cells with 5 mL of Hanks Balanced Salt solution to remove residual serum-containing media. Aspirate and discard the wash medium.

3. Add 2 mL of sterile 0.05% trypsin-EDTA to the washed cells. Gently rotate the flask to allow adherent cells to be coated with the trypsin-EDTA solution. Incubate × 3 min @ 37 °C.
4. Confirm that the cells are beginning to dislodge from the flask bottom and add 5 mL of complete DMEM. Gently agitate the flask and transfer cell suspension to a 15 mL sterile conical centrifuge tube.
5. Centrifuge the suspended cells (751 *g* × 5 min @ 4 °C). Resuspend by triturating the packed cell pellet in complete DMEM to a density of approximately 1×10^6 cells per mL.
6. Transfer 0.5 mL (~500,000 cells) to a new T75 cm^2 tissue culture flask containing 9.5 mL complete DMEM. Gently swirl the flask to evenly distribute the cells.
7. Culture at 37 °C under an humidified atmosphere of 95% air/5% CO_2.
8. Transformed TM cells can be stored long-term using the procedure outlined in Subheading 3.2.

3.4 Protein Prenylation Alters Rho GTPase Protein Stability

The effect of posttranslational prenylation on GTPase protein expression and stability in human TM cells is semiquantitatively determined using cell-permeable inhibitors that either disrupt endogenous availability of isoprenoids or prevent the transfer of isoprenoids onto native GTPases.

1. To determine whether altering endogenous isoprenoid availability effects monomeric GTPase protein expression, add to semi-confluent (~70%) replicate T75 cm^2 flasks of primary or transformed human TM cells (10 mL total volume) either 10 μL of a 10% solution of absolute ethanol (Veh, 0.01% final concentration) or 10 μL of a 10 mM working concentrate of activated lovastatin (Lov, 10 μM final concentration) in the absence or presence of 10 μM farnesyl pyrophosphate (FPP) or 10 μM geranylgeranyl pyrophosphate (GGPP) (Fig. 1, *see* **Note 5**).
2. To determine whether isoprenoid availability alters monomeric GTPase protein subcellular compartmentalization, pretreat (24 h) semi-confluent (~70%) replicate T75 cm^2 flasks of transformed human TM cells (10 mL total volume) with either 10 μL of a 10% solution of absolute ethanol (V, 0.01% final concentration) or 10 μL of a 10 mM working concentrate of activated lovastatin (10 μM) and subsequently chase (0–6h) pretreated cells with either media alone or with media supplemented with either 5 mM L-mevalonate, 10 μM GGPP, or 10 μM FPP + 10 μM GGTI-298 (Fig. 2, *see* **Note 6**).
3. To determine whether isoprenoids alter GTPase protein stability, pretreat (24 h) semi-confluent (~70%) replicate T75 cm^2

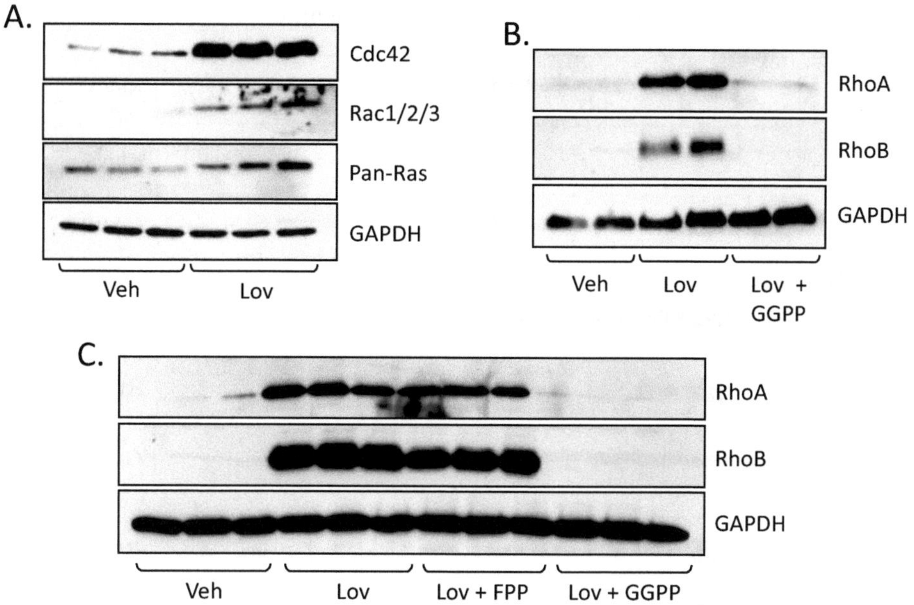

Fig. 1 Representative immunoblots of small monomeric GTPases present in lysates from (**a**, **c**) transformed or (**b**) primary human TM cells treated (24–48 h) in the absence (Veh, 0.01% ethanol) or presence of 10 μM activated lovastatin (Lov) without or with 10 μM farnesyl pyrophosphate (Lov + FPP) or 10 μM geranylgeranyl pyrophosphate (Lov + GGPP), as indicated. (Reproduced from ref. 1 with permission from Springer)

flasks of transformed human TM cells (10 mL total volume) with 10 μL of a 10 mM working concentrate of activated lovastatin (10 μM f.c.) and subsequently chase (0–24h) pretreated cells with fresh media containing 5 μM cycloheximide (a potent protein synthesis inhibitor) or with the same lovastatin-containing media supplemented with 5 μM cycloheximide in the absence (Control) or presence of 10 μM GGPP (Fig. 3).

4. To determine whether GGPP-facilitated decreases in Rho GTPase protein content occur as a result of proteasomal activation, pretreat (24 h) semi-confluent (~70%) replicate T75 cm^2 flasks of transformed human TM cells (10 mL total volume) with 10 μL of a 10 mM working concentrate of activated lovastatin (10 μM f.c.) and subsequently chase (24 h) pretreated cells with the same lovastatin-containing media supplemented with 10 μM GGPP ± 10 μM epoxomicin (a 20*S* proteasome inhibitor) (Fig. 4). Additional experiments are recommended to determine whether observed decreases in GGPP-facilitated Rho GTPase protein content are lysosomal-dependent (NH4Cl (0–30 μM or chloroquine 0–10 μM) or autophagy-dependent (0-10 mM 3-methyladenine or 0-10 μM wortmannin).

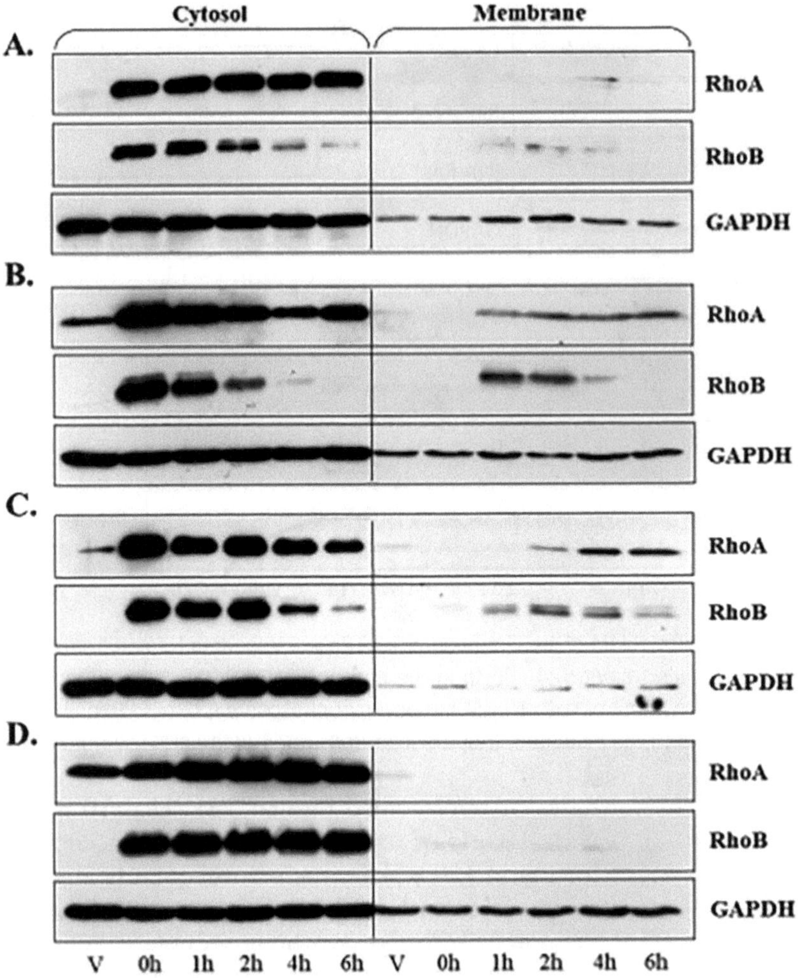

Fig. 2 Representative immunoblots of RhoA and RhoB GTPases expressed in soluble (cytosolic) or particulate (crude membrane) fractions prepared from pooled lysates of transformed human TM cells. Cell cultures ($n = 3$) were pretreated (24 h) with vehicle (V, 0.01% ethanol) or activated lovastatin (10 μM). Lovastatin-pretreated cells were washed and cultured for an additional 0-6 h, as indicated, in (**a**) media alone or in media supplemented with (**b**) L-mevalonate (5 mM), (**c**) GGPP (10 μM), or (**d**) FPP (10 μM) + GGTI-298 (10 μM). (Reproduced from ref. 14 with permission from Springer)

5. In each case, treated cells are harvested with 0.05% trypsin-EDTA (see **steps 3–5** in Subheading 3.2), wash the cell pellet once with fresh complete media. To determine the effect of added test agents on cell viability, gently mix an aliquot (15 μL) of harvested cells with an equal volume of 0.4% trypan blue dye (0.2% final concentration) for live-cell dye exclusion.
6. Prepare whole cell lysates by resuspending the washed cell pellet in 0.5 mL ultrapure water supplemented with a commercial cocktail of protease inhibitors. Store at −20 °C until use.

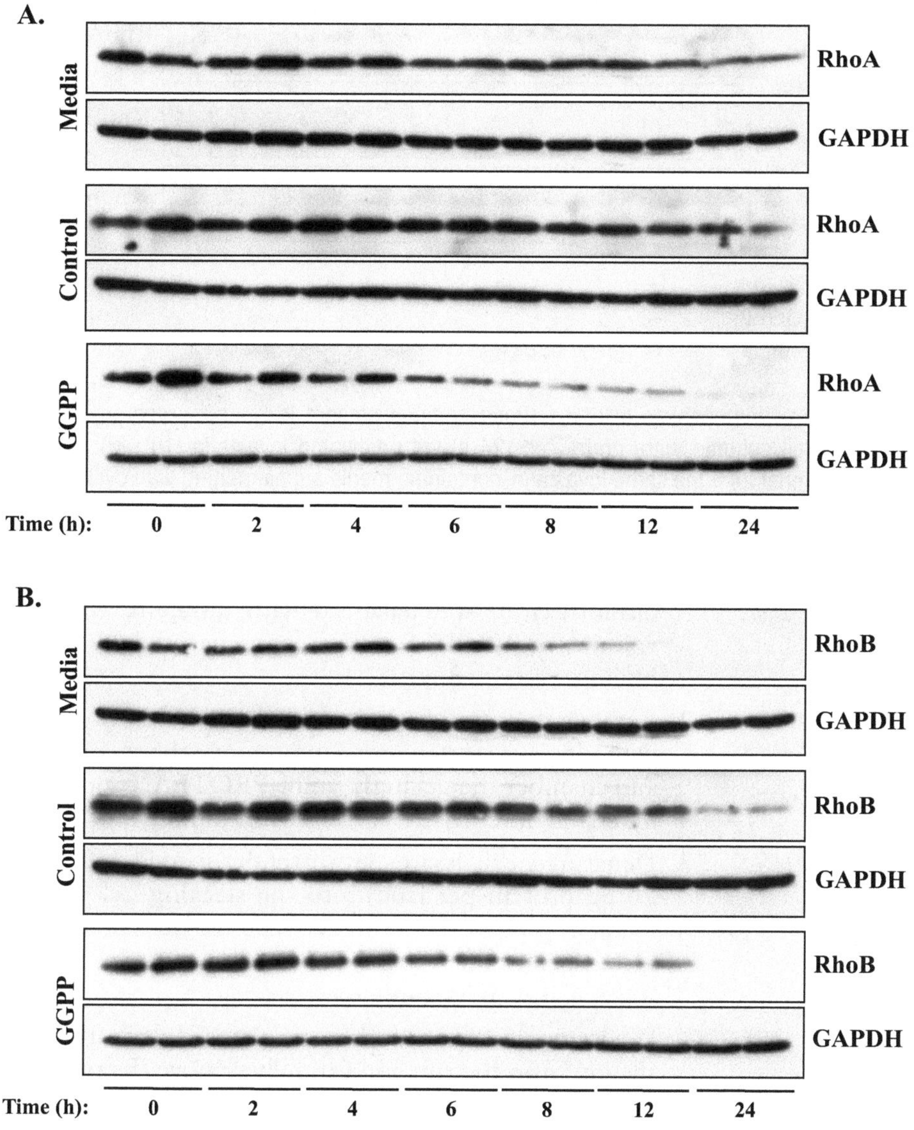

Fig. 3 Immunoblots of (**a**) RhoA and (**b**) RhoB GTPase proteins present in lysates prepared from transformed human TM cells. Cell cultures were pretreated (24 h) with activated lovastatin (10 μM) and subsequently chased with fresh media containing only cycloheximide (5 μM) or with the same lovastatin-containing media supplemented with cycloheximide (5 μM) in the absence (Control) or presence of GGPP (10 μM), as indicated. (Reproduced from ref. 14 with permission from Springer)

3.5 SDS-PAGE and Electrotransfer

Prior to electrophoresis, protein concentrations in cell lysates are determined using a commercially available protein assay kit (bicinchoninic acid, BCA). Here, proteins in cell lysates were resolved on pre-cast 4–20% denaturing gels using a mini-PROTEAN apparatus.

1. Cell lysates are thawed, probe sonicated on ice to homogeneity, and aliquots mixed with 20 μL 2× Laemmli sample buffer, 4 μL

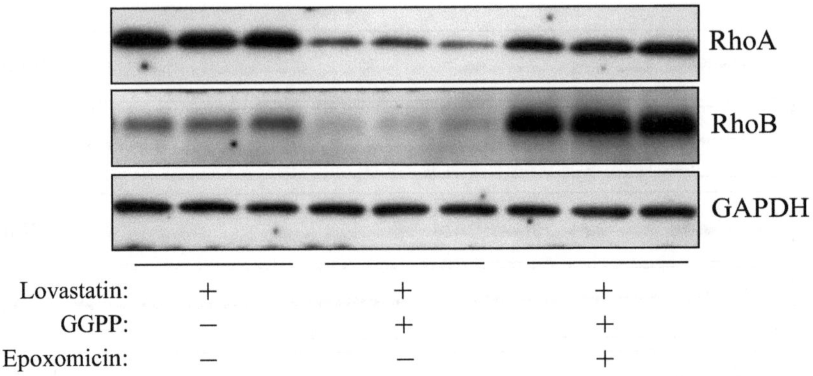

Fig. 4 Representative immunoblots of Rho GTPase proteins present in lysates prepared from transformed human TM cells. Cell cultures were pretreated (24 h) with activated lovastatin (10 μM) and subsequently chased for an additional 24 h the same lovastatin-containing media supplemented with cycloheximide (5 μM) in the absence or presence of GGPP (10 μM) and/or epoxomicin (10 μM), as indicated. (Reproduced from ref. 14 with permission from Springer)

dithiothreitol, and balanced with ultrapure water in a total volume of 40 μL. Samples are denatured by immersion in boiling water × 3 min.

2. Remove the comb from the pre-cast gel, rinse wells thoroughly with ultrapure water, score and remove the strip at the bottom, and chamber the gel in the mini-PROTEAN apparatus. Fill the inner and outer chamber wells with running buffer (*see* **Note 7**).
3. Denatured cell lysates are carefully loaded (12 μL per well, 20 μg protein per lane) into the stacking gel wells using gel loading pipet tips. Proteins resolved at room temperature across a constant voltage (100 V) alongside pre-stained protein molecular weight standards.
4. When the dye front reaches the exposed strip at the bottom of the gel, stop the run and carefully remove the plastic gel plates and rinse the gel with copious amounts of ultrapure water.
5. Slide the washed gel onto a glass support plate, cut away the stacking gel, and trim away unused side lanes and excess gel.
6. The trimmed gel is placed *front side* down on a filter pad pre-measured (gel size) and pre-soaked in electrotransfer buffer.
7. A pre-measured (gel-size) and pre-soaked piece of nitrocellulose membrane is positioned onto the gel surface (*see* **Note 8**).
8. Place another pre-measured (gel size) and pre-soaked filter pad onto the membrane. Remove any trapped air bubbles by rolling a clean test tube over the completed sandwich. Place the filter pad–gel membrane-filter pad sandwich in an electrotransfer cassette and submerge in electrotransfer buffer (*see* **Note 9**).
9. Electrotransfer the resolved proteins overnight at room temperature using a low constant current of 30 mAmps (*see* **Note 10**).

10. Disassemble the electrotransfer apparatus and using fine forceps gently peel the membrane off the gel. Immerse the membrane, protein side up, in ultrapure water (*see* **Note 11**).
11. Validate the quality and the efficacy of the electrotransfer by immersing the washed membrane in a solution of 0.2% Ponceau-S. Stain the membrane for no longer than 2 min.
12. Reduce background staining by dipping twice in ultrapure water and mark the lanes with a pencil.

3.6 Immunoblotting

1. Incubate the Ponceau-S stained membrane protein-side up in blocking buffer × 1 h at 37 °C with gentle rocking (*see* **Note 12**).
2. Rinse the blocked membrane with washing buffer and immerse protein-side up in 10 mL of fresh blocking buffer supplemented with a 1:1000 dilution of mouse anti-panRho monoclonal (clone 55), 1:200 dilution of mouse anti-RhoA monoclonal, or with a 1:200 dilution of rabbit anti-RhoB polyclonal antibody.
3. Incubate overnight at 4 °C with gentle rocking.
4. Rinse the immunostained membrane three times with 30 mL washing buffer for 5 min each and immerse protein-side up in 10 mL of fresh blocking buffer supplemented with appropriate horseradish peroxidase-conjugated secondary antibody (1: 2500 dilution of goat anti-mouse IgG or 1:10,000 dilution of goat anti-rabbit IgG) (*see* **Note 13**).
5. Incubate for 1 h at room temperature with gentle rocking.
6. Rinse the immunostained membrane three times with 30 mL washing buffer for 5 min each.
7. Remove the washed membrane and place protein-side up on tissue. Place the semi-dry membrane on a pre-cut piece of Saran Wrap and coat with an appropriate amount of freshly pre-mixed ECL reagent. Cover the membrane, place in an X-ray cassette.
8. In a dark room, place a piece of high performance chemiluminescence film over the Saran Wrap encased membrane, close the cassette to ensure good contact between film and membrane, and expose × 2–5 min depending on desired results.
9. Develop film using an auto-processor and semi-quantify relative changes in GTPase band densities. Normalize to GAPDH band densities present in the same sample on the same blot [1].
10. To confirm equal protein loading on the same membrane, immerse the developed immunoblot in 10 mL of Restore Western Blot stripping buffer for 10 min at room temperature, rinse three times with 30 mL washing buffer for 5 min each, and repeat steps 1–9 using a 1:10,000 dilution of rabbit anti-GAPDH as a primary antibody.

4 Notes

1. Increasing the percentage of Tween-20 (up to 0.2%) will reduce nonspecific background immunostaining, but at the costly expense of signal reduction as well.
2. Due to beige tone of the TM, it is very difficult to distinguish it from surrounding structures. Small stretches of the TM strip may break free and recoil around the tip of your forceps during the procedure. This is expected and is generally considered a good indicator that you are working with TM.
3. Depending on the clinical history of the donor, not all explants produce viable TM cells.
4. With increasing passage number, TM cell cultures become more reluctant to proliferate and may even change their phenotype.
5. In our hands, farnesyl or geranylgeranyl pyrophosphates were readily taken up by cultured TM cells by a mechanism that remains unclear.
6. GGTI-298 is a cell permeant inhibitor of geranylgeranyl transferase-I.
7. The gel should be immersed in running buffer up to, but not in communication with, the stacking gel.
8. Minimize moving the membrane once it is on the gel.
9. Care must be taken to be certain the membrane side of the sandwich is adjacent to the positive (red) pole and the gel side is adjacent to the negative (black) pole.
10. Proteins can be "hot"-transferred using a high current of 200 mAmps for 1 h if desired.
11. The orientation of the transfer should now be identical to the orientation used while loading the gel.
12. This will de-stain the membrane while blocking nonspecific binding sites.
13. Primary antibody solutions can be reused several times if desired. Add to the 10 mL of recovered solution 100 μL of a 2% sodium azide stock. Store the used solutions at 4 °C.

Acknowledgments

This work was supported, in part, by grants from the Department of Veterans Affairs, the Illinois Society for the Prevention of Blindness, and the Richard A. Perritt Charitable Foundation.

References

1. Stubbs EB Jr, Von Zee CL (2012) Prenylation of rho G-proteins: a novel mechanism regulating gene expression and protein stability in human trabecular meshwork cells. Mol Neurobiol 46(1):28–40
2. Zhang FL, Casey PJ (1996) Protein prenylation: molecular mechanisms and functional consequences. Annu Rev Biochem 65:241–269. https://doi.org/10.1146/annurev.bi.65.070196.001325
3. Gao J, Liao J, Yang GY (2009) CAAX-box protein, prenylation process and carcinogenesis. Am J Transl Res 1(3):312–325
4. Roberts PJ, Mitin N, Keller PJ, Chenette EJ, Madigan JP, Currin RO et al (2008) Rho family GTPase modification and dependence on CAAX motif-signaled posttranslational modification. J Biol Chem 283(37):25150–25163
5. Von Zee CL, Stubbs EB Jr (2011) Geranylgeranylation facilitates proteasomal degradation of rho G-proteins in human trabecular meshwork cells. Invest Ophthalmol Vis Sci 52(3):1676–1683
6. Lynen F (1967) Biosynthetic pathways from acetate to natural products. Pure Appl Chem 14(1):137–167. https://doi.org/10.1351/pac196714010137
7. Carruthers DN, Lee TS (2021) Diversifying isoprenoid platforms via atypical carbon substrates and non-model microorganisms. Front Microbiol 12:791089. https://doi.org/10.3389/fmicb.2021.791089
8. Kirby J, Keasling JD (2009) Biosynthesis of plant isoprenoids: perspectives for microbial engineering. Annu Rev Plant Biol 60:335–355. https://doi.org/10.1146/annurev.arplant.043008.091955
9. Kuzuyama T, Seto H (2012) Two distinct pathways for essential metabolic precursors for isoprenoid biosynthesis. Proc Jpn Acad Ser B Phys Biol Sci 88(3):41–52. https://doi.org/10.2183/pjab.88.41
10. Holstein SA, Hohl RJ (2004) Isoprenoids: remarkable diversity of form and function. Lipids 39(4):293–309. https://doi.org/10.1007/s11745-004-1233-3
11. Wright LP, Philips MR (2006) Thematic review series: lipid posttranslational modifications. CAAX modification and membrane targeting of Ras. J Lipid Res 47(5):883–891
12. Jeong A, Suazo KF, Wood WG, Distefano MD, Li L (2018) Isoprenoids and protein prenylation: implications in the pathogenesis and therapeutic intervention of Alzheimer's disease. Crit Rev Biochem Mol Biol 53(3):279–310. https://doi.org/10.1080/10409238.2018.1458070
13. Sever N, Song BL, Yabe D, Goldstein JL, Brown MS, DeBose-Boyd RA (2003) Insig-dependent ubiquitination and degradation of mammalian 3-hydroxy-3-methylglutaryl-CoA reductase stimulated by sterols and geranylgeraniol. J Biol Chem 278(52):52479–52490
14. Holstein SA, Wohlford-Lenane CL, Hohl RJ (2002) Isoprenoids influence expression of Ras and Ras-related proteins. Biochemistry 41(46):13698–13704
15. Holstein SA, Wohlford-Lenane CL, Hohl RJ (2002) Consequences of mevalonate depletion. Differential transcriptional, translational, and post-translational up-regulation of Ras, Rap1a, RhoA, AND RhoB. J Biol Chem 277(12):10678–10682
16. Von Zee CL, Richards MP, Bu P, Perlman JI, Stubbs EB Jr (2009) Increased RhoA and RhoB protein accumulation in cultured human trabecular meshwork cells by lovastatin. Invest Ophthalmol Vis Sci 50(6):2816–2823
17. Stubbs EB Jr (2023) Isoprenylation of monomeric GTPases in human trabecular meshwork cells. Methods Mol Biol 2625:217–230. https://doi.org/10.1007/978-1-0716-2966-6_19
18. Rao VR, Lautz JD, Kaja S, Foecking EM, Lukacs E, Stubbs EB Jr (2019) Mitochondrial-targeted antioxidants attenuate TGF-beta2 signaling in human trabecular meshwork cells. Invest Ophthalmol Vis Sci 60(10):3613–3624. https://doi.org/10.1167/iovs.19-27542
19. Keller KE, Bhattacharya SK, Borras T, Brunner TM, Chansangpetch S, Clark AF et al (2018) Consensus recommendations for trabecular meshwork cell isolation, characterization and culture. Exp Eye Res 171:164–173. https://doi.org/10.1016/j.exer.2018.03.001
20. Pang IH, Shade DL, Clark AF, Steely HT, DeSantis L (1994) Preliminary characterization of a transformed cell strain derived from human trabecular meshwork. Curr Eye Res 13(1):51–63

Chapter 11

α-Linolenic Acid Vesicles-Mediated Tau Internalization in Microglia

Subashchandrabose Chinnathambi

Abstract

In Alzheimer's disease, the synaptic loss is prominent due to the accumulation of Amyloid βeta (Aβ) protein in synapses, which affect neurotransmission, and thus ultimately causes neuronal loss. Tau, a microtubule-associated protein, is a vital protein of intracellular neurofibrillary tangles (NFTs) in AD. Along with the accumulation of aberrant proteins, glial cells, mainly astrocytes and microglia, play a major role in impairing neuronal network. Microglia have the ability to phagocytose Tau and rerelease in exosomes, which causes further spreading of Tau. Reduction in exosome synthesis can reduce spreading of Tau. Modulating microglia to clear the extracellular Tau seeds by its imported degradation would resolve the disease condition in Alzheimer's disease. In this study, we have shown the ability of α-linolenic acid (ALA) to inhibit the Tau aggregation and modulate their internalization property in microglial cells.

Key words α-Linolenic acid, Extracellualr Tau, Tau internalization, Tau microglia

1 Introduction

Tau is one of the microtubule-associated proteins that coordinates microtubule stabilization and has an important role in cell differentiation as well as polarization. The longest isoform of Tau (441 amino acids) consists of a projection domain, repeat domain along with the middle proline-rich domain [1–3]. The repeat domain is highly basic in nature due to the presence of lysine residues and contributes to major hydrophilic nature of Tau. The repeat domain acts as the core of PHFs formation, and hexapeptide motifs from repeat 2 and 3 contribute majorly to aggregation nucleation (Fig. 1a). Various anionic aggregation inducers like heparin, fatty acids were found to bind to repeat region, which helps to neutralize the charge and imitates PHF formation [4, 5]. Arachidonic acid as major abundant fatty acid in brain induces Tau assembly via charge compensation mechanism. α-Linolenic acid (ALA), a plant-based essential polyunsaturated

Marco Brotto and Sanjoy K. Bhattacharya (eds.), *Lipid Signalling: Methods and Protocols*, Methods in Molecular Biology, vol. 2816,
https://doi.org/10.1007/978-1-0716-3902-3_11,

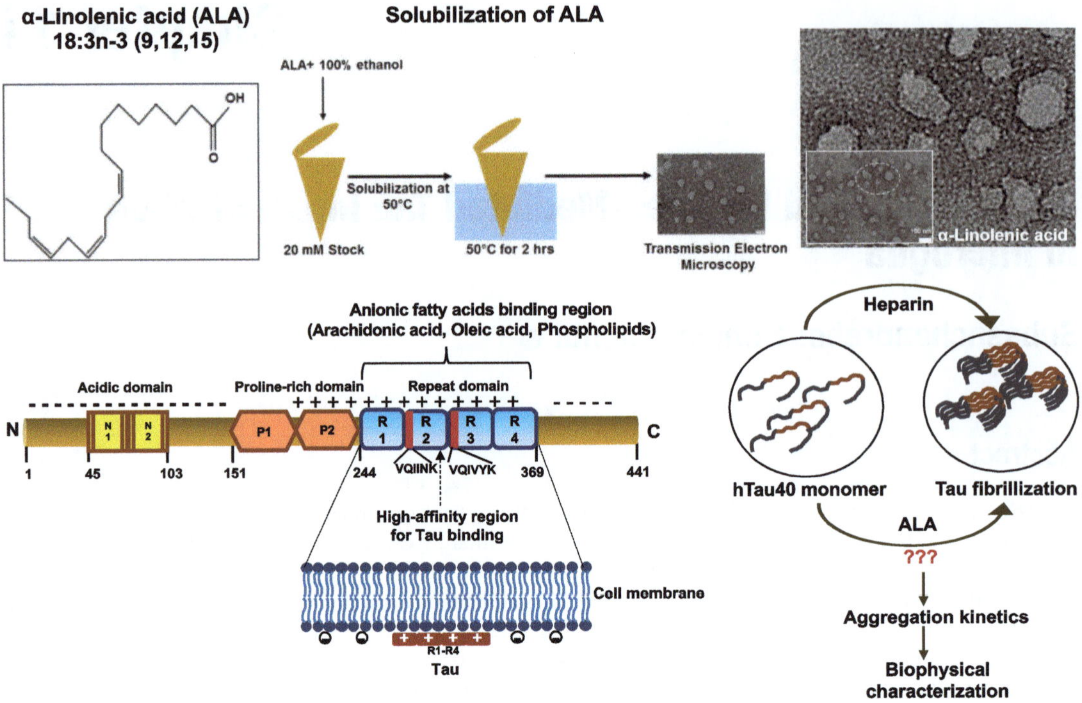

Fig. 1 Preparation of ALA vesicles. ALA is a 18c fatty acid with three double bonds and is a primary omega-3 fatty acid. The sample preparation of ALA was done by dissolving in 100% ethanol, and solubilization was carried out at 50° for 2 h. The vesicle formation of ALA was observed through HRTEM. The high positive charge imparted to Tau by repeat domain and poly proline region allows it to bind to polyanionic lipids and fatty acids from membrane and can cause its aggregation by charge compensation mechanism. Lipids have a high binding affinity for the various domains of Tau in this study to understand the ability of ALA to cause pathological fibrillization of Tau. Further, the biochemical and biophysical characterization of aggregation was carried out by using the ALA vesicles

omega-3 fatty acid (*n*-3 18:3), is a precursor of other important omega-3 fatty acids like DHA and EPA. ALA is a potential dietary nutrient to provide protection against coronary heart disease, brain injuries, etc. [6–9]. The necessary fatty acid for the brain function is DHA, which can be either synthesized from ALA or obtained from diet. The rate of synthesis of DHA from ALA is low, but evidence suggests that synthesis of DHA from ALA is sufficient to maintain the brain function [10]. Omega-3 fatty acid has neuroprotective properties and plays an important role in development and differentiation of neuronal cells. Docosahexaenoic acid (DHA-22:6n-3), Eicosapentaenoic acid (EPA-20:5n-3), and α-Linolenic acid (ALA 18 3n:3) are the most important long-chain fatty acids found in the brain and constitute a majority of the cell membrane. These fatty acids are incorporated in the cell membrane as long chains of phospholipids [11–14]. Omega-3 fatty acids are the vital factors to decide the polarity of microglia cells as M1 or M2. DHA tends to increase the M2 phenotype of microglia by changing the cell membrane composition and influencing the production of lipid mediators [15–18]. Omega-3 fatty acids also increase the phagocytosis of myelin debris, clear extracellular Aβ peptide in the AD brain environment, which is the major causes of neuroinflammation [19–21]. Free fatty acids are observed to induce self-assembly of both amyloid-β and Tau, whereas greater simulation is achieved with unsaturated fatty acids. To understand the effect of ALA on inducing assembly of Tau, Tau is exposed to ALA in the presence of buffer system and father allowed to aggregate for extended time. The aggregation kinetics can be studied by ThS fluorescence assay, SDS PAGE analysis, TEM analysis, and CD spectroscopy to understand the conformation changes in Tau protein upon aggregation [22–25]. Further, we aim to understand the effect of α-Linolenic acid, a precursor of DHA and EPA, on Tau internalization in microglial cells.

2 Materials

2.1 Cells

1. *N9* microglial cells
2. DH5α and BL21*

2.2 Chemicals and Antibodies

1. Protein expression and purification: LB media, Ampicillin, IPTG, NaCl, KCl, MgCl2, Na2HPO4, KH2PO4, MES, BES, CuSO4, glutaraldehyde, bovine serum albumin (BSA), ethylene glycol-bis (β-aminoethyl ether)- N,N,N′,N′-tetraacetic acid (EGTA), phenylmethylsulfonylfluoride (PMSF), bicinchoninic acid (BCA), 2 mM Dithiothreitol, methanol, protease inhibitor cocktail, isopropyl β-d-1-thiogalactopyranoside (IPTG),1 M NaCl.
2. SDS-PAGE reagents: Acrylamide, bis-acrylamide, tris, sodium dodecyl sulphate (SDS), ammonium persulfate (APS), tetramethylethylenediamine (TEMED), 6× protein loading dye, bio-rad dual color protein marker.
3. Chromatography materials: Cation exchange Sepharose fast flow column, Superdex 75 Hi load 16/600 column for size-exclusion chromatography, and 1× PBS.
4. Tau Aggregation Assay—Tau monomer (mg/mL), 25 mM NaCl, 1× Protease inhibitor cocktail (PIC), 0.01% Sodium azide, 1 mM DTT, Heparin (17,500 Da), BES pH 7.4, ThS fluorescent dye, Ammonium acetate, 384 black well micro plates, 2% Uranyl acetate, 400 mesh carbon-coated copper grids.
5. α-linolenic acid (ALA) vesicles preparation: α-linolenic acid (ALA), 100% molecular biology grade ethanol, and parafilm sheets.

6. Cell Culture reagents: N9 microglial cells, RPMI 1640 GlutaMax media, fetal bovine serum (FBS), trypsin-EDTA, 1× PBS buffer, penicillin-streptomycin, sterile centrifuge tubes, sterile petridishes, and multi-well plates.
7. Immunofluorescence assay reagents: 4% Paraformaldehyde solution in PBS, 1× PBS, Horse serum, Triton X-100, Moist tray, Parafilm sheet, 300 nM DAPI solution, Prolong antifade mounting media, glass slides, and sterile coverslips (12 & 18 mm).
8. Antibodies: Iba-1 Rabbit polyclonal antibody, K9JA Rabbit antibody, Phalloidin Alexa fluore-488 tagged, secondary antibody and control β-actin antibody.

2.3 Instruments

1. Laminar air flow with HEPA filter, CO_2 incubator at 5% CO_2 level.
2. Centrifuge, water bath and refrigerator.
3. AKTA pure chromatography system.
4. Optima MAX-XP ultracentrifuge (Beckman Coulter).
5. Avanti high-speed centrifuge.
6. Tecan plate reader Infinite 200 Pro.
7. SDS-gel electrophoresis unit (Bio-Rad).
8. Tecnai T20 transmission electron microscope 200 KeV (TEM).
9. FEI high-resolution transmission electron microscope 300 KeV (HR-TEM).
10. Zeiss Axio observer 7.0 with Apotome 2.0 microscope.

2.4 Softwares

1. ZEN 2.3 software.
2. ImageJ software.
3. SigmaPlot 10.
4. GraphPad Prism 8.0.1.

3 Methods

3.1 Expression and Purification of Tau

1. Use bacterial strains from previously stored glycerol stocks for culture inoculation.
2. Cultivate *E. coli* strains, DH5α and BL21* in LB broth for plasmid isolation and bacterial expression, respectively. After the plasmid transformation into BL21* bacterial strains, culture the cells on LB agar plates.
3. Inoculate the cells from isolated colonies in LB media in the presence of ampicillin (100 μg/mL) and incubate at 37 °C shaker incubator till the absorbance at 600 nm reaches 0.5.

4. Induce the protein expression with 0.5 mM IPTG. Further, incubate the culture for 4 hours before harvesting.
5. Pellet down the cells by centrifuging at 4000 rpm for 10 min at 4 °C.
6. Resuspend the cell pellet in a cell lysis buffer for isolating the protein. The Lysis buffer comprises 50 mM MES at pH 6.8, 1 mM EGTA, 2 mM $MgCl_2$, 5 mM DTT, 1 mM PMSF, and protease inhibitor cocktail. Lyse the cells at 15000 psi using a constant cell disruption system at 4 °C.
7. Heat the lysate at 90 °C in the presence of 0.5 M NaCl and 5 mM DTT to remove the contamination of proteins.
8. Perform dialysis for the collected supernatant with a buffer containing 50 mM NaCl to remove excess salt from the solution.
9. Carry-out the protein purification with two-step chromatography method. i.e., cation-exchange chromatography, size-exclusion chromatography.
10. Purify the dialyzed protein supernatant by cation-exchange chromatography using Sepharose fast flow column where elution will be performed with a gradient of 1 M NaCl.
11. Identify the Tau-containing fractions by SDS-PAGE, pool the fractions, concentrate, and further perform size-exclusion chromatography in a Superdex 75 Hi load 16/600 column by using 1× PBS and 2 mM DTT.
12. Confirm the protein quality by SDS-PAGE, pool the fractions and concentrate by using 5 kDa centrifugal filters.
13. Estimate the concentration of purified Tau by bicinchoninic acid assay (BCA) with1mg/ml BSA as a protein standard.
14. Aliquot and store the purified protein at −80 °C.

3.2 Preparation of ALA Vesicles

α-Linolenic acid (ALA), a plant-based essential polyunsaturated omega-3 fatty acid (n-3 18:3), is a precursor of other important omega-3 fatty acids, DHA and EPA (Fig. 1). The characterization of ALA micelles by transmission electron microscopy (TEM) and high-resolution TEM visualizes the vesicle-like structure formation of ALA (Fig. 1).

1. Prepare 20 mM of α-linolenic acid by dissolving in 100 μL of 100% Ethanol. (*see* **Note 1**).
2. Allow the ALA in ethanol solution to dissolve at 50 °C for 1–2 h (*see* **Note 2–3**). After 2 h of incubation at 50 °C, store the prepared ALA sample at 4 °C until the treatment (Fig. 1).
3. Observe the ALA vesicles under transmission electron microscopy (TEM) by preparing the grid using 40 μM ALA (working concentration).

4. Place 10 μL of 40 μM ALA on the parafilm and invert the carbon-coated TEM grid on the sample for 1.5 min with the help of tweezers. Soak the excess sample with Whatman 1 filter paper strips (*see* **Note 4–5**).
5. Lift the grid with tweezers and place on a drop of filtered MilliQ for washing twice for 30 s.
6. Finally, stain the grid with 2% Uranyl acetate for 5 min. Soak the excess of Uranyl acetate with Whatman 1 filter paper strips (*see* **Note 6**).
7. Observe ALA vesicles by TECNAI T20 120 kV Transmission electron microscope instrument. Capture images from 10,000× magnification to highest 1,000,000× magnification, which will help to visualize the minute details in the present structure. (*see* **Note 7–9**) [19].

3.3 Preparation of Tau Aggregates

The aggregation of Tau is driven majorly by PTMs and mutations that alter its charge and conformation. However, in vitro conditions mimic several cellular factors including reducing environment, ionic strength, inducer, etc., leading to Tau aggregation [15–17]. The full-length Tau has two cysteine residues that modulate kinetics and morphology of aggregates formed. The cysteine residues are involved in the formation of inter or intrachain disulfide bonds in Tau aggregates. The intrachain disulfide bonds formed during Tau aggregation alter the Tau assembly by forming compact Tau monomers. This results in increase of lag time for Tau aggregation. In the presence of DTT, interchain disulfide bonds are formed that accelerate Tau aggregates in the presence of inducer. To evaluate the ability of ALA in contributing to pathological aggregation of Tau, we expose Tau to ALA in the presence of buffer system and allow to aggregate for 120 hours [18–20].

1. Use heparin (17,500 Da) to induce aggregation of Tau with-Tau: heparin ratio 4:1 (*see* **Note 10**).
2. Incubate 20 μM Tau in 20 mM BES buffer containing 25 mM NaCl, 2 mM DTT, 5 μM heparin, 0.01% NaN_3 and protease inhibitor cocktail as per standard procedure established in our lab (*see* **Note 11**).
3. Incubate the reaction mixture at 37 °C for 120–150 h to understand the kinetics of Tau aggregation.
4. Characterize the Tau aggregates after 120 h of incubation by various biochemical and biophysical methods-like SDS-PAGE, ThS fluorescence, and TEM analysis (*see* **Note 12**).
5. To determine the role of ALA in Tau aggregation, incorporate ALA in the reaction mixture containing Tau for aggregation (Fig. 2).

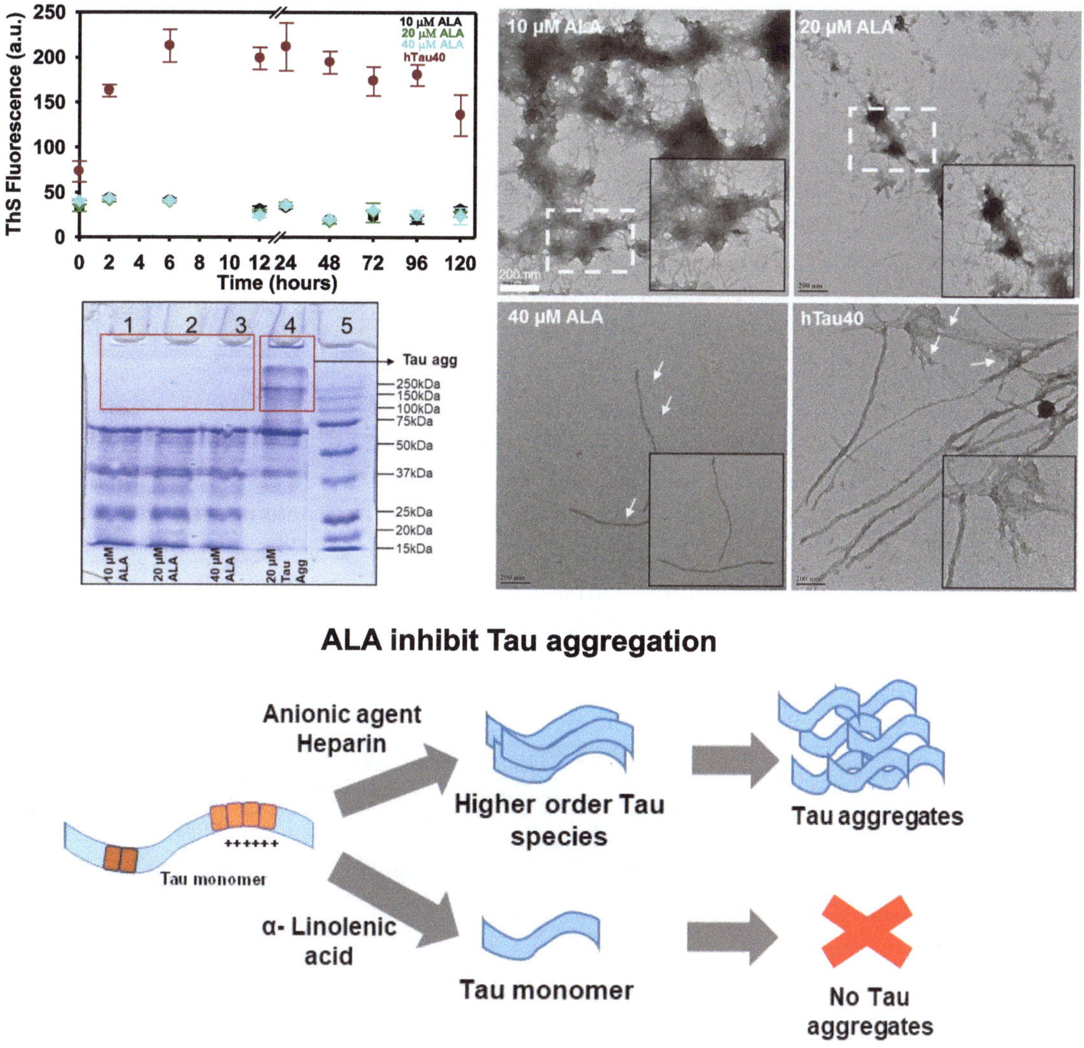

Fig. 2 ALA vesicles inhibit Tau aggregation in vitro. The ability of ALA to induce Tau aggregation was monitored with ThS fluorescence assay with 10, 20, and 40 μM of ALA for 120 h. Tau aggregates were analyzed by SDS-PAGE. The higher-order species are marked in red. The morphology of Tau fibril upon incubation with ALA and inducer, heparin was observed with TEM. The ALA vesicles inhibited the spontaneous assembly of Tau into aggregates, unlike other anionic unsaturated fatty acids and inducer

6. Add 10, 20, and 40 μM ALA at fixed Tau concentration (20 μM) in an assembly buffer to study its aggregation inducing/inhibiting capacity (*see* **Note 13**).
7. Incubate the reaction mixture at 37 °C and monitor the aggregation using fluorophores such as Thioflavin S (ThS), Thioflavin T (ThT), and 8-Anilinonaphthalene-1-sulfonic acid (ANS).

3.4 Culturing Microglia for Internalization of Tau

To study the ability of ALA to cause anti-inflammatory effects in microglia, observe the internalization and phagocytosis of extracellular Tau. Treat N9 microglial cells with 1 μM Tau monomer, 1 μM Tau aggregates along with 40 μM ALA as a test, with respective controls for comparison. In the immunofluorescence analysis, detect the K9JA Tau (red) and Iba-1 (green) intensity in microglia cells after exposure of Tau and ALA. (Fig. 3).

1. Grow N9 microglial cells in RPMI media supplemented with 10% heat-inactivated serum, penicillin-streptomycin antibiotic solution and glutamine in a T25 flask or 60 mm dish to maintain the culture.
2. Passage the cells using 0.25% trypsin-EDTA solution after washing with PBS, on attaining 90% confluency or more.
3. For immunofluorescence, seed 25,000 cells of N9 microglia on 12 mm coverslip in 24-well plate supplemented with 10% FBS and 1% penicillin-streptomycin.
4. During treatment of hTau40 monomer/aggregates and ALA, cells should be supplemented with 0.5% serum-deprived RPMI

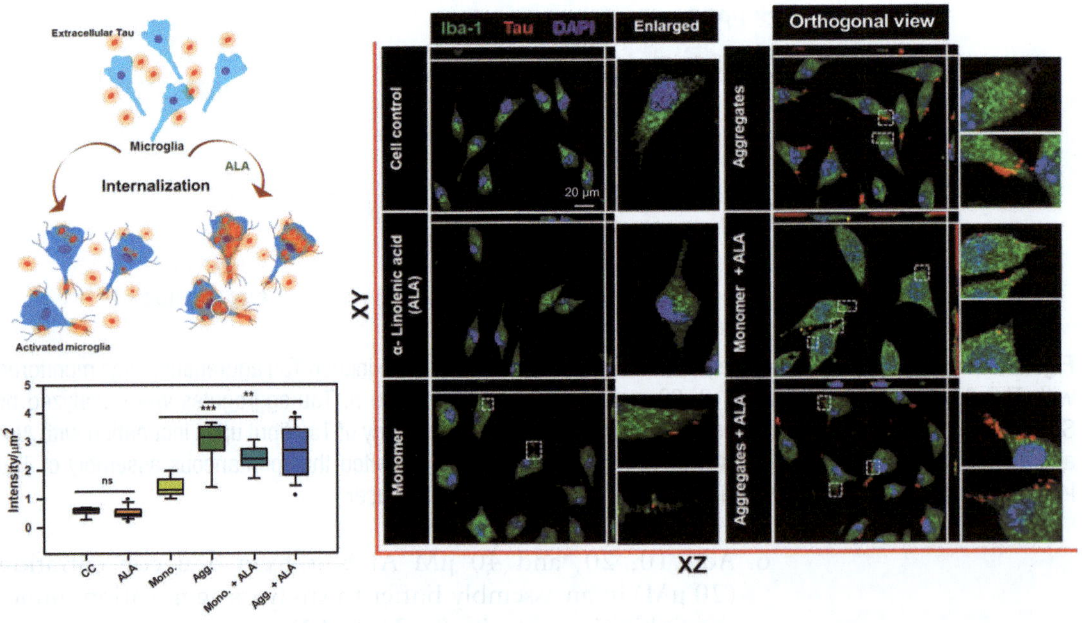

Fig. 3 Extracellular Tau internalization escalated by ALA in microglia. The microglial internalization of extracellular Tau in the presence of ALA can help in Tau clearance in AD. The internalization was studied by using immunofluorescence microscopy. Further, we have employed various analysis of the images and 3D subsectioning of the image and presented as a orthogonal view to understand the presence of protein inside the cell at x and y plane. Iba-1 is one of the calcium-binding proteins, which recognizes the marker for microglia/macrophages, and thus, it has actin remodeling activity, phagocytosis membrane ruffling, and various other cytoskeleton activities in migratory cells

media. Treat the cells with hTau40 monomer, aggregates, and ALA for 24 hours (*see* **Note 14**).

5. Fix the cells with 4% paraformaldehyde solution upon treatment for 20 min at room temperature then wash with 1× PBS (*see* **Note 15**).
6. Before staining, permeabilize the cells using 0.2% Triton X-100 in 1× PBS buffer (PBSt) for 15 min followed by blocking with 5% horse serum in PBSt buffer for 1 h at room temperature (*see* **Note 16**).
7. Prepare primary antibody/phalloidin solution at appropriate dilutions in blocking buffer for cell staining and incubate coverslips for overnight at 4 °C in a moist chamber (*see* **Note 17**).
8. The next day, wash the cells with 1× PBS buffer thrice for 10 min each.
9. Incubate the coverslips with secondary antibody conjugated with Alexa fluor anti-rabbit 555 (1:500) diluted in blocking buffer at 37 °C for 1 h.
10. Further, wash the cells thricewith 1× PBS buffer and counterstain with DAPI (300 nM) solution for nuclear staining (*see* **Note 18**).
11. Mount the coverslips on a glass slide with 5–10 μL prolong antifade mounting media.
12. Place the coverslips at RT overnight for curing before scanning in the microscope.
13. Observe the cells of different treatment groups under a 63× oil immersion in Axio observer 7.0 Apo-tome 2.0 Carl Zeiss microscope.

4 Notes

1. The stock of ALA should always be stored at 4 °C, and the tubes or the bottle should have opened minimum times, since there are high possibilities of peroxidation of fatty acids.
2. The Eppendorf tubes used for preparation of ALA should be tightly covered with parafilm, since the solubilization takes place at 50 °C and there are possibilities of concentrating the samples.
3. Briefly, ALA was dissolved in 100% molecular biology grade ethanol and solubilized at 50 °C in the stock concentration of 20 mM.
4. The working concentration of ALA 40 μM was decided according to previous studies, and the final concentration of ethanol

in the culture was maintained below 0.5% to avoid solvent-mediated toxicity.

5. ALA-based vesicles are freshly prepared before each and every experiment. This would allow the intact structure of ALA-based vesicles.
6. The carbon grids should be kept in a desiccator or dried properly before imaging to get the proper contrast. The wet grids give hazy structures that cannot be focussed with TEM.
7. The higher and lower magnification would help to understand the entire field of interest and the morphological differences in ALA vesicles.
8. The fatty acid preparation is very sensitive, and they undergo peroxidation and degradation very rapidly; hence, the preparation should be fresh before every experiment. To store the prepared lipids for longer time, alternatively, they can be dried with nitrogen purging and stored at −80 °C. This would keep the vesicles intact for longer time.
9. For the comparison purpose, the samples of every reaction mixtures were analyzed at 0 h as well by SDS-PAGE analysis. This would help to distinguish the higher-order aggregates formation after the incubation period of 120 h.
10. Tau is a highly soluble and natively unfolded protein, which does not form any kind of precipitates, but on the other hand, Tau form aggregates if not handled carefully.
11. Any biochemical or biophysical characterization of Tau will be conducted prior to the pelleting assay. At the end of pelleting, only supernantant was collected without touching the bottom of the tube. There will be a small amount of pellet that can nucleate the aggregation process.
12. The higher-order aggregates of Tau were analyzed by SDS-PAGE since Tau forms SDS stable aggregates. The increased molecular weight of aggregates can be visualized as protein bands at higher molecular weight.
13. The number of microglial cells for the various internalization experiments are standardized as per the laboratory protocol. The various concentrations of ALA vesicles have been standardized in microglial cells for the efficacy of their toxicity.
14. During the exposure, the ALA and Tau species will be first mixed in an Eppendorf tubes before directly adding to the seeded cells. The sudden inclusion of various compounds may affect the cells adhesion, and the microglial cells will respond differently.

15. The pH of 4% PFA solution should be maintained properly at 7.4, and the fixed cells should be washed properly with PBS for PFA removal.
16. The percentage of Triton-X100 should be maintain at 0.2%, increased percentage might disrupt the cell integrity.
17. While undergoing primary or secondary antibody staining during immunofluorescence processing, the dish containing coverslips should be maintained in humid environments with the help of tissues or filter papers and MilliQ water. Since antigen-antibody reactions are always performed in a humid condition, this also helps in retaining the antibody solutions intact and prevent its drying.
18. The coverslips should allow to dry after DAPI treatment as it creates background while scanning the coverslips.

Acknowledgments

We are grateful to Chinnathambi's lab members for their scientific discussions, helpful suggestions, and critical reading of the manuscript. The author greatly acknowledge Smita Desale for helping in the preparation of figures and for helpful discussions.

Funding

This project is supported in part by grant from India ENIGMA Initiative for Global Aging & Mental Health grant R01AG060610 and under National Institute of Aging/National Institute of Health (NIA/NIH), USA.

References

1. Desale SE, Chinnathambi S (2020) Role of dietary fatty acids in microglial polarization in Alzheimer's disease. J Neuroinflammation 17: 1–14
2. Kopf A, Renkawitz J, Hauschild R, Girkontaite I, Tedford K, Merrin J, Thorn-Seshold O, Trauner D, Häcker H, Fischer K-D (2020) Microtubules control cellular shape and coherence in amoeboid migrating cells. J Cell Biol 219:e201907154
3. Bloom GS (2014) Amyloid-β and Tau: the trigger and bullet in Alzheimer disease pathogenesis. JAMA Neurol 71(4):505–508
4. Desale SE, Chinnathambi S (2021) Phosphoinositides signaling modulates microglial actin remodeling and phagocytosis in Alzheimer's disease. Cell Commun Signal 19:1–12
5. Desale SE, Dubey T, Chinnathambi S (2021) α-Linolenic acid inhibits Tau aggregation and modulates Tau conformation. Int J Biol Macromol 166:687–693
6. Desale SE, Chinnathambi S (2021) α-Linolenic acid induces clearance of Tau seeds via actin-remodeling in microglia. Mol Biomed 2:1–14
7. Das R, Balmik AA, Chinnathambi S (2020) Phagocytosis of full-length Tau oligomers by actin-remodeling of activated microglia. J Neuroinflammation 17:1–15
8. Desale SE, Chinnathambi S (2021) α–Linolenic acid modulates phagocytosis and

endosomal pathways of extracellular Tau in microglia. Cell Adhes Migr 15:84–100
9. Chen G-f, Xu T-h, Yan Y, Zhou Y-r, Jiang Y, Melcher K (2017) Amyloid beta: structure, biology and structure-based therapeutic development. Acta Pharmacol Sin 38(9): 1205–1235
10. Desale SE, Chinnathambi S (2020) Role of dietary fatty acids in microglial polarization in Alzheimer's disease. J Neuroinflammation 17(1):1–14
11. Sallaberry CA, Voss BJ, Majewski J, Biernat J, Mandelkow E, Chi EY (2021) Tau and membranes: interactions that promote folding and condensation. Front Cell Dev Biol 21 (9):725241
12. Murphy DA, Courtneidge SA (2011) The 'ins' and 'outs' of podosomes and invadopodia: characteristics, formation and function. Nat Rev Mol Cell Biol 12:413–426
13. Ballestrem C, Wehrle-Haller B, Hinz B, Imhof BA (2000) Actin-dependent lamellipodia formation and microtubule-dependent tail retraction control-directed cell migration. Mol Biol Cell 11:2999–3012
14. Zech T, Calaminus SD, Machesky LM (2012) Actin on trafficking: could actin guide directed receptor transport? Cell Adhes Migr 6:476–481
15. Desale SE, Chidambaram H, Chinnathambi S (2021) G-protein coupled receptor, PI3K and Rho signaling pathways regulate the cascades of Tau and amyloid-β in Alzheimer's disease. Mol Biomed 2(1):1–18
16. Desale SE, Chidambaram H, Chinnathambi S (2022) α-Linolenic acid induces extracellular Tau internalization and its activation in microglial cells. In: Tau protein-II, methods and protocols, Methods in molecular biology. Humana Press, New York
17. Desale SE, Chidambaram H, Chinnathambi S (2022) Biochemical and biophysical characterization of Tau and α-linolenic acid vesicles in vitro. In: Tau protein-II, methods and protocols, Methods in molecular biology. Humana Press, New York
18. Chidambaram H, Desale SE, Chinnathambi S (2022) Interaction of Tau with G-protein coupled purinergic P2Y12 receptor by molecular docking and molecular dynamic simulation. In: Tau protein-II, methods and protocols, Methods in molecular biology. Humana Press, New York
19. Chidambaram H, Desale SE, Chinnathambi S (2022) Purinergic receptor P2Y12-mediated Tau internalization in microglia. In: Tau protein-II, methods and protocols, Methods in molecular biology. Humana Press, New York
20. Desale, S. E., & Chinnathambi, S. (2023). α-Linolenic acid-induced podosome-like structure formation and spontaneous higher-order assemblies ameliorate microglial migration in Tauopathy. https://doi.org/10.1101/2020.04.15.042143
21. Desale, S. E., & Chinnathambi, S. (2023). Microtubule dependent polarization and migration of microglia conserved by α-linolenic acid and extracellular Tau-mediated stimulation. https://doi.org/10.21203/rs.3.rs-2695599/v1
22. Mandelkow E-M, Mandelkow E (2012) Biochemistry and cell biology of Tau protein in neurofibrillary degeneration. Cold Spring Harb Perspect Med 2(7):a006247
23. Gustke N, Trinczek B, Biernat J, Mandelkow E-M, Mandelkow E (1994) Domains of Tau protein and interactions with microtubules. Biochemistry 33(32):9511–9522
24. Chi SW, Kim DH, Lee SH, Chang I, Han KH (2007) Pre-structured motifs in the natively unstructured preS1 surface antigen of hepatitis B virus. Protein Sci 16(10):2108–2117
25. King ME, Ahuja V, Binder LI, Kuret J (1999) Ligand-dependent Tau filament formation: implications for Alzheimer's disease progression. Biochemistry 38(45):14851–14859

Chapter 12

Imaging Phospholipase D Activity with Clickable Alcohols via Transphosphatidylation

Shyamal P. Waghwala, Taylor Ngo, and Ainsely Verkaik

Abstract

Phospholipase D (PLD) is an enzyme with many functions, one of which is the synthesis of phosphatidic acid (PA), a molecule with a myriad of effects on various organ systems and processes. These numerous roles make it hard to understand the true action of PA in cellular and bodily processes. Imaging PLD activity is one way to better understand the synthesis of PA and start to elucidate its function. However, many of the current imaging techniques for PLD come with limitations. This chapter presents a thorough methodology of a new imaging technique for PLD activity with clickable alcohols via transphosphatidylation (IMPACT) and Real-Time IMPACT (RT-IMPACT) that takes advantage of clickable chemistry to overcome current limitations. Using strain-promoted azide-alkyne cycloaddition (SPAAC), inverse electron-demand Diels-Alder (IEDDA), and the synthesis of various organic compounds, this chapter will explain a step-by-step procedure of how to perform the IMPACT and RT-IMPACT method(s).

Key words Phospholipase D, Transphosphatidylation, IMPACT, SPAAC, Clickable alcohols

1 Introduction

Phospholipase D (PLD) is a protein that serves to hydrolyze glycerophospholipids, such as phosphatidylcholine (PC), to create a headgroup and phosphatidic acid (PA), a lipid second messenger with pleiotropic effects [1]. Initially, it was thought that PA was created primarily from other enzymes such as phospholipase C (PLC) and diacylglycerol; however, studies later determined that PLD was responsible for the creation of PA [2, 3]. There are many subtypes of PLD including PLD1 and PLD2, which are involved in generating PA as previously discussed. Other subtypes like PLD3 and PLD4 accumulate in a cell's endoplasmic reticulum with PLD3 potentially having a role in Alzheimer's and PLD4 playing a role in triggering autoimmune conditions. Lastly, PLD6 has been shown to cleave mitochondrial cardiolipin to PA and cleave long RNAs during spermatogenesis [4]. For the purposes of this protocol, we

Marco Brotto and Sanjoy K. Bhattacharya (eds.), *Lipid Signalling: Methods and Protocols*, Methods in Molecular Biology, vol. 2816,
https://doi.org/10.1007/978-1-0716-3902-3_12,

will be focusing on PLD1 and PLD2 due to their ability to generate PA. This function of generating PA represents a major role that these proteins play in human biology. As a lipid second messenger, PA's pleiotropic effects include various secretory processes like hormone, cytokine, and neurotransmitter release [5]. PA can also affect cell survival signals through PLD activity, such as mTOR and MAPK signaling [5]. Additionally, specific PLD1 and PLD2 isoforms have been found to have importance in vesicular trafficking, Golgi function, and cytoskeletal reorganization [1]. As has been discussed, PLD normally creates PA from PC by using water as the nucleophile; however, early studies found that PLD can alternatively use glycerol or other primary alcohols as the nucleophile. These alternatives result in a transphosphatidylation reaction producing a stable substrate that does not interact with lipid phosphatases—a reaction exploited for the technique discussed in this protocol [1, 6].

The pleiotropic effects of PLD and PA make it difficult to understand the exact effect that these enzymes have on different cells within the human body. Imaging PLD is one way to better elucidate the locations in which this enzyme acts along with its associated biological context. The following are a couple techniques that are used to image PLD activity [4]. One such method is to use liposomes with ^{3}H-PC to perform an in-vitro analysis of activators and inhibitors of PLD1 and PLD2 separately from cellular regulation. In this process, ^{3}H-Palmitic Acid can be used due to its product ^{3}H-Ptd-alcohol being identifiable via TLC; however, the time this process takes may result in alterations to the PA. Fluorescent sensors can also be used to tag PA-binding domains. Though these methods seem promising, each has inherent drawbacks lowering their ideality. These drawbacks include its rapid turnover, the presence of other enzymes creating PA, and disproportionate radioactive labeling [4]. Additionally, the techniques that measure transphophatidylation using TLC or LC-MS cause a loss of spatial information [7]. Due to these complications, many of the primary approaches to imaging PLD activity require ex-vivo approaches [8].

The outlined alternative, Imaging PLD Activity with Clickable Alcohols using Transphosphatidylation (IMPACT), provides solutions to some of these problems. The rationale behind this reaction is to use clickable, bioorthogonally tagged, primary alcohols allowing for easy tagging with fluorophores [9]. By measuring these fluorophores, single cell and subcellular measurements can be made. It has been found that both alkynyl and azido primary alcohols allow for PLD transphosphatidylation and subsequent detection with Copper-catalyzed Azide-alkyne Cycloaddition (CuAAC) or Strain-Promoted Azide-Alkyne Cycloaddition (SPAAC) and a fluorophore [9]. These techniques allow for single cell measurement of PLD activity. To measure sub-cellular concentrations of PLD, Real-Time IMPACT (RT-IMPACT) can be used, which uses the Inverse Electron-Demand Diels-Alder (IEDDA) reaction between a Trans-Cyclooctene (TCO) and tetrazine (Tz) instead of the SPAAC reaction [10]. This reaction happens at a much faster rate and does not require a wash out step, thus helping to eliminate the challenge of rapid turnover faced by other imaging modalities [10]. Additionally, specific inhibitors for PLD1 and PLD2 (as will be specified in the methods) are used to ensure that the PLD signaling is truly from PLD and not another enzyme [10].

2 Materials

2.1 For Single Cell Quantification

1. 24-well plate.
2. 750 nM 5-fluor-2-indolyl deschlorohalopemide (FIPI) pan-PLD inhibitor.
3. 250 nM PLD1 isoform-selective inhibitor, VU0359595.
4. 350 nM PLD2 isoform-selective inhibitor, VU0364739.
5. 100 nM phorbol 12-myristate 13-acetate (PMA).
6. 1 mM 3-azido-1-propanol (AzProp).
7. Phosphate-buffered saline (PBS).
8. Fetal bovine serum (FBS).
9. 1 μM BCN-BODIPY or BODIPY-cyclooctene.
10. Tyrode's HEPESbuffer: 135 mM NaCl, 5 mM KCl, 1.8 mM $CaCl_2$, 1 mM $MgCl_2$, 1 mg/mL glucose, 1 mg/mL bovine serum albumin, 20 mM HEPES, pH 7.4.
11. Trypsin.

For bulk measurement:

1. 60-mm dish.
2. Multiple 1.5 mL Eppendorf tubes.
3. Multiple 0.45 μm filters.
4. Ice.
5. PMA.
6. AzProp.
7. 125 μL 20 mM acetic acid.
8. 250 μL methanol.
9. 250 μL chloroform.
10. Stream of nitrogen gas.
11. 250 μL of chloroform:methanol:water mixture (73:23:1 v:v:v).
12. 2.5 μM of BCN-BODIPY.

13. 7 μL of degassed chloroform.
14. 30 μL of reaction master mix:
    - Reaction master mix (for one reaction) contains: 100 μg Alk-QA, 6 μL of 10 mM [acetonitrile]$_4$CuBF$_4$ in degassed acetonitrile and 24 μL of degassed ethanol.

RT-IMPACT including reagent synthesis (The following synthetic reagents are used as needed with concentrations ranging between 100 mM and 1 μM):

1. Glycidol.
2. Tert-Butyl-(chloro)-diphenylsilane (TBDSCI).
3. Imidazole.
4. Dichloromethane (DCM) ×4.
5. Allyl magnesium chloride.
6. Dilithium tetrachlorocuprate (Li_2CuCl_4).
7. Tetrahydrofuran (THF).
8. Chloromethyl methyl ether (MOMCl).
9. N,N-Diisopropylethylamine (DIEA).
10. 4-Dymethylaminopyridine (DMAP).
11. Stannic chloride ($SnCl_4$).
12. Allyltrimethylsilane (allyl-$SiMe_3$).
13. Denzylidene-bis(tricycloheylphosphine).
14. Dichlororuthenium (Grubbs first-generation catalyst).
15. Tetra-n-butylammonium fluoride (TBAF, n-Bu_4NF).
16. 254 nm light.
17. Methyl benzoate ($PhCO_2Me$).
18. Diethyl ether (Et_2O).
19. Silver nitrate impregnated silica gel ($AgNO_3 \cdot SiO_2$).
20. 1,3,5,7-Tetramethyl-8-(4-cyanophenyl-)BODIPY (p-cyanophenyl-BODIPY).
21. 50 mol% zinc trifluoromethanesulfonate (zinc triflate).
22. Hydrazine (N2H4).
23. Acetonitrile (MeCN).
24. Two 35-mm glass dishes.
25. DMEM.
26. 4 mL Tyrode's HEPES buffer.
27. PMA.
28. Red-fluorescent ER-tracker (Thermo Fischer) if tracking PM to ER trafficking.

3 Methods

3.1 Single Cell Quantification [10]

1. Day before experiment, seed 125,000 cells per well on a 24-well plate.
2. Next day, negative control samples are treated with one of the following specific inhibitors for 30 min prior to labeling:
   (a) PLD1/2 dual inhibitor, FIPI, 750 nM.
   (b) PLD1 isoform-selective inhibitor, VU0359595, 250 nM.
   (c) PLD2 isoform-selective inhibitor, VU0364739, 350 nM.
3. Incubate half-cells (including both those with and without inhibitor) with 100 nM phorbol 12-myistate 13-acetate (PMA) (an agonist for the growth of endogenous PLD) for 20 min followed by AzProp, 1 mM, with or without the inhibitor, serving as a positive control.
4. Rinse the cells three times with PBS.
5. Subject the cells to a strain-promoted azide-alkyne cycloaddition (SPAAC) using 1 μM BCN-BODIPY (BODIPY-cyclooctyne) in Tyrode's-HEPES buffer for 10 min at 37 °C.
6. Rinse the cells two times with PBS.
7. Incubate the cells at 37 °C in Tyrode's-HEPES buffer for 10–15 min (*see* **Note 1**).
8. Lift the cells with trypsin and transfer to 96-well plate.
9. Rinse the cells twice with cold PBS containing FBS (0.5% v/v).
10. Resuspend the cells in 50 μL cold PBS containing FBS (0.5% v/v).
11. Analyze the cells using flow cytometry (*see* **Note 2**).

3.2 Bulk Measurements of PLD Activity [10]

1. Seed 500,000 cells on a 60-mm dish on the day prior to experimentation.
2. On the day of experimentation, incubate the cells with PMA, 100 nM, and AzProp (3-Azido-1-propanol), 1 mM for approximately 20 min.
3. Place the cells on ice, aspirate the media and rinse with cold PBS to remove excess alcohol.
4. Add a solution of 100 μL cold PBS, 125 μL 20 mM acetic acid, and 250 μL methanol to stop the labeling.
5. Scrape and transfer the cells to a 1.5 mL Eppendorf tube.
6. Add 250 μL of chloroform to the tube.
7. Vortex the mixture for 1 min followed by centrifugation at 13,300×*g* until a pellet is formed.

8. Transfer the bottom organic layer to a new tube.
9. Add 250 μL to the original tube (not the organic layer).
10. Vortex and centrifuge (13,300×*g*) this new mixture.
11. Combine the organic layers (bottom of both tubes) and dry under a stream of nitrogen gas.
12. Follow the protocol in next section depending on if using HPLC (recommended) or mass spectrometry.

3.3 If Using Normal-Phase HPLC Separation [10]

1. Dissolve the dried organic layers in 37 μL of a mixture with chloroform:methanol:water (73:23:3 v:v:v) and 2.5 μM of BCN-BODIPY.
2. Incubate this mixture for 16 h in the dark at 42 °C.
3. Dilute the sample with 113 μL chloroform:methanol:water (73:23:3 v:v:v).
4. Filter the sample with a 0.45 μm filter.
5. Analyze using normal phase HPLC.

3.4 If Using Mass Spectrometry [10] (See Note 3)

1. Add 7 μL of degassed chloroform to the lipid residue in the Eppendorf tube.
2. Add 30 μL of reaction master mix (details in materials).
3. Briefly flush tube with argon or nitrogen gas.
4. Place the tube in a 42 °C water bath (preventing the top half of the tube from being submerged) for 5 h.
5. Dilute the reactions with 113 μL of chloroform:methanol: water (73:23:3 v:v:v) mixture.
6. Filter using a 0.45 μm filter.
7. Analyze using mass spectrometry.

The following procedures should only be followed if intending to do RT-IMPACT.

3.5 Synthesizing Oxo-TCO-C1 for RT-IMPACT [11, 12]

1. Starting with glycidol, add tert-Butyl-(chloro)-diphenylsilane (TBDSCI) as an alcohol protecting agent along with imidazole and dichloromethane (DCM). Let the reaction proceed at room temperature for 16 h.
2. Add allyl magnesium chloride, dilithium tetrachlorocuprate (-Li_2CuCl_4), and tetrahydrofuran (THF), and let the reaction proceed at −30 °C for 1 h.
3. Treat the newly created substance with chloromethyl methyl ether (MOMCl), N,N-Diisopropylethylamine (DIEA), 4-dymethylaminopyridine (DMAP), and DCM. Let the reaction proceed at room temperature for 20 h.

4. Combine the new reagent with stannic chloride ($SnCl_4$), DCM, and the organosilicon, allyltrimethylsilane (allyl-$SiMe_3$) to induce the Sakurai reaction. Letting the reaction run at room temperature for 2 h will generate butenyl ether.
5. To this butenyl ether, add DCM and benzylidene-bis (tricycloheylphosphine), dichlororuthenium (Grubbs first-generation catalyst) and reflux for 2.5 h to perform a ring-closing metathesis; this will produce cis-oxocene.
6. To this newly generated cis-oxocene, add Tetra-n-butylammonium fluoride (TBAF, n-Bu_4NF) and let the reaction proceed for 8 h at room temperature to unprotect the alcohol.
7. Next, perform photoisomerization with 254 nm light with methyl benzoate ($PhCO_2Me$), diethyl ether (Et_2O) and run on a silver nitrate impregnated silica gel ($AgNO_3 \cdot SiO_2$) [13]. This will produce a mixture of diastereomers that can be used as Oxo-TCO-C1 (trans-5-oxocene-C1) (*see* **Note 4**).

3.6 Synthesizing Tz-BODIPY for RT-IMPACT [14]

1. Starting with p-cyanophenyl-BODIPY, add 50 mol% zinc trifluoromethanesulfonate (zinc triflate), hydrazine (N_2H_4), and acetonitrile (MeCN) and run the reaction at 60 °C. This will generate a H-Tetrazine and a methyl-tetrazine derivative.
2. To isolate the methyl-tetrazine derivative, use a clean fluorescence baseline on HPLFC (*see* **Note 5** for storage and working concentrations).

3.7 Performing RT-IMPACT [10, 12]

1. Seed the cells in a 35 mm glass plate overnight.
2. Use (s)-oxoTCO-C1 and Tz-BODIPY as created in the previous two steps as the transphosphatidylation substrates (*see* **Notes 4** and **5** for working concentrations).
3. If stimuli of PLD are novel/unknown, first use flow cytometry (refer to Subheading 3.1) to optimize concentration, time window, and signal-to-noise ratio for optimal PLD imaging (*see* **Note 6**).
   (a) For this, use 3 mM Oxo-TCO with PMA in DMEM (or other buffer) for 5 min at 37 °C.
4. If the stimulus is already known, use fluorescence microscopy (epi-microscopy or confocal microscopy) to capture steady-state imaging as demonstrated in **steps 5** through **9**.
5. Use the concentration of oxoTCO and PMA determined from previous flow cytometry step (*see* **Note 7**).
6. Followed by this addition, aspirate the solution and rinse for 1 min with DMEM at 37 °C (*see* **Notes 8** and **9**).
7. Treat the cells with 0.5 μM Tz-BODIPY for 1 min.

8. Aspirate the BODIPY solution and replace it with 1–2 mL of Tyrode's HEPES buffer (*see* **Note 9**).
9. Image using appropriate microscopy tool. Pay special attention to ideal microscope settings.
10. **Steps 11** through **19** demonstrate real-time imaging.
11. Seed the cells overnight on 35-mm glass plate.
12. Incubate the cells with appropriate concentration of PMA and oxo-TCO-Cl (from **step 5**) for approximately 5 min (*see* **Note 10**).
13. If looking at plasma membrane to endoplasmic reticulum trafficking, add red-fluorescent ER-tracker.
14. Aspirate and rinse for 1 min with DMEM at 37 °C (*see* **Note 8** and **9**).
15. Add 100 μL of Tyrode's-HEPES buffer to the center of the dish.
16. Mount the dish on microscope and locate a region of 10–20 cells for imaging (*see* **Note 11**).
17. Start a time-lapse movie with the settings from the steady-state experiment from **step 9** using focus control (*see* **Note 12**).
18. If looking at plasma membrane to endoplasmic reticulum trafficking, add red-fluorescent ER-tracker.
19. Dropwise, but rapidly add 100 μL of 1 μM (in PBS) Tz-BODIPY to the center of the glass well.

4 Notes

1. The time and temperature of the rinse-out step are critical. Duration may have to be catered to specific cell type.
2. If flow cytometry is unavailable, fluorescence microscopy can be used in which 200,000 cells should be seeded on a 35-mm glass bottom imaging dish. Additionally, rather than using trypsin to lift cells, 1 mL of Tyrode's-HEPES buffer should be added to the dish before imaging.
3. This technique uses alkyne-quarternary ammonium salt—N,N, N-trimethylproparygylamine bromide (Alk-QA)—via CuAAC to increase LC-MS detection.
4. Working and storage concentrations for Oxo-TCO-Cl [11, 12]:
   - Working solution concentration: 3 mM in DMEM (or other buffer) that should be used within 30 min.

- Storing solution concentration: 3 M in DMSO corresponding to 5:7 v/v ratio of oxoTCO to DMSO stored at −20 °C.

5. Working and storage concentrations for Tz-BODIPY [14]:
   - Working solution concentration: 1 μM in PBS, which can last for 4 h.
   - Tz-BODIPY: 1 mM in DMSO in 3 μL aliquots at −80 °C.
6. Ideal signal-to-noise ratio should be below 2.
7. As an example, for HeLA cells, the following concentrations were used: 100 nM PMA and 3 mM oxoTCO for 5 min at 37 ° C in 100 μL DMEM.
8. Time for rinse step may be adjusted accordingly.
9. Ensure that the entire dish is covered to minimize evaporation effects.
10. This time will change depending on specific cell lines. For HeLa cells, it is reported to be 5 min.
11. Work quickly to minimize the movement of the product.
12. Typical time lapse is taken at a 3 s interval for 3 min. For plasma membrane to endoplasmic reticulum trafficking, an 8 s interval for 4 min may be required.

References

1. McDermott MI, Wang Y, Wakelam MJO, Bankaitis VA (2020) Mammalian phospholipase D: function, and therapeutics. Prog Lipid Res 78:101018. https://doi.org/10.1016/j.plipres.2019.101018
2. Bocckino SB, Blackmore PF, Wilson PB, Exton JH (1987) Phosphatidate accumulation in hormone-treated hepatocytes via a phospholipase D mechanism. J Biol Chem 262(31):15309–15315
3. Pai JK, Siegel MI, Egan RW, Billah MM (1988) Phospholipase D catalyzes phospholipid metabolism in chemotactic peptide-stimulated HL-60 granulocytes. J Biol Chem 263(25):12472–12477
4. Philip F, Ha EE, Seeliger MA, Frohman MA (2017) Measuring phospholipase D enzymatic activity through biochemical and imaging methods. Methods Enzymol 583:309–325. https://doi.org/10.1016/bs.mie.2016.09.041
5. Tanguy E, Wang Q, Moine H, Vitale N (2019) Phosphatidic acid: from pleiotropic functions to neuronal pathology. Front Cell Neurosci 13:2. https://doi.org/10.3389/fncel.2019.00002
6. Bermudez V, Tenconi PE, Giusto NM, Mateos MV (2022) Canonical phospholipase D isoforms in visual function and ocular response to stress. Exp Eye Res 217:108976. https://doi.org/10.1016/j.exer.2022.108976
7. Chiu DC, Baskin JM (2022) Imaging and editing the phospholipidome. Acc Chem Res 55(21):3088–3098. https://doi.org/10.1021/acs.accounts.2c00510
8. Bumpus TW, Baskin JM (2016) A Chemoenzymatic strategy for imaging cellular phosphatidic acid synthesis. Angew Chem Int Ed Engl 55(42):13155–13158. https://doi.org/10.1002/anie.201607443
9. Bumpus TW, Baskin JM (2017) Clickable substrate mimics enable imaging of phospholipase D activity. ACS Cent Sci 3(10):1070–1077. https://doi.org/10.1021/acscentsci.7b00222
10. Bumpus TW, Liang D, Baskin JM (2020) IMPACT: imaging phospholipase d activity with clickable alcohols via transphosphatidylation. Methods Enzymol 641:75–94. https://doi.org/10.1016/bs.mie.2020.04.037
11. Lambert WD, Scinto SL, Dmitrenko O, Boyd SJ, Magboo R, Mehl RA et al (2017)

Computationally guided discovery of a reactive, hydrophilic trans-5-oxocene dienophile for bioorthogonal labeling. Org Biomol Chem 15(31):6640–6644. https://doi.org/10.1039/c7ob01707c

12. Liang D, Wu K, Tei R, Bumpus TW, Ye J, Baskin JM (2019) A real-time, click chemistry imaging approach reveals stimulus-specific subcellular locations of phospholipase D activity. Proc Natl Acad Sci USA 116(31):15453–15462. https://doi.org/10.1073/pnas.1903949116

13. Royzen M, Yap GP, Fox JM (2008) A photochemical synthesis of functionalized trans-cyclooctenes driven by metal complexation. J Am Chem Soc 130(12):3760–3761. https://doi.org/10.1021/ja8001919

14. Carlson JC, Meimetis LG, Hilderbrand SA, Weissleder R (2013) BODIPY-tetrazine derivatives as superbright bioorthogonal turn-on probes. Angew Chem Int Ed Engl 52(27):6917–6920. https://doi.org/10.1002/anie.201301100

Chapter 13

Synthesis of Azo Analogs for Investigating Phosphatidic Acid-Mediated Signaling

Jawad Muayad and Maria D. Cabrera Gonzalez

Abstract

Phosphatidic acid (PA) is a key signaling lipid that plays a crucial role in regulating various cellular processes. Studies have shown that azobenzene-containing PA analogues can be used as an all-chemical strategy for light-mediated control of PA signaling. These photoswitchable lipids offer a solution to the limitations of traditional bulk dosing methods by allowing for light- and shape-dependent interactions with protein effectors and lipid-metabolizing enzymes. This chapter describes how to synthesize AzoPA and dAzoPA.

Key words Azo-analogs, Phosphatidic acid, Photoswitchable

1 Introduction

Phosphatidic acid (PA) is a key signaling lipid that plays a crucial role in regulating various cellular processes. One way to study PA-mediated signaling is using azobenzene-containing PA analogues [1]. These photoswitchable lipids offer an all-chemical strategy for light-mediated control of PA signaling. Traditionally, membrane composition can be perturbed by incubating cells with natural or unnatural lipids through a procedure known as bulk dosing [1]. However, this approach has its limitations as the delivered lipid will always be in the "on" state [1] Photoswitchable lipids bearing azobenzene groups offer a solution to this problem. These unnatural lipids are capable of photoinduced isomerization between the thermally stable trans isomer and the cis isomer induced by 365 nm light. The trans isomer mimics a saturated tail while the cis isomer mimics an unsaturated tail. This allows for light- and shape-dependent interactions with protein effectors and lipid-metabolizing enzymes, enabling control over their bioactivity using UV light. In addition, unsaturated forms of PA can activate mitogenic signaling through mTOR, a central regulator of cell growth and proliferation [1] Specifically, cis-dAzoPA, but not

Marco Brotto and Sanjoy K. Bhattacharya (eds.), *Lipid Signalling: Methods and Protocols*, Methods in Molecular Biology, vol. 2816, https://doi.org/10.1007/978-1-0716-3902-3_13,

trans-dAzoPA, has been found to stimulate mTOR signaling in cells. Additionally, cis forms of AzoPA and dAzoPA can selectively repress the Hippo pathway, promoting growth in a manner that depends on the conformation of the hydrophobic azobenzene groups [1].

Overall, these photoswitchable PA analogs provide a powerful tool for probing PA-mediated signaling in cells. Overall, these photoswitchable PA analogs provide a powerful tool for probing PA-mediated signaling in cells. In this chapter, we will guide you through the detailed synthesis of AzoPA and dAzoPA, beginning with the preparation of R(-)-2,3-O-Isopropylidene-sn-glycerol. The synthesis process involves several critical steps, including the protection of hydroxyl groups, the introduction of azobenzene moieties, and the careful control of reaction conditions to ensure the formation of the desired isomers. Subsequent purification and characterization steps will also be detailed, ensuring the production of high-purity and correctly configured AzoPA and dAzoPA for experimental use [2, 3].

2 Materials

2.1 Preparation of S-2,3-O-isopropylidene-1-O-stearoyl-sn-glycerol

1. R(-)-2,3-O-Isopropylidene-sn-glycerol (2.85 g, 22.0 mmol, 1.0 equiv.)
2. 4-(dimethylamino)pyridine (DMAP) (268 mg, 2.20 mmol, 0.1 equiv.)
3. Dry CH_2Cl_2 (75 mL; 25 mL)
4. Dry NEt_3 (6.16 mL, 4.48 g, 44.0 mmol, 2.0 equiv.)
5. Stearoyl chloride
6. H2O (28 mL)
7. Aqueous HCl (2 M, 50 mL)
8. Saturated aqueous $NaHCO_3$ (2 × 50 mL)
9. Brine solution (2 × 50 mL)
10. Anhydrous Na_2SO_4
11. Flash silica gel chromatography column materials: 150 g of SiO_2, Hexane, EtOAc (20:1 ratio).

2.2 Preparation of 1-O-stearoyl-3-O-triethylsilyl-sn-glycerol

1. S-2,3-O-Isopropylidene-1-O-stearoyl-sn-glycerol (2) (500 mg, 1.25 mmol, 1.0 equiv.)
2. Dry 1,2-dichloroethane (DCE) (20 mL)
3. N,N-Diisopropylethylamine (DIPEA) (1.20 mL, 6.88 mmol)
4. Triethylsilyl trifluoromethanesulfonate (TESOTf) (496 mg, 0.425 mL, 1.88 mmol, 1.5 equiv.; 169 mg, 0.145 mL, 0.5 equiv.)

5. EtOAc
6. Aqueous hydrochloric acid (0.1 M, 50 mL)
7. Brine solution (50% saturated, 2 × 50 mL)
8. THF (20 mL)
9. Aqueous Na_2CO_3 solution (10%, 10 mL)
10. I_2 (610 mg, 2.4 mmol, 1.95 equiv.; 300 mg, 1.2 mmol, 1.0 equiv.)
11. Saturated aqueous $Na_2S_2O_3$ (50 mL)
12. Water (2 × 50 mL)
13. Anhydrous sodium sulfate
14. Flash silica gel chromatography column materials: 50 g of SiO_2, Hexane, EtOAc (20:1 ratio)

2.3 Preparation of 2-O-(4-(4-((4-butylphenyl)diazenyl) phenyl)butanoyl)-1-O-stearoyl-3-otriethylsilyl-sn-glycerol

1. FAAzo-419 (1.08 g, 3.3 mmol, 2.0 equiv.)
2. N-(3-dimethylaminopropyl)-N′-ethylcarbodiimide (EDC) (773 mg, 4.38 mmol, 3.0 equiv.)
3. DMAP (20.2 mg, 166 μmol, 0.1 equiv.)
4. Dry CH_2Cl_2 (50 mL; 30 mL; 200 mL)
5. Water (2 × 100 mL)
6. Brine solution (100 mL)
7. Flash silica gel chromatography column materials: 170 g of SiO_2, Hexane, EtOAc (30:1 ratio)

2.4 Preparation of 2-O-(4-(4-((4-butylphenyl)diazenyl) phenyl)butanoyl)-1-O-stearoyl-sn-glycerol (PhoDAG-1)

1. S-triethylsilyl-sn-glycerol (4) (500 mg, 0.641 mmol, 1.0 equiv.)
2. CH_2Cl_2 (5 mL; 25 mL; 200 mL)
3. $FeCl_2{\cdot}6H_2O$ (5 mM in 25 mL 3:1 MeOH:CH_2Cl_2)
4. EtOAc (200 mL)
5. Water (2 × 200 mL)
6. Na_2SO_4
7. Flash silica gel chromatography column with SiO_2 (3:1 hexanes:EtOAc)

2.5 Preparation of AzoPA and dAzoPA

1. PhoDAG-1 (photoswitchable analogue of diacylglycerol)
2. Di-tert-butyl-N,N-diisopropyl phosphoramidite
3. TFA (Trifluoroacetic acid)
4. FAAzo4
5. AzoLPA33 (di-tert-butylprotected)

3 Methods

3.1 Preparation of S-2,3-O-isopropylidene-1-O-stearoyl-sn-glycerol

1. Dissolve R(-)-2,3-O-Isopropylidene-sn-glycerol and DMAP in dry CH_2Cl_2 (75 mL) under an argon atmosphere.
2. Add dry NEt_3 and cool the reaction to 0 °C.
3. Dissolve stearoyl chloride in dry CH_2Cl_2 (25 mL) and slowly add to the previously prepared solution.
4. Allow the reaction to slowly warm to room temperature and stir for 2 h.
5. Slowly add H_2O (28 mL) and stir rapidly for 10 min.
6. Separate the phases, and wash the organic phase with aqueous HCl, followed by saturated aqueous $NaHCO_3$ and brine solutions.
7. Dry the organic phase over anhydrous Na_2SO_4 and filter.
8. Purify the filtrate by flash silica gel chromatography to yield S-2,3-O-isopropylidene-1-O-stearoyl-sn-glycerol (2, 6.99 g, 81%) as an off-white solid.

3.2 Preparation of 1-O-stearoyl-3-O-triethylsilyl-sn-glycerol

1. Dissolve S-2,3-O-Isopropylidene-1-O-stearoyl-sn-glycerol in dry DCE (20 mL) under an argon atmosphere.
2. Add N,N-Diisopropylethylamine and TESOTf (496 mg, 0.425 mL, 1.88 mmol, 1.5 equiv.) at room temperature and stir the reaction at 90 °C.
3. After 1 h, add a second portion of TESOTf (169 mg, 0.145 mL, 0.5 equiv.) and continue stirring at 90 °C for 2.5 h.
4. Upon consumption of the starting material as determined by TLC analysis, cool the solution to room temperature and dilute with EtOAc.
5. Wash the solution with aqueous hydrochloric acid and brine solutions.
6. Concentrate the organic phase under reduced pressure and dissolve the resulting oil in THF (20 mL).
7. Add an aqueous Na_2CO_3 solution followed by I_2 (610 mg) to this solution and stir rapidly for 2.5 h at room temperature.
8. Add a further portion of I_2 (300 mg) and continue stirring at room temperature for 1 h.
9. Dilute the solution with EtOAc (50 mL) and wash with saturated aqueous $Na_2S_2O_3$ (50 mL), water (2 × 50 mL), and brine solutions (50 mL).
10. Dry the organic phase over anhydrous sodium sulfate and filter.

11. Concentrate the filtrate under reduced pressure and purify the resulting oil using flash column chromatography to yield S-triethylsilyl-sn-glycerol as a colorless liquid.

3.3 Preparation of 2-O-(4-(4-((4-butylphenyl)diazenyl)phenyl)butanoyl)-1-O-stearoyl-3-Otriethylsilyl-sn-glycerol

1. Dissolve FAAzo-419, EDC and DMAP in dry CH_2Cl_2 (50 mL) under an argon atmosphere.
2. Stir this solution at room temperature for 15 min.
3. After cooling to 0 °C, slowly add S-triethylsilyl-sn-glycerol in dry CH_2Cl_2 (30 mL).
4. Warm the solution to room temperature and stir overnight under an argon atmosphere.
5. Dilute the solution with CH_2Cl_2 (200 mL) and wash with water (2 × 100 mL) and brine solutions (100 mL).
6. Filter the solution and concentrate the filtrate under reduced pressure.
7. Purify the resulting red oil using flash silica gel chromatography to yield 2-O-(4-(4-((4-butylphenyl)diazenyl)phenyl)butanoyl)-1-O-stearoyl-3-O-triethylsilyl-sn-glycerol (4, 944 mg, 72%) as a red oil.

3.4 Preparation of 2-O-(4-(4-((4-butylphenyl)diazenyl)phenyl)butanoyl)-1-O-stearoyl-sn-glycerol (PhoDAG-1)

1. Dissolve 2-O-(4-(4-((4-Butylphenyl)diazenyl)phenyl)butanoyl)-1-O-stearoyl-3-O-triethylsilyl-sn-glycerol in CH_2Cl_2 (5 mL).
2. Add this solution to a solution of $FeCl_2{\cdot}6H_2O$.
3. Stir this solution at room temperature for 30 min.
4. Dilute the solution with EtOAc (200 mL) and wash it with water.
5. Dry the organic phase over Na_2SO_4.
6. Filter the mixture and concentrate the filtrate under reduced pressure.
7. Purify the resulting oil using flash silica gel chromatography (3: 1 hexanes:EtOAc) to yield PhoDAG-1 as an orange solid. (*See* **Note 1**)

3.5 Preparation of AzoPA and dAzoPA

1. Synthesize AzoPA by phosphorylating PhoDAG-1 with di-tert-butyl-N,N-diisopropyl phosphoramidite to yield compound 1.
2. Oxidize compound 1 to produce AzoPA.
3. Deprotect AzoPA using TFA.
4. Synthesize dAzoPA by starting with di-tert-butylprotected AzoLPA33.
5. Acylate AzoLPA33 with FAAzo4 to produce dAzoPA.
6. Deprotect dAzoPA using TFA.

4 Notes

1. Short reaction times and quick chromatography are essential to avoid acyl chain migration.

References

1. Chiu D, Baskin JM (2022) Imaging and editing the phospholipidome. Acc Chem Res 55(21): 3088–3098. https://doi.org/10.1021/acs.accounts.2c00510
2. Tei R, Morstein J, Shemet A et al (2021) Optical control of phosphatidic acid signaling. ACS Cent Sci 7(7):1205–1215. https://doi.org/10.1021/acscentsci.1c00444
3. Frank JA, Yushchenko DA, Hodson DJ et al (2016) Photoswitchable diacylglycerols enable optical control of protein kinase C. Nat Chem Biol 12(9):755–762. https://doi.org/10.1038/nchembio.2141

Chapter 14

Method to Assess the Intracellular Fate and Bioavailability of Clusterin Using Live Cell Confocal Microscopy Imaging

Avinash Soundararajan and Padmanabhan Paranji Pattabiraman

Abstract

Clusterin, also known as apolipoprotein J, is an ATP-independent holdase chaperone protein. Clusterin is involved in various functions including protein quality control and lipid transport. Though clusterin is secreted upon stress, the intracellular fate of clusterin after a stress response is not well understood. The protocol described here utilizes clusterin tagged to fluorescent proteins like green fluorescent protein and red fluorescent protein to understand the intracellular fate of clusterin.

Key words Human trabecular meshwork, Clusterin, Secretory protein, Signal peptide, Red fluorescent protein, Green fluorescent protein

1 Introduction

Within the intricate landscape of cellular biology, the secretory pathway emerges as a dynamic conduit responsible for orchestrating the synthesis, folding, and precise delivery of an extensive repertoire of cellular proteins. Central to this journey is the endoplasmic reticulum (ER), which regulates protein translation, folding, and posttranslational modification that includes addition of sugar chains. These modified proteins are transported to the Golgi apparatus by a coated vesicle using the Coat Protein Complex II (COPII). Proteins are then sorted into the regulated secretory pathway in the trans Golgi network and packaged into specialized secretory vesicles that carry them to extracellular locations [1, 2]. Conventionally, secretory proteins contain a unique short amino acid sequence called signal peptides in the N-terminal region. This sequence directs the protein into the secretory pathway [3]. Protein secretion is essential for various functions including cell growth and cellular response [4].

Secretory proteins leave the cell to reach the extracellular space where they can bind to other proteins, the extracellular matrix

Marco Brotto and Sanjoy K. Bhattacharya (eds.), *Lipid Signalling: Methods and Protocols*, Methods in Molecular Biology, vol. 2816, https://doi.org/10.1007/978-1-0716-3902-3_14,

(ECM), or membrane receptors [5]. Engagement with the membrane receptors can trigger an outside-in signaling or can be taken into the cells through endocytosis [6]. Therefore, it becomes essential to evaluate the bioavailability of secretory proteins. In our previous publication, we evaluated the bioavailability of a secretory protein, clusterin, under conditions of pathological stress [7].

Clusterin, also known as apolipoprotein J, serves as a stress response secretory ATP-independent chaperone protein [8–12]. Chaperone proteins aids in protein folding in cell under stress and physiological conditions. Besides its chaperone function, clusterin serves as a lipid transporter due to its lipoprotein nature [13, 14]. Clusterin's diverse functionality extends to potential binding interactions with membrane receptors, lipids, complement factors, and ECM proteins [15–17]. Despite these interactions, the precise post-binding functions of clusterin are not well understood. An important goal of our lab is to study the clusterin receptors and their significance in the regulation of intraocular pressure (IOP). Being aware of this significance, it becomes increasingly imperative to scrutinize clusterin levels and its precise intracellular localization in response to pathological stress. The pathological stressors include elevated pressure, growth factors, and steroids like dexamethasone. Our recent study showed that clusterin loss elevated IOP and thereby understanding the fate and bioavailability of clusterin can help determine the mechanistic basis of IOP increase under clusterin loss [7].

Therefore, we have utilized an experimental protocol aimed at determining the bioavailability of clusterin in live cells. This methodology holds the potential for broad applications of any secretory protein beyond clusterin, providing a valuable framework for the assessment of the fate of secretory proteins and potential intracellular dynamics.

2 Materials Required

1. Cellvis 35 mm 4-chambered glass bottom dish (#D35C4-20-1.5-N).
2. Gelatin Solution: Prepare 2% gelatin in milliQ water and autoclave it, then store it at 4 °C.
3. Dulbecco's modified Eagle's medium (DMEM) for cell maintenance.
4. Penicillin-Streptomycin-Glutamine (PSQ) (Gibco, #10378-016) for antibiotic supplementation in DMEM.
5. Fetal bovine serum (FBS) to prepare serum media (add 10% FBS to DMEM).
6. Normal trabecular meshwork (NTM) cell line (NTM5).

7. Cell culture incubator (37 °C, humidified, and 5% CO2-controlled).
8. Lipofectamine 3000 for transfection (Invitrogen, #L3000-015).
9. OptiMEM (Gibco, #31985-070) devoid of antibiotics.
10. Zeiss LSM 700 confocal microscope attached to a Tokai Hit Stage Top Incubator.

3 Methodology

3.1 Culturing NTM Cells

1. NTM cells from fully confluent 10 cm plate is trypsinized, suspended in OptiMEM (Gibco, #31985-070) and counted.
2. Around 500–1000 NTM cells are plated on 2% gelatin-coated Cellvis 35 mm four-chambered glass bottom dishes.
3. Cells are maintained in DMEM supplemented with 10% FBS and PSQ at 37 °C in a 5% CO_2-controlled incubator for 2 days. Now the cells are ready for transfection.

3.2 Transfection

1. Before transfection, replace the conditioned media with Opti-MEM and maintain in the incubator until transfected.
2. Transfect NTM5 cells with clusterin tagged to red fluorescent protein and green fluorescence protein (CLU-RG) or signal peptide tagged to red fluorescent protein and green fluorescence protein (SP-RG) plasmid as a control using Lipofectamine 3000 as per the manufacturer's protocol or previously published work [7]. These plasmids were kindly provided by Dr. Itakura [18].
3. Prepare Solution A containing lipofectamine 3000 in Opti-MEM and Solution B containing either CLU-RG or SP-RG plasmid and P3000 in OptiMEM. The volume is determined based on the amount of plasmids used as per the manufacturer's protocol.
4. Gently mix Solution B with Solution A and incubate undisturbed for 15 min.
5. Slowly add the incubated mixture to the cells.
6. After 8 h, replace the medium with fresh DMEM without serum.

3.3 Live-Cell Imaging

1. Forty-eight hours post-transfection, perform live-cell time-lapse imaging using any confocal imaging microscope equipped with a stage top incubator. We use a Zeiss LSM 700 confocal microscope equipped with a Tokai Hit Stage Top Incubator. The cell culture dish should be maintained at

37 °C inside the moisture chamber until the imaging is completed.

2. Begin with 20× objective to identify the transfected cells based on the fluorescence (either GFP or RFP).
3. Select the cell for imaging and capture the baseline image at 63× objective or as per requirement.
4. Record the cell's position in the stage for continuous analysis.
5. Perform any necessary treatments and continuously image the cell for changes, recording RFP and GFP fluorescence signals.

4 Interpretation

CLU-RG and SP-RG plasmids produce both green and red fluorescence, typically colocalizing. However, under acidic conditions, such as within lysosomal vesicles, the Clusterin and GFP undergo lysosomal enzymatic degradation, sparing RFP, which is highly resistant to the acidic environment [7, 18]. Therefore, an increased accumulation in RFP puncta indicates clusterin degradation.

5 Conclusion

In conclusion, this methodology for assessing clusterin bioavailability and intracellular localization within live cells aids in explaining the intricate roles of clusterin in diverse cellular contexts. In our case, we identified that transforming growth factor β2 (TGFβ2), a profibrotic growth factor that can elevate IOP, stress on ER and protein synthesis machinery can shuttle clusterin for lysosomal degradation [7]. Thus, we believe that this loss of functional clusterin is a potential mechanism for TGFβ2-induced ocular hypertension. By leveraging the insights gained from this methodology, researchers can study the fate of clusterin in the cell under different stress environments and extend its application to the investigation of other secretory proteins.

Acknowledgments

The authors thank the National Eye Institute (R01EY029320); Ralph W. and Grace M. Showalter Award; Research Support Funds Grant (RSFG); Cohen Pilot Grants in Macular Degeneration Research; Research to Prevent Blindness (RPB) Pilot grant and Challenge grant from RPB to IU.

References

1. Barlowe CK, Miller EA (2013) Secretory protein biogenesis and traffic in the early secretory pathway. Genetics 193(2):383–410. https://doi.org/10.1534/genetics.112.142810
2. Gomez-Navarro N, Miller E (2016) Protein sorting at the ER-Golgi interface. J Cell Biol 215(6):769–778. https://doi.org/10.1083/jcb.201610031
3. Karamyshev AL, Tikhonova EB, Karamysheva ZN (2020) Translational control of secretory proteins in health and disease. Int J Mol Sci 21(7). https://doi.org/10.3390/ijms21072538
4. Praznik A, Fink T, Franko N, Lonzarić J, Benčina M, Jerala N et al (2022) Regulation of protein secretion through chemical regulation of endoplasmic reticulum retention signal cleavage. Nat Commun 13(1):1323. https://doi.org/10.1038/s41467-022-28971-9
5. Farhan H, Rabouille C (2011) Signalling to and from the secretory pathway. J Cell Sci 124(2):171–180. https://doi.org/10.1242/jcs.076455
6. Carraway KL, Carraway CAC, Carraway KL (1998) Intracellular membrane trafficking, secretion/exocytosis and endocytosis. In: Carraway KL, Carraway CAC, Carraway KL (eds) Signaling and the cytoskeleton. Berlin, Heidelberg, Springer, pp 211–242
7. Soundararajan A, Wang T, Ghag SA, Kang MH, Pattabiraman PP (2022) Novel insight into the role of clusterin on intraocular pressure regulation by modifying actin polymerization and extracellular matrix remodeling in the trabecular meshwork. J Cell Physiol 237(7):3012–3029. https://doi.org/10.1002/jcp.30769
8. Poon S, Easterbrook-Smith SB, Rybchyn MS, Carver JA, Wilson MR (2000) Clusterin is an ATP-independent chaperone with very broad substrate specificity that stabilizes stressed proteins in a folding-competent state. Biochemistry 39(51):15953–15960
9. Wyatt AR, Yerbury JJ, Wilson MR (2009) Structural characterization of clusterin-chaperone client protein complexes. J Biol Chem 284(33):21920–21927. https://doi.org/10.1074/jbc.M109.033688
10. Rohne P, Prochnow H, Wolf S, Renner B, Koch-Brandt C (2014) The chaperone activity of clusterin is dependent on glycosylation and redox environment. Cell Physiol Biochem 34(5):1626–1639. https://doi.org/10.1159/000366365
11. Wyatt AR, Yerbury JJ, Berghofer P, Greguric I, Katsifis A, Dobson CM, Wilson MR (2011) Clusterin facilitates in vivo clearance of extracellular misfolded proteins. Cell Mol Life Sci 68(23):3919–3931. https://doi.org/10.1007/s00018-011-0684-8
12. Geraghty NJ, Satapathy S, Wilson MR (2022) The emerging roles of extracellular chaperones in complement regulation. Cells 11(23):3907
13. Baralla A, Sotgiu E, Deiana M, Pasella S, Pinna S, Mannu A et al (2015) Plasma clusterin and lipid profile: a link with aging and cardiovascular diseases in a population with a consistent number of centenarians. PLoS One 10(6):e0128029. https://doi.org/10.1371/journal.pone.0128029
14. Ishikawa Y, Akasaka Y, Ishii T, Komiyama K, Masuda S, Asuwa N et al (1998) Distribution and synthesis of apolipoprotein J in the atherosclerotic aorta. Arterioscler Thromb Vasc Biol 18(4):665–672. https://doi.org/10.1161/01.ATV.18.4.665
15. Yuste-Checa P, Bracher A, Hartl FU (2022) The chaperone clusterin in neurodegeneration—friend or foe? BioEssays 44(7):2100287. https://doi.org/10.1002/bies.202100287
16. Foster EM, Dangla-Valls A, Lovestone S, Ribe EM, Buckley NJ (2019) Clusterin in Alzheimer's disease: mechanisms, genetics, and lessons from other pathologies. Front Neurosci 13:13. https://doi.org/10.3389/fnins.2019.00164
17. Satapathy S, Wilson MR (2021) The dual roles of clusterin in extracellular and intracellular proteostasis. Trends Biochem Sci 46(8):652–660. https://doi.org/10.1016/j.tibs.2021.01.005
18. Itakura E, Chiba M, Murata T, Matsuura A (2020) Heparan sulfate is a clearance receptor for aberrant extracellular proteins. J Cell Biol 219(3). https://doi.org/10.1083/jcb.201911126

Chapter 15

Detailed Protocol for Solid-Phase Extraction for Lipidomic Analysis

Marian N. Aziz, Leticia Brotto, Ahmed S. Yacoub, Kamal Awad, and Marco Brotto

Abstract

Developing robust analytical techniques is a vital phase to facilitate understanding the roles and impacts of various omic profilings in cellular functions. The comprehensive analysis of various biological molecules within a biological system requires a precise sample preparation technique. Solid-Phase Extraction (SPE) has proven to be an indispensable method in lipidomic analysis, providing an uncomplicated and user-friendly technique for extraction and purification of lipid components from complex biological matrices. Of all the factors influencing the reliability and success of SPE, column or adsorbent materials, flow rate, and storage conditions are paramount in terms of their significance. In this chapter, we will discuss in detail the SPE steps for lipidomic analysis in biofluid samples (serum and plasma) and muscle tissues.

Key words SPE for lipidomic, SPE for muscle, SPE for plasma, SPE for serum, Sample preparation

1 Introduction

Solid-Phase Extraction, also known as SPE, is one of the most widely employed and preeminent sample preparation techniques used in the field of Bioanalytics [1]. This liquid-solid extraction technique provides a rapid and efficient method for the extraction of analytes from a complex matrix [2]. Profiling lipids in biological samples is imperative for abstaining dependable lipid profiles, and the preparation of these samples remains highly labor-intensive. Conventional preparation methods are excessively time-consuming and highly prone to errors [3, 4]. For example, liquid-liquid extraction (LLE), a fundamental method of sample extraction in the field of chromatography, stands out as one of the widely employed methods of sample extraction [5]. However, LLE is preferred for comprehensive lipidomic analysis, SPE finds its primary utility in targeted methods when the isolation or fractionation of particular lipid classes is imperative [6, 7]. Nevertheless, it is worth noting

Marco Brotto and Sanjoy K. Bhattacharya (eds.), *Lipid Signalling: Methods and Protocols*, Methods in Molecular Biology, vol. 2816, https://doi.org/10.1007/978-1-0716-3902-3_15,

that SPE can exhibit variability, time-intensiveness, and difficulties in automation. By adhering to the guidelines and insights provided in this chapter, which are informed by our laboratory's expertise and design, you can enhance the reliability and success of the SPE process.

Solid-Phase Extraction employs a low-resolution chromatographic process to effectively isolate and separate analytes from the liquid matrix [8]. This extraction method uses the disparity in affinity between an analyte and interferents present in the sample matrix, for a solid or sorbent phase [9]. This discernible affinity between the analyte and interferents facilitates the selective separation of the desired analyte from the interferents, culminating in the accumulation of the analyte concentration at the polar surface of the solid. Hence, choosing the appropriate SPE sorbent material is crucial. The sorbent should have the right affinity for the target analytes and be compatible with the sample matrix. And the selection of a lipid SPE cartridge is governed by the dual mechanism involving both size exclusion and hydrophobic interaction [3, 10].

SPE utilizes a fast and highly efficient four-step process to attain a pure sample. The initial step consists of conditioning the adsorbent, commonly referred to as solvation. During this phase, the pertinent sites of the absorbent are activated to facilitate the absorption of the analyte. Following the conditioning phase is loading. During this phase, the sample is introduced into the column at a suitable flow rate allowing it to optimize the retention of the analyte to the stationary phase. Subsequently, the column undergoes a washing process to eliminate all impurities and other matrix constituents. Finally, a solvent that has a heightened affinity for the analyte is then used to acquire a concentrated form of the analyte. In this chapter, we demonstrate the application of solid-phase extraction technique and describe the standard procedures employed for the extraction of plasma or serums.

2 Materials

2.1 Reagents and Materials

1. 1.5 mL Low-Retention microcentrifuge tubes.
2. 15 mL centrifuge tubes.
3. Formic Acid Solutions:
   (a) 1.0% Formic Acid (v/v) in water (Prepare by adding 1 mL of 100% Formic Acid to 99 mL of nanopure water).
   (b) 0.1% Formic Acid (v/v) in water (Prepare by adding 20 mL of 1.0% Formic Acid to 180 mL of nanopure water).
4. Ethanol Solution:

   (a) 15% Ethanol (v/v) in water (Prepare by mixing 30 mL of absolute ethanol with 170 mL of nanopure water).

5. Methanol Solution:
   (a) 80% Methanol (v/v) in water (Prepare by mixing 160 mL of Methanol with 40 mL of nanopure water).
6. Store all solvents in a 4 °C refrigerator.
7. Pipettes (10 μL, 100 μL, 1 mL, serological 5 mL) and pipette tips.
8. Pipette waste holder.
9. Plasma/serum samples (ensure no less than 50 μL for each sample) and pre-weighed muscle tissue (minimum 20 mg) before the preparation step.
10. Racks designed for 1.5 mL microcentrifuge tubes.
11. Refrigerated Microcentrifuge.
12. SPE cartridge (Polymeric Reversed Phase, 10 mg/1 mL).
13. SPE Vacuum Manifold.
14. Internal Standard (IS) mixture stock, prepared using methanol, and stored at −80 °C (Contains all internal standards for lipid mediator of targets).

2.2 Materials Required for Muscle Tissue Analysis

1. Muscle tissue samples aliquoted into 50–100 mg per sample, stored in a −80 °C freezer.
2. Stainless steel disposable scalpel, with a size 15 being highly recommended.
3. Scissors and forceps.
4. Stainless steel beads for homogenization.
5. TissueLyser II Homogenizer.

3 Methods

3.1 Solid Phase Extraction

In our laboratory, we utilize an SPE Vacuum Manifold that can handle up to 12 samples in one operation. However, when working with multiple groups and a sample size equal to or exceeding three ($n \geq 3$), we choose to process four samples in each run. This approach ensures that all three replicates within each group undergo the same processing conditions.

1. Prepare two units of ice buckets.
2. Put on your gloves.
3. Retrieve the frozen plasma/serum samples and arrange them according to their ID numbers as listed on your worksheet. Place all the samples inside the ice bucket. Maintaining your

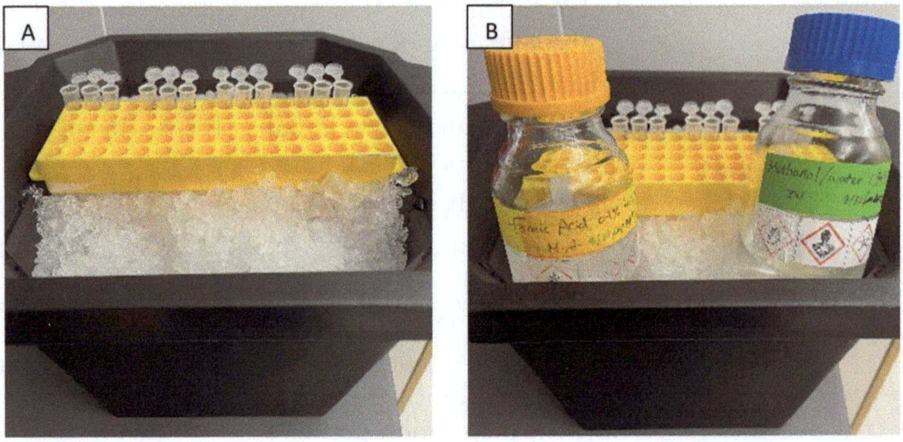

Fig. 1 (**a**) 1.5 mL tubes arranged in groups of three tubes per set, with all samples placed on the rack and stored on ice. (**b**) Solutions and aliquots kept on ice

samples within the ice bucket always is crucial to prevent any alterations in lipid profiling.

4. Label all the 1.5 mL microcentrifuge tubes (low retention) based on the sample numbers. For instance, label the first tube as #1, corresponding to sample ID #1 (or use the sample information from the worksheet).
5. Organize the 1.5 mL tubes in sets of three tubes per sample ($n \geq 3$) and place them inside the ice container (see Fig. 1a).
6. Retrieve the solutions and aliquots from the refrigerator (0.1% Formic Acid +80% Methanol) and the Internal Standard Mixture (IS) from the −80 °C freezer. Keep all of them on ice (refer to Fig. 1b).

3.2 Activation of SPE Cartridge

1. Retrieve the solutions and aliquots from the 4 °C refrigerator (formic acid + methanol) and take the ID mixture from the −80 °C freezer. Place them on ice.
2. Affix the necessary number of SPE cartridges to the Vacuum Manifold, ensuring that the valves are in the closed position.
3. Label the side of the cartridges with your sample number or code (#1 corresponding to sample ID #1).
4. Add 1 mL of room temperature 100% Methanol to each SPE cartridge.
5. Open the valves to activate the cartridge without applying vacuum.
6. Add 1 mL of ice-cold 0.1% formic acid to each SPE cartridge (*See* **Note 1**).

3.3 Procedures

1. Label 1.5 mL microcentrifuge tubes (low retention) with #1, corresponding to sample ID #1 (or use the sample ID), and keep them on ice.
2. Turn ON the centrifuge and set the speed to 1600 units and the temperature to 4 °C for 10 min. It's important to do this before proceeding to **step 7** (as described further) to allow the system to cool down.
3. Add 1 mL of ice-cold 80% methanol (v/v) in water to the 1.5 mL tubes.
4. Transfer 50 μL of plasma/serum into the 1.5 mL tubes.
5. Add 5 μL of IS mixture stock solution (0.5 μg/mL) to each sample or to tissue homogenates (refer to Subheading 3.4 for solid sample preparation).
6. Vortex each sample for 10 s.
7. Place the samples on the shaker located inside a dark refrigerator (to prevent degradation or oxidation reactions) and set it to run for 10 min at speed 5.
8. After 10 min, remove the samples from the dark refrigerator and keep them all on ice.
9. Arrange the tubes in the Refrigerated Centrifuge.
10. Centrifuge them at 16,000×*g*, at 4 °C, for 10 min.
11. Label 15 mL centrifuge tubes with sequential numbers, starting with #1, corresponding to sample ID #1. Keep all the labeled tubes on ice.
12. Remove the samples from the centrifuge and place them in the ice.
13. Transfer the supernatant from each sample into the labeled 15 mL centrifuge tubes.
14. Add 4 mL of ice-cold 0.1% formic acid to each of the centrifuge tubes.
15. Gently agitate the tubes slowly three times.
16. Load 1 mL of each sample into an activated SPE cartridge.
17. Open the valve and allow the liquid to pass through in a dropwise manner.
18. If necessary, apply vacuum, but ensure that the flow rate is not too fast, as this may reduce the efficiency of mediator species capture on the cartridge.
19. Close the vacuum.
20. After loading the sample, rinse the cartridge with 1 mL of ice-cold 0.1% formic acid.
21. Open the vacuum to rinse the cartridge.
22. Close both the vacuum and the valve.

23. Add 1 mL of a 15% ethanol aliquot prepared from a 100% Absolute Ethanol stock to rinse the cartridge.
24. Open the valve and use vacuum to rinse the cartridge slowly, allowing the liquid to flow dropwise.
25. Maintain the vacuum at 10–20 Hg pressure for 15 min to remove as much remaining water from the cartridge as possible. (*See* **Note 2**).
26. Label 1.5 mL low-retention sterile microcentrifuge tubes with the sample ID and any other pertinent information (e.g., serum, date, your initials).
27. Close the vacuum after 15 min.
28. Remove the SPE Cartridge from the Manifold.
29. Take out the waste bucket from the Manifold and transfer its contents to an acid waste container. Rinse the waste bucket with water.
30. Replace the waste bucket with the Manifold Sample Holder. Ensure that the Manifold Sample Holder is correctly positioned to match the curved edge inside the manifold.
31. Insert the 1.5 mL labeled microcentrifuge tubes onto the Holder.
32. Elute with 400 μL of a 100% Methanol aliquot.
33. Open the valve.
34. Gradually increase the vacuum to 5 Hg, maintaining it for 2–3 min. It's preferable to avoid applying excessive vacuum at this stage, but this may vary depending on the nature of your biological matrix.
35. Close the vacuum.
36. Discard all used cartridges.
37. Transfer the tubes containing your extracted and purified lipids to a rack.
38. Evaporate the Methanol from your samples using a Centrifugal Vacuum Concentrator.
    - To configure the machine:
      (a) Verify that the machine is powered ON.
      (b) Set the time to a range between 50 min to 1 h.
      (c) Ensure that the brake is set to OFF.
      (d) Leave the temperature setting as it is (--).
      (e) Select the mode as V-AQ.
      (f) Open the tubes and arrange them inside the Centrifugal Vacuum Concentrator in a sequence that ensures the machine's balance.

(g) Place all the dried samples in a −80 °C freezer for future LC-MS/MS analysis. Label them outside the freezer box with the sample type as indicated in the worksheet, the date, and your initials.

3.4 Solid Tissue Preparation (Muscle of Other Soft Tissues)

1. Prior to assessing the frozen muscles:
   (a) To prevent any potential contamination between samples, wash both the scissors and forceps with ice-cold 80% Methanol after each sample and once more after the final sample.
   (b) For LC–MS/MS analysis, it is generally recommended that a tissue sample should weigh between 50 and 100 mg, with a minimum acceptable weight of 20 mg. Therefore, if the tissue samples exceed 100 mg, it becomes necessary to section them appropriately before extraction. This can be achieved using a disposable scalpel, specifically a size 15 scalpel.
2. Measure the frozen muscle's weight in milligrams (mg) and document it in a spreadsheet. Subsequently, insert the recorded weight into the 2 mL microcentrifuge tube already containing the bead.

   Steps to consider before cutting the portion of frozen muscle:
   (a) To prevent cross-contamination, make sure to clean the scissors with ice-cold 80% Methanol after each sample and after the final sample.
   (b) Proceed to cut the allocated frozen muscle tissue into small fragments directly within the tubes using scissors.
   (c) Place a single stainless-steel bead into each 2 mL microcentrifuge tube.
3. To homogenize the minced tissue using the TissueLyser II Homogenizer and configure the machine, follow these steps:
   (a) Activate the machine by switching it ON.
   (b) Ensure that the memory function is enabled.
   (c) Set the frequency to 30/s (30.0).
   (d) Adjust the time setting to 2.00 min (or 0.30 s for each of the 8 intervals).
   (e) Lift the transparent lid to open it.
   (f) Acquire the two "Qiagen-sample holders."
   (g) Place six tubes in the central slots of each holder, ensuring that you align the pieces correctly by matching the outer "buttons" in parallel and orienting the black lid toward the caps of the tubes (See Fig. 2).

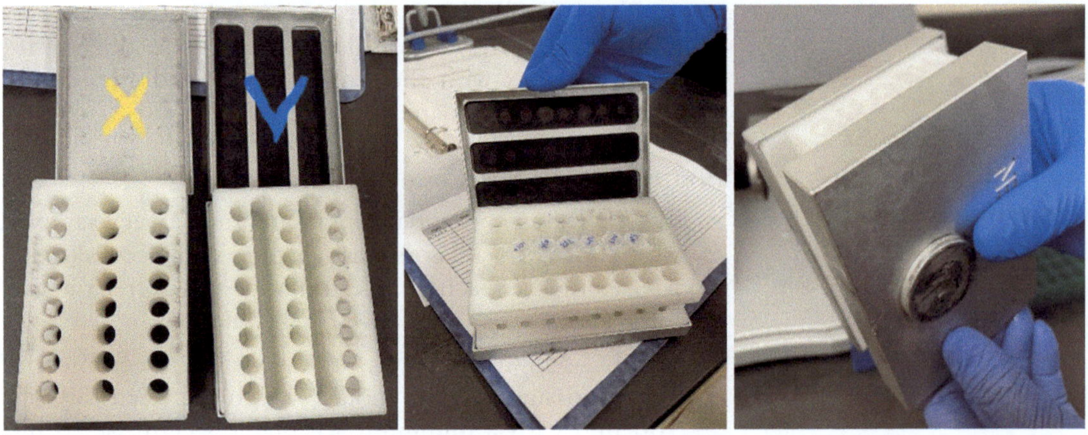

Fig. 2 Sample holders for TissueLyser II Homogenizer

(h) The outer "buttons" should be positioned within the bracket, ensuring that they are securely fastened by tightening the screws.

(i) Verify that the holders are correctly aligned, both parallel and in a vertical position (*See* Fig. 2).

(j) Secure the system by locking it before initiating the run, and then unlock it when you need to remove the "Qiagen sample holders" from the system.

(k) Close the lid and commence the operation.

(l) Pause the machine after each 30-s run interval, allowing for a 20-s pause to prevent overheating.

(m) Transfer all the samples directly onto the ice bucket and wait for 1 min.

(n) Return the samples to the "Qiagen-sample holders" and repeat the homogenization process, running it for 30 s each time, repeating four times for a total of 2 min.

(o) Maintain the samples on ice to prepare them for the subsequent Solid-Phase Extraction (SPE) step.

(p) Add 5 μL of the Internal Standard (IS) to each tissue homogenate, and gently mix them by simpla agitation.

(q) Put the samples onto the shaker and let them sit for 1 h (with a maximum duration of 2 h). Set the shaker to a speed of 5 during this time.

(r) Follow the procedure outlined in Subheading 3.3, starting at **step 5**, and apply the same steps for cartridge activation as mentioned in Subheading 3.2. (*See* **Note 3**).

4 Notes

1. When the formic acid is running low, turn off the valve, ensuring that a small amount of solution remains on the filter. Never allow the filter to dry out at this stage.
2. Note that before stopping the drying process, ensure that there are no water droplets on the tube walls.
3. When the formic acid is running low, turn off the valve, ensuring that a small amount of solution remains on the filter. Never allow the filter to dry out at this stage.

References

1. Stone J (2017) Chapter 3 – Sample preparation techniques for mass spectrometry in the clinical laboratory. In: Nair H, Clarke W (eds) Mass spectrometry for the clinical laboratory. Academic Press, San Diego, pp 37–62
2. Naviglio D et al (2019) Rapid solid-liquid dynamic extraction (RSLDE): a powerful and greener alternative to the latest solid-liquid extraction techniques. Foods 8. https://doi.org/10.3390/foods8070245
3. Apffel A, Zhao L, Sartain MJ (2021) A novel solid phase extraction sample preparation method for lipidomic analysis of human plasma using liquid chromatography/mass spectrometry. Metabolites 11. https://doi.org/10.3390/metabo11050294
4. Addepalli RV, Mullangi R (2021) A concise review on lipidomics analysis in biological samples. ADMET DMPK 9(1):1–22
5. Zhang J, Hu B (2013) Liquid-liquid extraction (LLE). In: Separation and purification technologies in biorefineries. Wiley, Chichester, pp 61–78
6. Cajka T, Fiehn O (2016) Toward merging untargeted and targeted methods in mass spectrometry-based metabolomics and lipidomics. Anal Chem 88(1):524–545
7. Li M et al (2014) Analytical methods in lipidomics and their applications. Anal Chem 86(1):161–175
8. Pavkovich AM, Bell DS (2019) Extraction| QuEChERS. In: Worsfold P et al (eds) Encyclopedia of analytical science, 3rd edn. Academic Press, Oxford, pp 84–88
9. Ścigalski P, Kosobucki P (2020) Recent materials developed for dispersive solid phase extraction. Molecules 25. https://doi.org/10.3390/molecules25214869
10. Maranata GJ, Surya NO, Hasanah AN (2021) Optimising factors affecting solid phase extraction performances of molecular imprinted polymer as recent sample preparation technique. Heliyon 7(1):e05934

Chapter 16

Analysis of Lipid GPCR Molecular Interactions by Proximity Labeling

Xian Zhao and Hideru Obinata

Abstract

G-protein-coupled receptors (GPCRs) are hepta-helical transmembrane proteins that mediate various intracellular signaling events in response to their specific ligands including many lipid mediators. Although analyses of GPCR molecular interactions are pivotal to understanding diverse intracellular signaling events, affinity purification of interacting proteins by a conventional co-immunoprecipitation method is challenging due to the hydrophobic nature of GPCRs and their dynamic molecular interactions. Proximity labeling catalyzed by a TurboID system is a powerful technique for defining the molecular interactions of target proteins in living cells. TurboID and miniTurbo (a modified version of TurboID) are engineered biotin ligases that biotinylate neighboring proteins in a promiscuous manner. When fused with a target protein and expressed in living cells, TurboID or miniTurbo mediates the biotin labeling of the proteins with close proximity to the target protein, allowing efficient purification of the biotinylated proteins followed by a shot-gun proteomic analysis. In this chapter, we describe a step-by-step protocol for the labeling of GPCR neighboring proteins by TurboID or miniTurbo, purification of the biotin-labeled proteins, and subsequent sample preparation for proteomic analysis. We utilized S1PR1 as a model GPCR, a receptor for a bioactive lipid molecule sphingosine 1-phosphate (S1P) that plays various roles in physiological and pathological conditions. This analysis pipeline enables the mapping of interacting proteins of lipid GPCRs in living cells.

Key words Proximity labeling, TurboID, GPCR, Sphingolipid, Sphingosine 1-phosphate

1 Introduction

Sphingolipids constitute a specialized group of lipids essential to the composition of the plasma membrane and are identified as potent signaling and messenger molecules in many cell types [1]. Among them, sphingosine 1-phosphate (S1P) is an important bioactive metabolite involved in various physiological and pathological processes [2]. Dysregulation of S1P signaling is one of the crucial factors for cancer initiation, progression, and resistance to anticancer treatments [3]. S1P exerts its biological functions by acting on its receptors S1PR1-5, which belong to a G-protein-

Marco Brotto and Sanjoy K. Bhattacharya (eds.), *Lipid Signalling: Methods and Protocols*, Methods in Molecular Biology, vol. 2816,
https://doi.org/10.1007/978-1-0716-3902-3_16,

coupled receptor (GPCR) family. GPCRs are expressed on the cell surface, and many undergo internalization after ligand binding. The subcellular localization of GPCRs and the molecular identities of their interacting proteins help shape GPCRs' biological functions [4]. Conventional approaches (e.g., biochemical fractionation and affinity purification) for mapping the interactome of GPCRs are challenging due to low throughput, modification or overexpression of effector proteins, and low temporal resolution [5].

In recent years, enzyme-catalyzed proximity labeling technique has been developed as a complementary approach to spatial mapping at the molecular level in living cells [6]. In a proximity labeling system, a promiscuous labeling enzyme is fused to the target protein, and proximity labeling is carried out by the enzyme that catalyzes the conversion of an inert small-molecular substrate into a highly reactive but short-lived diffusible intermediate. This reactive species diffuses from the enzyme active site and covalently labels neighboring biomolecules (around 10–20 nm) in a promiscuous manner [7]. Since the small-molecular substrate for labeling is biotin moiety in many cases, the biotinylated proteins can be selectively enriched by affinity purification using streptavidin or tamavidin beads [8]. As the biotinylation is carried out in living cells while molecular complexes and cellular membranes are intact, the spatial relationships and interaction networks are preserved in their native state.

In the past several years, various labeling enzymes have been engineered, including ascorbate peroxidase 2 (APEX2), BirA, and TurboID, each with their own strengths and limitations [9–12]. APEX2, an enhanced ascorbate peroxidase, was engineered from a plant ascorbate peroxidase [9]. In the presence of hydrogen peroxide, APEX2 rapidly catalyzes exogenous biotin-phenol to produce highly reactive biotin-phenoxyl intermediates, which then covalently react with adjacent proteins at electron-rich side chains (primarily tyrosine). APEX2 has rapid labeling kinetics (less than 1 min of labeling time), enabling the detection of biological processes with high temporal resolution [10]. However, the potential oxidative stress from the presence of hydrogen peroxide and the low membrane permeability of biotin-phenol both limit its application in living organisms.

By contrast, BirA in the BioID system [11], derived from *Escherichia coli* biotin ligase, does not require toxic reagents and simply uses soluble substrate biotin while cells provide ATP. BirA can adenylate biotin to generate reactive biotin–adenosine monophosphate, which diffuses from its active site and reacts with lysine side chains on proximal proteins to achieve promiscuous protein biotinylation [11]. Although the BioID system simply requires an addition of exogenous nontoxic biotin, allowing its application in vivo, the very low activity (more than 18 h of labeling time) limits its application to defining the interactome mapping within the timescale of minutes to a few hours.

TurboID and its modified enzyme miniTurbo systems are improved versions of the BioID system by directed evolution for efficient labeling to overcome the slow labeling kinetics of BirA [12]. With the addition of biotin in cell culture, TurboID and miniTurbo can label neighboring proteins in 10 min, which broadens the scope of dynamic processes that can be probed [12]. TurboID is about twofold more active than miniTurbo and exhibits labeling activity even in the absence of a high concentration of exogenous biotin. On the other hand, miniTurbo has a lower affinity for biotin and doesn't show labeling activity until a high concentration of exogenous biotin is provided, making it preferable for precise temporal control of the labeling window. Additionally, miniTurbo is 20% smaller than TurboID, which may minimize the disturbance to the trafficking and functions of the fusion protein [12].

miniTurbo has been reported to be less stable than TurboID likely due to removal of its N-terminal domain, resulting in lower expression levels in some cell lines [13]. However, we were able to stably express miniTurbo and keep its enzyme activity in human umbilical vein endothelial cells (HUVECs) [14], suggesting that the stability of miniTurbo may vary in different cell lines. To identify interacting proteins of GPCRs, we have compared the biotinylated proteins catalyzed by TurboID and miniTurbo fused to S1PR1. We prioritized labeling specificity and chose miniTurbo as the labeling enzyme for subsequent proteomic analysis. The biotinylated proteins were captured by tamavidin beads, followed by on-beads digestion by LysC and trypsin, then applied to a shotgun proteomic analysis by nano-scale high-performance liquid chromatography-tandem mass spectrometry (nano LC-MS/MS) (Fig. 1).

The localized biotin labeling by the TurboID system gives a spatial resolution of protein interaction. Nonetheless, due to the complex molecular environment in cells, several hundreds of proteins in proximity to the target protein can be labeled, including nonspecific proteins. Therefore, it is essential to regulate the expression of the TurboID-tagged target protein at the similar level as the endogenous target protein. It is also necessary to place the proper control as a reference to distinguish the specifically interacting proteins from the bystander proteins. In our recently published study on S1PR1 [14], we placed the S1PR1-TurboID/-miniTurbo constructs under the control of a Tet-On system to regulate the expression levels. We also utilized S1PR1 mutant type (S1PR1-TM4) as a reference fusion construct, which shows intracellular localization in contrast to the cell-surface localization of wild-type S1PR1 (S1PR1-WT). We compared the proteins identified by S1PR1-WT with those by S1PR1-TM4, aiming the

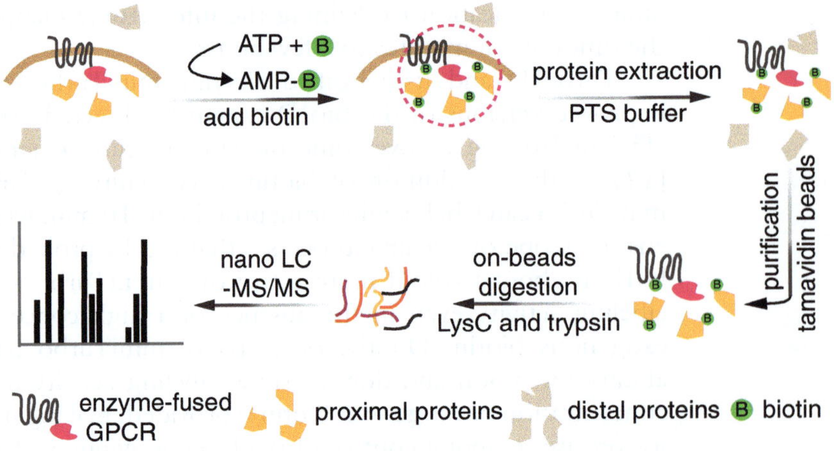

Fig. 1 Workflow for performing a TurboID (miniTurbo) proteomic analysis. The biotin ligase is fused to the C terminus of the G-protein-coupled receptor (GPCR). In the presence of exogenous biotin, the enzyme biotinylates proteins in proximity (<20 nm) to the receptor in an unbiased manner. Proteins are extracted by PTS buffer, then biotinylated proteins are purified by Tamavidin 2-REV magnetic beads, followed by on-beads digestion and mass spectrometry analysis

identification of a subset of proteins that were preferentially biotinylated in S1PR1-WT.

2 Materials

For the best results, all solutions should be prepared with mass-spec grade ultrapure water. Depending on the cell lines used, additional reagents such as media additives may be required. The materials described here are for standard procedures using HUVECs with expression of TurboID-tagged S1PR1 (WT or TM mutant) under the control of the Tet-On system.

2.1 TurboID and miniTurbo Fusion Protein Expression

1. Cell line of interest: human embryonic kidney (HEK) 293T cells and HUVECs (*see* **Note 1**).
2. Growth medium: Dulbecco's modified Eagle's medium (DMEM) containing 4.5 g/L glucose, L-glutamine, sodium pyruvate and sodium bicarbonate supplemented with 10% fetal bovine serum and 1% Penicillin–Streptomycin (P/S) for HEK293T cells, and Endothelial Cell Growth Medium-2 (EGM-2, Lonza) supplemented with 10% fetal bovine serum for HUVECs.
3. Mammalian expression plasmids for TurboID and miniTurbo fusion proteins: TurboID (Addgene, #107169) and miniTurbo (Addgene, #107170) fragments were amplified by PCR and subcloned into pLVX-TetOne-Puro vectors (Takara Bio) which had been inserted with S1PR1-WT or -TM4 fragment,

to fuse the TurboID or miniTurbo at the C-terminal end of S1PR1 (*see* **Note 2**).

4. Lentiviral packaging plasmids: pVSV-G, pMDL/pPRE and pRSV-REV (Addgene).
5. Fibronectin solution: 10 μg/mL in PBS (*see* **Note 3**).
6. Doxycycline: 1 mg/mL stock solution is prepared in H_2O and stored at −20 °C. Working solution is freshly prepared by dilution of the stock solution in EGM-2.

2.2 Labeling Reaction and Lysis of the Cells

1. Biotin: 100 mM stock solution is prepared in DMSO and stored at −20 °C.
2. Tris-HCl: 1 M stock solution is prepared in H_2O with pH adjusted by hydrochloric acid to the required values.
3. 20× protease inhibitor cocktail (PIC) stock solution: One tablet of cOmplete protease inhibitor cocktail, EDTA free (Roche) is dissolved in 2.5 mL of H_2O. Working solution is prepared by dilution of the stock solution in the lysis buffer.
4. Cell lysis buffer: 50 mM Tris-HCl (pH 8.0), 100 mM NaCl, 2 mM EDTA, 10 mM 2-glycerophosphate, 1 mM Na_3VO_4, 5 mM NaF, 1% Triton X-100, 0.5% Fos-Choline and 1× PIC.
5. Phase transfer surfactant (PTS) buffer: 100 mM Tris-HCl (pH 9.0 or 7.4), 12 mM sodium deoxycholate, 12 mM sodium lauroyl sarcosinate and 1 × PIC (*see* **Note 4**).
6. Protein assay kit: BCA protein assay kit (ThemoFisher Scientific) or equivalent protein assay kits for determination of protein concentrations.
7. 4× Sample loading buffer for SDS-PAGE: 200 mM Tris-HCl (pH 6.8), 4% SDS, 30% glycerol, 20% 2-mercaptoethanol and 0.04% bromophenol blue.

2.3 Immunoblotting

1. Tris-buffered saline containing 0.1% Tween 20 (TBS-T): 20 mM Tris-HCl (pH 7.6), 137 mM NaCl and 0.1% Tween-20 (v/v).
2. 5% (w/v) skim milk: dissolve 5 g of skim milk powder in 100 mL TBS-T and store at 4 °C for no more than 1 week.
3. Primary antibody: Rabbit monoclonal anti-S1PR1 antibody (Cell Signaling Technology) is diluted 1:2000 in 5% skim milk for immunoblotting.
4. Secondary antibody: anti-rabbit HRP-conjugated secondary antibody is diluted 1:2000 in 5% skim milk.
5. HRP-conjugated streptavidin: HRP-conjugated streptavidin (ThermoFisher Scientific) is diluted 1:5000 in 5% skim milk.

2.4 On-Beads Digestion

1. Tamavidin 2-REV-coated magnetic beads suspension: Tamavidin beads are available from Wako pure Chemicals (*see* **Note 5**).
2. Ammonium bicarbonate: 50 mM solution is freshly prepared in H_2O.
3. Dithiothreitol solution (DTT): 100 mM DTT solution is freshly prepared in 50 mM ammonium bicarbonate.
4. Iodoacetamide solution: 550 mM iodoacetamide in 50 mM ammonium bicarbonate, prepared immediately before use.
5. LysC solution: LysC (Promega) is dissolved in H_2O to make 0.5 μg/μL stock solution, aliquoted and stored at −80 °C.
6. Trypsin solution: Trypsin Gold, Mass Spectrometry Grade (Promega) is dissolved in 50 mM acetic acid to make 0.5 μg/μL stock solution, aliquoted and stored at −80 °C. Do not refreeze more than two times to avoid auto-proteolysis.

2.5 Peptides Purification

1. Trifluoroacetic acid (TFA): HPLC grade.
2. Solid-phase extraction column: StageTip, GL-tip SDB is available from GL Sciences.
3. StageTip solvents: water/acetonitrile (95/5) with 0.1% TFA (Solvent A) and water/acetonitrile (20/80) with 0.1% TFA (Solvent B).
4. Peptide resuspension buffer: 0.1% formic acid in water.

3 Methods

3.1 Generation of Cell Lines Stably Expressing S1PR1-TurboID/miniTurbo

1. Transfect HEK293T cells with the fusion construct plasmids and lentivirus packaging plasmids using a calcium phosphate method (*see* **Note 6**).
2. Collect the lentiviral-containing medium 72 h after transfection, filtrate through a 0.45 μm filter, make aliquots and store at −80 °C until use.
3. Seed HUVECs at early passage (P2 or P3) onto fibronectin-coated 6-well plates. Allow the cells to reach 40–50% confluency. Infect HUVECs with packaged lentivirus for the expression of the fusion constructs (S1PR1-WT/-TM4-TurboID or S1PR1-WT/-TM4-miniTurbo in our case).
4. After incubation for 48 h, transfer the HUVECs to 100 mm culture dishes with growth medium containing 2 μg/mL puromycin for more than a week for stable expression (*see* **Note 7**).
5. Once all non-infected cells have died and infected cells reached confluence, propagate the cells to the desired number and cryopreserve in appropriate aliquots in liquid nitrogen until use.

3.2 Characterization of the Expression and Labeling Activity of the Fusion Proteins

1. Thaw HUVECs expressing TurboID- or miniTurbo-fused proteins and seed onto fibronectin-coated 6-well plates. Allow the cells to recover and grow to around 70% confluency.
2. Replace the medium with fresh growth medium containing various concentrations of doxycycline and incubate for 24 h to induce the expression of the fusion proteins (*see* **Note 8**).
3. Replace the medium with fresh growth medium containing 500 μM biotin to initiate labeling (*see* **Note 9**).
4. Stop the labeling reaction by aspirating the medium and washing the cells with ice-cold PBS.
5. Add the cell lysis buffer to each well, collect the cells by scraping into 1.5 mL microcentrifuge tubes, then disrupt the cells by sonication (*see* **Note 10**).
6. Remove cell debris by centrifugation at 10,000×*g* for 10 min at 4 °C, and transfer the supernatant to fresh microcentrifuge tubes. Protein concentration is determined by the BCA protein assay kit (ThemoFisher Scientific) or equivalent protein assays.
7. Add the 4× Sample loading buffer for SDS-PAGE to the 3 volumes of the cell lysate, and boil the samples for 5 min at 98 °C.
8. Load the equal amounts of proteins (10–20 μg), and separate the proteins on an SDS-polyacrylamide gel (*see* **Note 11**).
9. Electro-transfer the proteins onto PVDF membranes (*see* **Note 12**).
10. Block the membranes with 5% skim milk in TBS-T for 1 h at RT.
11. For detection of TurboID- or miniTurbo-fused S1PR1 (WT/TM4) expression, incubate the membranes with rabbit anti-S1PR1 antibody in 5% skim milk overnight at 4 °C. Wash the membrane three times with TBS-T for 10 min each. Incubate with secondary antibody for 1 h at RT.
12. For detection of biotinylated proteins, incubate the membranes with streptavidin-HRP in 5% skim milk overnight at 4 °C.
13. Wash the membranes three times with TBS-T for 10 min each. Incubate with ECL western blotting substrate for 1 min and image the membranes.
14. Choose the optimal concentration of doxycycline and the incubation time for biotin labeling for proteomic analysis according to the results (*see* **Note 13**, Figs. 2 and 3).

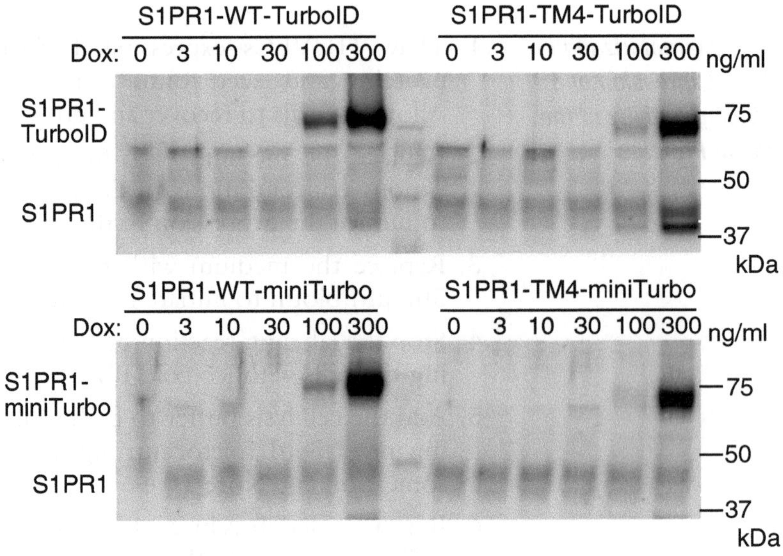

Fig. 2 Western blot analysis of S1PR1-WT/-TM4-TurboID and S1PR1-WT/-TM4-miniTurbo expressions in HUVECs induced by increasing concentrations of doxycycline (Dox) for 24 h. The cells were lysed, and the enzyme-tagged S1PR1 was identified by anti-S1PR1 antibody. Adapted from reference [14] with a slight modification

3.3 Preparation of Biotinylated Proteins for Proteomic Analysis

1. Thaw HUVECs expressing the fusion proteins and seed onto 100 mm fibronectin-coated culture dishes. Allow the cells to recover and grow to confluence (*see* **Note 14**).
2. Transfer one million cells onto each 100 mm fibronectin-coated culture dish and incubate for 24 h (*see* **Note 15**).
3. Replace the medium with fresh growth medium containing the optimized concentration of doxycycline, and incubate for 24 h to induce the expression of the fusion proteins.
4. Replace the medium with fresh growth media containing 500 μM biotin and incubate for desired time.
5. Stop the labeling reaction by aspirating the medium and washing the cells with ice-cold PBS once.
6. Add 1 mL of PTS buffer (pH 9.0) to each dish, collect the cells by scraping into a 15 mL conical tube, then disrupt the cells by sonication.
7. Remove the cell debris by centrifugation at 10,000×*g* for 10 min at 4 °C and transfer the supernatant to new tubes. Protein concentration is determined by the BCA protein assay kit.
8. Take 200 μg of protein from each cell lysate into a 1.5 mL microcentrifuge tube. Make a replicate for a no digestion control, if necessary.

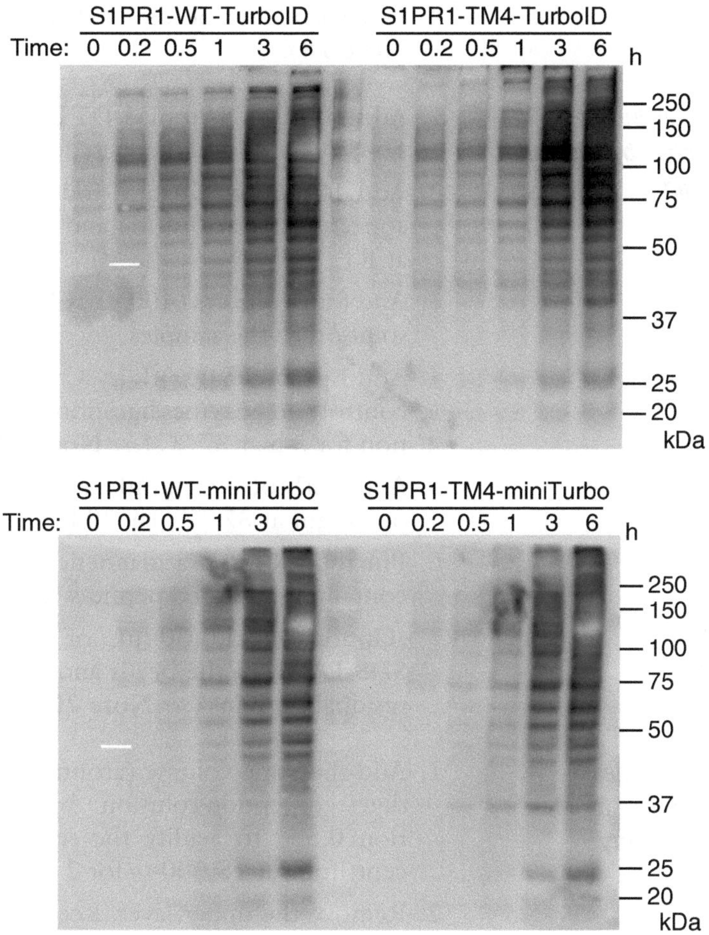

Fig. 3 Western blot analysis of the biotinylated proteins by S1PR1-WT/-TM4-TurboID and S1PR1-WT/-TM4-miniTurbo in HUVECs. The cells were treated with 100 ng/mL doxycycline for 24 h, then incubated with 500 μM biotin for indicated times. The cells were lysed by the cell lysis buffer, and the biotinylated proteins were visualized by streptavidin-HRP. Adapted from reference [14] with a slight modification

9. Add PTS buffer (pH 7.4) to make final volume of 1 mL in each tube (*see* **Note 16**).
10. Add 30 μL of Tamavidin beads and incubate overnight at 4 °C with continuous rotation (*see* **Note 17**).
11. Separate the beads from the solution by using a magnetic stand.
12. Wash the beads with 1 mL of PTS buffer (pH 7.4) with gentle vortex for 30 s.
13. Repeat the wash step five times.

14. Resuspend the beads in 90 μL of PTS buffer (pH 9.0) (*see* **Note 18**).

3.4 On-Beads Digestion by LysC and Trypsin

1. Add 10 μL of 100 mM DTT to the samples (final concentration: 10 mM) and incubate for 30 min at RT.
2. Add 10 μL of 550 mM iodoacetamide to the samples (final concentration: 50 mM) and incubate in the dark for 30 min at RT.
3. Add four volumes (400 μL) of 50 mM ammonium bicarbonate solution to the samples.
4. Add 1 μL of 0.5 μg/μL LysC to one tube. Keep another tube as control (no enzyme digestion), and incubate with gentle agitation for 3 h at 37 °C (*see* **Note 19**).
5. Add 1 μL of 0.5 μg/μL trypsin to the samples and incubate overnight at 37 °C (16–20 h).
6. Put the tubes on a magnetic stand, and transfer the solution containing digested peptides into new tubes.
7. (Optional) Add 30 μL of 2× Sample loading buffer for SDS-PAGE to the beads and boil for 5 min at 95 °C to elute residual proteins (*see* **Note 20** and Fig. 4).

3.5 Peptides Purification for Mass Spectrometry

1. Add the same volume (around 500 μL) of ethyl acetate to the digested peptide solution. Add 5 μL of TFA (final concentration 0.5%) to acidify the solution. Vortex for 1 min, then centrifuge at 15,000×*g* for 2 min at RT.
2. Remove the upper layer. Keep the middle and lower layer (*see* **Note 21**).
3. Vacuum evaporate the aqueous lower layer with a centrifugal evaporator.
4. Resuspend the peptides in 100 μL Solvent A.
5. Desalt the samples using a StageTip.
   - 5.1. Conditioning: Add 20 μL of Solvent B to a StageTip and centrifuge at 1500×*g* for 2 min at RT. Discard the flowthrough.
   - 5.2. Equilibration: Add 20 μL of Solvent A and centrifuge at 1500×*g* for 2 min at RT. Discard the flowthrough.
   - 5.3. Loading: Add the samples (100 μL) into the StageTip and centrifuge at 1500×*g* for 4 min. Discard the flowthrough.
   - 5.4. Wash: Add 20 μL of Solvent A and centrifuge at 1500×*g* for 2 min at RT, twice.
   - 5.5. Elution: Transfer the StageTip into a fresh tube. Add 40 μL of Solvent B and centrifuge at 1500×*g* for 2 min at RT.

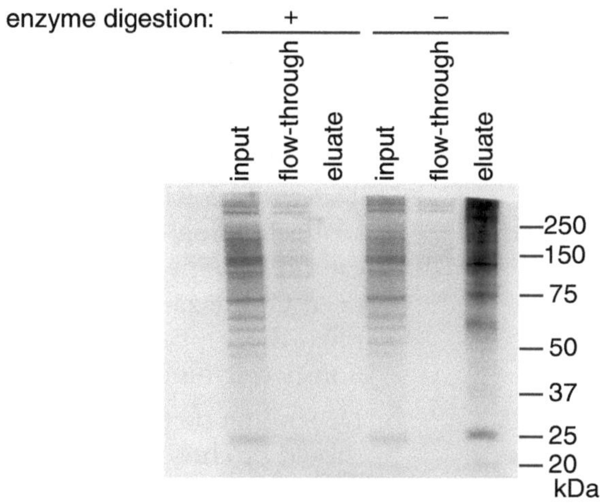

Fig. 4 Western blot analysis of biotinylated proteins in the course of purification by tamavidin beads. Input: whole cell lysate; flow-through: supernatant after the incubation with tamavidin beads for overnight; eluate: eluates from tamavidin beads in the 2× Sample loading buffer for SDS-PAGE after boiling, with or without LysC and trypsin digestion

6. Vacuum evaporate the desalted samples with a centrifugal evaporator and store the samples at −80 °C until use.
7. Resuspend the peptides in 20 μL of the Peptide resuspension buffer.
8. Transfer the peptide solution to autosampler vials.
9. Apply the samples to the nanoLC-MS/MS analyses (*see* **Note 22**).
10. Perform protein identification with an algorithm at your choice (*see* **Note 23**).
11. Data analysis of the proteomic results and subsequent biological validation are described elsewhere [14].

4 Notes

1. We use HEK293T cells for lentivirus packaging and HUVECs for expressing the fusion proteins.
2. The expression levels of the fusion proteins are controlled by a doxycycline-induced Tet-On expression system, pLVX-TetOne-Puro vector.
3. For the culture of HUVECs, it is recommended to coat the cultureware with fibronectin to ensure the attachment of the cells.

4. PTS buffer increases the solubility of hydrophobic proteins and enhances the activity of trypsin [15]. After trypsin digestion, sodium deoxycholate and sodium lauroyl sarcosinte can be removed by adding a water-immiscible organic solvent prior to LC-MS/MS analysis.
5. Tamavidin 2-REV is a novel avidin-like tetramer with reversible biotin-binding capability [8]. Streptavidin also can be used to capture biotinylated proteins for subsequent on-beads digestion. Optimization of bead volume may be required to maximize the capture of biotinylated proteins while minimizing nonspecific binding to the beads.
6. Depending on the cell lines used, proper transfection methods should be chosen. Calcium phosphate method is good enough for us to transfect for the production of lentivirus. Lentivirus system is good for stable expression in primary cells including HUVECs.
7. We generate cell lines stably expressing the fusion constructs to achieve optimal expression levels for maximizing labeling activity with less variation in signals among biological replicates. The concentration of antibiotics for selection depends on cell type and needs to be determined.
8. To control the expression level of the fusion protein, we suggest conducting the titration of doxycycline concentrations. We have tested 0–300 ng/mL of doxycycline (Fig. 2) and chosen 100 ng/mL for subsequent assays.
9. Labeling time should be optimized according to the purpose.
10. To avoid sample heating, sonication should be conducted on ice with appropriate breaks.
11. We use 4–12% gradient gels to observe the wide molecular range of proteins. Run two replicate gels, if necessary, one for detecting the expression of the fusion proteins and the other for detecting biotinylated proteins.
12. Nitrocellulose membrane can be used instead.
13. In our case, we chose 100 ng/mL doxycycline to achieve the expression of the fusion protein at the similar levels as the endogenous target protein (Fig. 2), and 3 h of the biotin-labeling time for subsequent proteomic analysis (Fig. 3).
14. Optimized conditions from the small-scale assays should be scaled up accordingly for the large-scale proteomics experiments.
15. Biological replicates should be placed properly for each experimental condition.

16. While lysis efficiency is better in PTS buffer (pH 9.0), binding of biotinylated proteins to Tamavidin beads is facilitated under the neutral pH [8].
17. The amount of Tamavidin beads should be optimized.
18. After this step, low-protein-binding tubes should be used to reduce the loss of peptides.
19. Thaw the stock solution of LysC on ice immediately before use. Same for trypsin.
20. This step is for western blot analysis to check the successful digestion by LysC and trypsin, compared to the no digestion control (Fig. 4).
21. The upper layer contains mostly ethyl acetate and PTS, while peptides remain in the lower layer.
22. We conduct nanoLC-MS/MS analysis with an Eksigent Ekspert NanoLC 425 system coupled to a TripleTOF 6600 mass spectrometer. The peptide mixture is separated by an ODS column (Eksigent ChromXP-C18-CL, 3 μm, 120 Å, 0.075 mm I.D. × 150 mm L) with 2–30% acetonitrile gradient containing 0.1% formic acid for 60 min.
23. We use the Paragon algorithm search engine using a ProteinPilot software (Sciex).

References

1. Quinville BM, Deschenes NM, Ryckman AE et al (2021) A comprehensive review: sphingolipid metabolism and implications of disruption in sphingolipid homeostasis. Int J Mol Sci 22(11):5793
2. Cartier A, Hla T (2019) Sphingosine 1-phosphate: lipid signaling in pathology and therapy. Science 366(6463):eaar5551
3. Riboni L, Abdel Hadi L, Navone SE et al (2020) Sphingosine-1-phosphate in the tumor microenvironment: a signaling hub regulating cancer hallmarks. Cells 9(2):337
4. Yang D, Zhou Q, Labroska V et al (2021) G protein-coupled receptors: structure- and function-based drug discovery. Signal Transduct Target Ther 6(1):7
5. Low TY, Syafruddin SE, Mohtar MA et al (2021) Recent progress in mass spectrometry-based strategies for elucidating protein-protein interactions. Cell Mol Life Sci 78(13): 5325–5339
6. Gingras AC, Abe KT, Raught B (2019) Getting to know the neighborhood: using proximity-dependent biotinylation to characterize protein complexes and map organelles. Curr Opin Chem Biol 48:44–54
7. Xu Y, Fan X, Hu Y (2021) In vivo interactome profiling by enzyme-catalyzed proximity labeling. Cell Biosci 11(1):27
8. Takakura Y, Sofuku K, Tsunashima M (2013) Tamavidin 2-REV: an engineered tamavidin with reversible biotin-binding capability. J Biotechnol 164(1):19–25
9. Lam SS, Martell JD, Kamer KJ et al (2015) Directed evolution of APEX2 for electron microscopy and proximity labeling. Nat Methods 12(1):51–54
10. Paek J, Kalocsay M, Staus DP et al (2017) Multidimensional tracking of GPCR signaling via peroxidase-catalyzed proximity labeling. Cell 169(2):338–349 e311
11. Roux KJ, Kim DI, Raida M et al (2012) A promiscuous biotin ligase fusion protein identifies proximal and interacting proteins in mammalian cells. J Cell Biol 196(6):801–810
12. Branon TC, Bosch JA, Sanchez AD et al (2018) Efficient proximity labeling in living cells and organisms with TurboID. Nat Biotechnol 36(9):880–887
13. May DG, Scott KL, Campos AR et al (2020) Comparative application of BioID and

TurboID for protein-proximity biotinylation. Cells 9(5):1070

14. Zhao X, Kiyozuka K, Konishi A et al (2023) Actin-binding protein filamin B regulates the cell-surface retention of endothelial sphingosine 1-phosphate receptor 1. J Biol Chem 299(7):104851
15. Masuda T, Tomita M, Ishihama Y (2008) Phase transfer surfactant-aided trypsin digestion for membrane proteome analysis. J Proteome Res 7(2):731–740

Chapter 17

Downregulation of ATP8B2 to Assess Plasmalogen Distribution and Far1 Expression in Primary Trabecular Meshwork Cells

Ishan Chaplot, Carolina Cruz-Wegener, Maria D. Cabrera Gonzalez, and Sanjoy K. Bhattacharya

Abstract

The trabecular meshwork (TM) from primary open-angle glaucoma (POAG) cases has been found to contain decreased levels of intracellular plasmalogens. Plasmalogens are a subset of lipids involved in diverse cellular processes such as intracellular signaling, membrane asymmetry, and protein regulation. Proper plasmalogen biosynthesis is regulated by rate-limiting enzyme fatty acyl-CoA reductase (Far1). ATPase phospholipid transporting 8B2 (ATP8B2) is a type IV P-type ATPase responsible for the asymmetric distribution of plasmalogens between the intracellular and extracellular leaflets of the plasma membranes. Here we describe the methodology for extraction and culturing of TM cells from corneal tissue and subsequent downregulation of ATP8B2 using siRNA transfection. Further quantification and downstream effects of ATP8B2 gene knockdown will be analyzed utilizing immunoblotting techniques.

Key words ATP8B2, Plasmalogens, Asymmetrical membrane distribution, Cell culture, Metabolomics, Primary open-angle glaucoma, Far1, Trabecular meshwork

1 Introduction

ATPase phospholipid transporting 8B2 (ATP8B2) has been increasingly studied in recent years with relation to lipid metabolism and membrane composition. ATP8B2, a type IV P-type ATPase (P4-ATPase), is involved in the asymmetric distribution of plasmalogens between the extracellular and cytoplasmic leaflets of the cellular membranes [1]. Plasmalogens are a subclass of glycerophospholipids containing a vinyl-ether bond at the sn-1 position of the glycerol backbone. The vinyl-ether bond substitutes one of the two acyl bonds in a conventional phospholipid [2]. Plasmenyl-ethanolamine (PlsE) and plasmenyl-choline (PlsC) are the most common plasmalogens, differentiated by groups on the sn-3 position [2]. The functionality of plasmalogens with the plasma

Marco Brotto and Sanjoy K. Bhattacharya (eds.), *Lipid Signalling: Methods and Protocols*, Methods in Molecular Biology, vol. 2816, https://doi.org/10.1007/978-1-0716-3902-3_17,

membrane has been studied in different contexts, such as plasmalogen importance for protein regulation, cell growth, intracellular signaling, and membrane asymmetry [3]. Fatty acyl-CoA reductase (Far1) is regarded as the rate-limiting enzyme involved in plasmalogen biosynthesis. Additionally, a feedback mechanism between Far1 and plasmalogens exists to regulate cellular plasmalogen quantity by targeting Far1 enzyme stability and not Far1 transcription frequency [4, 5]. Far1 is degraded when there is an overabundance of cellular plasmalogens and is upregulated during cellular plasmalogen deficiency [3]. As a result, Far1 quantities can be utilized in a laboratory setting to observe plasmalogen distribution. Protein kinase B (AKT) is a protein pertaining to the family of serine-threonine kinases, which are fundamental regulators for various cellular processes including metabolism, proliferation, and survival [6]. Given the diversification of AKT functions, studies have suggested the potential role of AKT in homeostatic control of intraocular pressure of glaucomatous patients and its interactions with intracellular plasmalogen levels [7]. Reduced plasmalogen concentration as a result of downregulated or absent ATP8B2 has been found to subsequently downregulate the activation of AKT via phosphorylation of its serine-473 (Ser473) and threonine-308 (Thr308) domains [8, 9]. Despite this, the exact role plasmalogens exhibit biologically is still evolving. However, ATPase mechanisms are certainly involved in the intermembrane translocation of plasmalogens and may have potential implications in primary open-angle glaucoma (POAG). Decreased levels of intracellular plasmalogens found in POAG patients may deduce that ATP8B2 may be involved in the distribution of plasmalogens in lipid membranes of trabecular meshwork (TM) cells. This can be attributed to the proposed ATP8B2 mechanism, which involves flippase activity with an affinity to phospholipids, including plasmalogens. By modulating the knockdown of ATP8B2 to determine plasmalogen levels, we can confirm that it plays a role in lipid distribution and may support the potential role in decreased polyunsaturated fatty acids (PUFA)-carrying PlsCs throughout POAG progression [2, 10].

This chapter provides the protocols necessary to analyze the relationship between ATP8B2 and plasmalogens in TM cells. TM cells will be extracted and cultured utilizing consolidated cell culture methods. Subsequently, the ATP8B2 gene will be knocked using small interfering RNA (siRNA) transfection methodology for comparison to control groups. Quantitative and qualitative analyses using reverse transcription-quantitative polymerase chain reaction (RT-qPCR) and immunoblotting will be necessary for a conclusive study.

2 Materials

2.1 Trabecular Meshwork (TM) Cell Culturing

2.1.1 Preparation of Gelatin for Well-Plate

1. Gelatin from bovine skin (Sigma-Aldrich G9391).
2. Six-well plate.
3. 18 mm coverglass.
4. 70% EtOH.

2.1.2 Retrieval of Corneal Tissue Samples, Cell Culturing, and Storage

1. Donor human corneas retrieved from corneal transplantations; communicated and arranged with local eye bank, corneal surgeon or pathology department.
2. Optisol-GS medium (Bausch & Lomb Incorporated).
3. Eppendorf tube with 70% ethanol.
4. Eppendorf tube with PBS.
5. Empty Eppendorf tube.
6. Bacterial trypsin enzyme +1 mL of TM-containing Eppendorf tube.
7. Sterile coverglass 18 mm. Submerge coverglass in 70% EtOH (use autoclaved ddH2O), 10 min. Remove ethanol, allow coverglass to air dry in hood w/plate partially covered.
8. 5% CO_2 incubator set at 37 °C.
9. Fine-curved vannas scissors.
10. Curved fine forceps.
11. Straight fine forceps.
12. Large blunt scissors.
13. TM Media preparation of 50 mL in 4 separate tubes: 39 mL Opti-MEM, 10 mL fetal bovine serum (FBS), 500 μL GlutaMax (100×) (Gibco Cat. #35050061), 500 μL penicillin-streptomycin.

2.1.3 Passaging of TM Cells

1. 0.05% trypsin/0.5 mM EDTA.
2. In 50 mL of DMEM add 5.5 mL of FBS, 550 μL of GlutaMax (100×), 550 μL of Na pyruvate, and 550 μL of penicillin-streptomycin. Multiple bottles should be prepared for passaging.
3. TM Cell Media.
4. PBS buffer.
5. Gelatin-coated well plate with isolated samples.

2.2 siRNA-Mediated Knockout of ATP8B2

1. Human ATP8B2 siRNA SMARTpool (Horizon discovery).
2. ON-TARGET*plus* GAPD Control Pool (Positive Control) (Horizon discovery).

3. ON-TARGET*plus* Non-targeting Control Pool (Negative Control) (Horizon Discovery).
4. 5× siRNA buffer (Horizon Discovery).
5. 96-well plate.
6. Antibiotic free growth medium.
7. Lipofectamine® RNAiMAX Reagent.
8. Opti-MEM® Medium.

2.3 RT-PCR

2.3.1 Homogenization of TM Cell Samples for RNA Isolation

1. Qiagen RNeasy Plus Mini Kit (Cat. #74034).
2. Microtubes and Caps.
3. Vortex Mixer.
4. Centrifuge.

2.3.2 Quantification of RNA Using NanoDrop Spectrophotometer

1. NanoDrop Spectrophotometer.
2. RNA sample.
3. Blanking solution.

2.3.3 cDNA Synthesis

1. Thermo Scientific Maxima First Strand cDNA synthesis Kit (Cat. #K1671).
2. Centrifuge.

2.3.4 RT-qPCR

1. Thermo Scientific Luminaris Color HiGreen High ROX qPCR Master Mix (Cat. #K0363).
2. Thermocycler.

2.4 Immunoblotting to Assess Far1 Expression

2.4.1 Protein Extraction for Cells in Suspension

1. TM Cell suspension samples.
2. Ice-cold PBS.
3. Microcentrifuge tubes.
4. Ice bath.
5. Ice-cold lysis buffer.

2.4.2 Electroblotting and Analysis for ATP8B2

1. SDS Page (ThermoFisher).
2. 0.45 μm Nitrocellulose membrane.
3. StartingBlock Blocking Buffer (TBST Format, ThermoFisher Cat. #37543).
4. 5% non-fat dry milk.
5. Coomassie® brilliant blue R-250, purple powder C.I. 42660, (VWR Cat. #IC19068280). Dissolve the Coomassie Brilliant Blue R in methanol prior to addition of acetic acid and water; 50% (v/v) methanol, 0.05% (w/v) Coomassie Brilliant Blue R-250, 10% acetic acid, 40% Water. (VWR).
6. Shaker.

7. Primary antibody: Polyclonal Rabbit anti-Human FAR1 Antibody (aa31-80, WB). Unit size: 100 μL (0.5 mg/mL), (LS Bio Cat #LS-C81505-GOS11).
8. Secondary antibodies: *For Phospho-AKT1 (Thr308) & Phospho-AKT1 (Ser473):* Goat anti-Rabbit IgG (H + L) Secondary Antibody, HRP 2 mL (ThermoFisher cat. #31460).
9. 200 mL Clarity™ Western ECL Substrate (Bio-Rad).
10. Imaging Software for analysis.

2.5 Immunoblotting for Assessing AKT Phosphorylation

2.5.1 Protein Extraction for Cells in Suspension

1. TM Cell suspension samples.
2. Ice-cold PBS.
3. Microcentrifuge tubes.
4. Ice bath.
5. Ice-cold lysis buffer.

2.5.2 Western Blotting

1. 0.45 μm Nitrocellulose membrane (Bio-Rad).
2. 1× Transfer Buffer (Semi-Dry): 48 mM Tris base, 39 mM glycine, 20% methanol, pH adjusted to 8.3.
3. Prepare 1 L of 10× TBS by dissolving 88 g of NaCl and 24 g of tris base in 900 mL distilled, with pH adjusted to 7.6 with 12 N HCl. Add 100 mL for a 1 L volume. To make 1× TBS dilute 100 mL 10× TBS with 900 mL distilled water.
4. Tris-buffered saline 0.1% Tween® 20 detergent (TBST) from 10× TBS stock: Dilute 100 mL of 10× TBS with 900 mL of distilled water and add 1 mL tween 20.
5. 5% non-fat dry milk added to prepared 1× TBST.
6. Bel-Art™ SP Scienceware™ Fritware™ Porous Polyethylene Sheets (Fisher Scientific, Catalog No. 22-164384).
7. Whatman® qualitative filter paper, Grade 1 (Sigma Aldrich).
8. Cytiva EPS 3501 High Voltage Power Supplies for Electrophoresis Units.
   (a) Power supply subject to variation according to laboratory availability; must have capacity to reach currents of 200–250 mA and 100 V.
9. Primary antibodies:
   (a) Phospho-AKT1 (Thr308) Recombinant Rabbit Monoclonal Antibody (AktT308-G12), 200 μL (ThermoFisher cat. #MA5-37083).
   (b) Phospho-AKT1 (Ser473) Monoclonal Antibody (14-6), 100 μL (ThermoFisher cat. #44-621G).

10. Secondary antibodies: *For Phospho-AKT1 (Thr308) & Phospho-AKT1 (Ser473):* Goat anti-Rabbit IgG (H + L) Secondary Antibody, HRP 2 mL (ThermoFisher cat. #31460).
11. Clarity™ Western ECL Substrate for immunoblot development.
12. Imaging Software for analysis.

3 Methods

3.1 TM Cell Culturing

3.1.1 Preparation of Gelatin for Well-Plate (See Note 1)

1. Retrieve prewarmed 2% gelatin and dissolve with 100 mL of sterile ddH2O.
2. Sterilize by autoclaving the gelatin with the following conditions: 121 °C, 15 psi for 30 min for complete reuse pension of the gelatin.
3. In a six-well plate, coat each well with 1 mL gelatin. Place the coated plate in a 37 °C incubator for 15 min.
4. After incubation, remove the gelatin from the well plate. Rinse the gelatin with sterile 1× PBS or TM cell media before seeding cells.
5. Rinse an 18 mm coverglass with 70% EtOH, then allow for coverglass to air dry entirely for use.

3.1.2 Retrieval of Corneal Tissue Samples, Cell Culturing, and Storage

1. Suitable corneal tissue samples (*see* **Note 2**) retrieved from local eye banks must be stored in Optisol-GS medium at 2–8 °C for 14 days prior to its expiration date for transplantation. Trabecular meshwork isolation completed approximately 1–2 days prior to expiration.
2. Disinfect all surfaces of fume hood and microscope, as well surgical instruments and tubes, with 70% ethanol solution.
3. Place the corneal tissue sample in sterile Petri dish. Bathe the sample in 70% ethanol for 1 min.
4. Transfer the sample to a new Petri dish and douse the corneal tissue sample in sterile PBS solution, ensuring that the samples do not dry out.
5. Carefully remove the uveal structures (iris, ciliary body, and choroid) from donor corneal tissue samples using vannas scissors. Use forceps to secure sample posteriorly and bisect it near the equator utilizing a single-edged razor blade. Place the remnants on a new sterile Petri plate. Extracted uveal components in the new petri plate may be discarded into biohazardous waste after completing full extraction.
6. Isolate the primary trabecular meshwork cells from corneal tissue with curved fine jewelers under direct examination with

dissection microscope, extracted as continuous strips to ensure better-quality visualization and analysis.

7. Prepare the isolated TM strip(s) for plating by trimming into 2–3 mm fragments with vannas scissors.
8. Wash the TM strips 3–4× with PBS to remove any extraneous uveal components.
9. Place the TM strip(s) into a new Petri dish and pour a thin layer of PBS to maintain moisture of the sample.
10. Centrifuge the sample(s) at 1500 × *g* for 5 min, then remove supernatants through aspiration with a 200 μL pipette. For a higher yield of supernatant, cutoff the tip of the p1000 pipette diagonally.
11. Increase the pipette dial to approximately 380 μL and pre-aspirate with PBS. Then, aspirate the TM tissue and transfer the aspirated supernatants and PBS to the dry gelatin-coated well plate.
12. Deposit the TM supernatant onto the well plate, place the sterilized coverslip onto the tissue. The coverslip will serve to anchor the isolated sample; therefore, ensure that the sample is not freely floating around the coverslip.
13. Remove excess PBS and resuspend the cell pellets in 1 mL of prepared TM cell media (*see* **Note 3**).
14. Maintain the cell culture in humidified air containing 5% CO_2 at 37 °C during the process of cell migration onto the plate over the course of 2–4 weeks.

3.1.3 Passaging of TM Cells (See ***Note 4****)*

1. Disinfect all surfaces of fume hood and materials entering the fume food with 70% ethanol solution.
2. Allocate and retrieve the TM samples in the CO_2 incubator. Transfer the well plate into fume hood following thorough sterilization.
3. Aspirate the media utilizing a pipette gun and dispose of old media in a waste beaker.
4. Rinse the TM cells thoroughly with PBS.
5. Add trypsin EDTA and allow the cells to incubate for 10 min.
6. Add 100 mL of prepared, appropriate media (designated ratios according to passage number), allowing for trypsin deactivation.
7. Procedures according to passage number.
   (a) P0
      (i) *Media:* Opti-MEM/20% FBS, 1% PS, 1% GlutaMAX, 1% Na pyruvate
   (b) P1

(i) *Media:* Opti-MEM/20% FBS, 1% PS, 1% GlutaMAX, 1% Na pyruvate

(ii) *Passage:* 3 wells of 6-well plate

(iii) *Cryopreserve:*1 well of 6-well plate

(c) P2

(i) *Media:* DMEM/10% FBS, 1% PS, 1% GlutaMax, 1% Na pyruvate

(ii) *Passage:* 3× 10 cm plates

(iii) *Cryopreserve:* 2 × 10 cm plates

(d) P3

(i) *Media:* DMEM/10% FBS, 1% PS, 1% GlutaMax, 1% Na pyruvate

(ii) *Passage:* 9 × 10 cm

(e) P4

(i) *Media:* DMEM/10% FBS, 1% PS, 1% GlutaMax, 1% Na pyruvate

3.1.4 Biweekly Changing TM Media for Confluency

1. Warm up the prepared TM media at 37 °C in a water bath for a minimum of 30 min.
2. Transfer the incubating cell cultures and pre-warmed media to a fume hood for media change.
3. Using a pipette, carefully aspirate the current cell culture media, paying close attention as to not accidentally aspirate TM strips. Dispose of old media in a waste beaker.
4. Using a pipette gun, transfer 100 mL of pre-warmed media to each well of the well plate.
5. Following media change for each well, re-incubate the cell culture with newly changed media to CO_2 37 °C incubator.

3.2 siRNA-Mediated Knockout of ATP8B2

There will be four conditions, each repeated three times for a total of 12 wells:

(1) ATP8B2 gene knockdown. (2) Negative Control (Scramble), (3) Positive Control (GAPDH), (4) Mock

1. 1.4 × 10^4 trabecular meshwork cells should be plated in a growth medium that is antibiotic-free and contains serum. If cells are adherent, this should be done 24 h before transfection. Cells should be plated and observed to have 60–80% confluency at the time of transfection. This should be performed and observed for each well.
2. 10 μM stock siRNA must be produced from the original 5 nmol sirNA using 5× siRNA buffer dilution (*see* **Note 5**).
3. Two dilutions should be performed:

(a) Dilute 0.36 μL of siRNA (10 μM) or 0.36 μL of controls into 30 μL Opti-MEM® Medium. This dilution is performed only three times, one time for each condition except for Mock (*see* **Note 6**).

(b) Dilute 4.32 μL Lipofectamine® RNAiMAX Reagent in 119.96 μL Opti-MEM® Medium. This dilution is performed for all four conditions at once.

4. Combine 30 μL of the diluted siRNA with 30 μL of the diluted Lipofectamine® RNAiMAX Reagent for a 1:1 ratio. This should be done for each of the three conditions except for Mock. The mock condition will use 30 μL of Opti-MEM® Medium with no siRNA or control and 30 μL of the diluted Lipofectamine® RNAiMAX Reagent.
5. Mix gently with pipette, then incubate the four mixtures for 5–10 min at room temperature.
6. Add 10 μL of siRNA-lipid complex mixture to cells in each well in a dropwise manner. Each condition has three replicates, making a total of 30 μL for each condition. The mock condition does not receive any siRNA, only transfection reagents. Mix gently with a rocking plate.
7. Cells should be incubated for 24–72 h at 37 °C before analysis of transfected cells (*see* **Note 7**).

3.3 RT-PCR

3.3.1 Homogenization of TM Cell Samples for RNA Isolation

1. Cells grown in monolayer: Cell-culture medium should be removed with pipette, and 300 μL Buffer RLT plus from the Qiagen RNeasy Plus Mini Kit should be added to each well.
2. The lysates should be transferred to microtubes and sealed (*see* **Note 8**).
3. Homogenize the lysates through vortexing microtubes for 30 s.
4. Transfer the homogenized lysate to the provided gDNA Eliminator spin column that is in the 2 mL collection tube.
5. For 30 s at ≥8000× *g*, centrifuge the samples. Then, store the flow-through and discard the column.
6. Add 350 μL or 600 μL of 70% ethanol to the flow-through depending on the cell count. Use pipette to mix and do not centrifuge (*see* **Note 9**).
7. Immediately transfer a maximum 700 μL of sample along with any precipitate, to the RNeasy spin column within the supplied 2 mL collection tube.
8. Close the tube, and for 15 s at ≥8000× *g*, centrifuge the sample. The flow-through can be discarded.
9. To the RNeasy Mini spin column within a 2 mL collection tube, add 700 μL Buffer RW1.

10. Centrifuge at ≥8000× *g* for 15 s. Flow-through can be discarded.
11. To the RNeasy spin column, add 500 μL Buffer RPE.
12. Close the tube and Centrifuge at ≥8000× *g* for 15 s. Flow-through can be discarded.
13. Again, to the RNeasy spin column, 500 μL Buffer RPE is to be added. Centrifuge for 2 min at ≥8000× *g* (*see* **Note 10**).
14. The RNeasy spin column can be relocated to a new 1.5 mL collection tube. To the spin column, add 30–50 μL RNase-free water directly.
15. Close the tube and centrifuge for 1 min at ≥8000× *g* for final elution of RNA.

3.3.2 Quantification of RNA Using Nanodrop Spectrophotometer

1. Start the NanoDrop software and linked spectrophotometer.
2. Choose among DNA, RNA, or protein for intended use.
3. Carefully lift the horizontal arm of the spectrophotometer.
4. The dark spot on the pedestal under the arm is for sample placement. Slowly clean the pedestal with a wipe (lint free).
5. Pipette 1–2 μL of blanking solution and lower the arm.
6. Press on "Measure blank."
7. Lift the arm and clean with a wipe (lint free).
8. **Steps 5–7** can be repeated with the RNA solution instead of blanking solution.

3.3.3 DNA Synthesis

1. Thaw, mix, and centrifuge the kit components. Then, keep on ice (*see* **Note 11**).
2. On ice, add the following components in chronological order for a total volume of 10 μL in an RNase-free tube on ice: 1 μL 10× dnDNase Buffer, 1 μL dsDNase, 1 μg Template RNA (Dependent on RNA concentration from NanoDrop). Spectrophotometer, Fill the rest of the tube with nuclease-free water.
3. Mix the sample gently and then centrifuge.
4. Incubate the sample at 37 °C for 2 min in a water bath. Then place the sample on ice, centrifuge momentarily, and place on ice again.
5. In the same tube from **step 1**, add the following components: 4 μL 5× Reaction Mix, 2 μL Maxima enzyme mix, and 4 μL nuclease-free water.
6. Mix the sample gently and then centrifuge.

7. Two controls are performed using reverse transcriptase minus (RT-) negative control and no template control (NTC) (*see* **Note 12**).
8. Incubate the tube for 10 min at 25 °C, then incubate the tube for 15 min at 50 °C.
9. Heat the sample at 85 °C for 5 min to stop the reaction (*see* **Note 13**).

3.3.4 RT-qPCR

1. Thaw the solutions. Then, vortex and centrifuge (*see* **Note 14**).
2. Proper calculations have to be completed for correct qPCR quantities (*see* **Note 15**).
3. Create a reaction mix by adding forward primer and reverse primer, 2× Master Mix, preferred cDNA quantity, and water for a total of 20 μL. Mix the reactions gently and centrifuge if necessary. Do not vortex. Place the samples in the thermocycler.

3.4 Immunoblotting for Assessing Far 1 Expression

Immunoblotting procedure will be applied for TM cell samples of each condition described in Subheading 3.2.

3.4.1 Protein Extraction for Cells in Suspension (See Note 16)

1. Centrifuge the TM cell suspension at 2000× *g*, 5–7 min, 4 °C. Collect the cell pellet and dispose of the supernatant.
2. Add ice-cold PBS to the cell pellet and wash via centrifuging at 2000× *g*, 5–7 min, 4 °C.
3. Add ice-cold lysis buffer to the cell pellet and agitate the mixture in microcentrifuge tube for 30 min, 4 °C.
4. Centrifuge the tubes at 16,000× *g*, 20 min, at 4 °C.
5. Transfer and collect the supernatant into a fresh tube. Discard the pellet.
6. Place the tube with supernatant on an ice bath.

3.4.2 Western Blotting

1. Prepare the blotting apparatus by thorough rinse with DI/high-purity water.
2. Fill the transfer tank with 1× transfer buffer. Submerge the gel cassette holder, fiber pads, and polyethylene sheet in the transfer buffer.
3. Disassemble the gel apparatus and remove the stacking gel carefully. Trim a small piece of the lower left-hand corner of the gel for gels loaded from left to right.
4. Equilibrate the gel in a glass dish with 250 mL of 1× transfer buffer based on thickness of the gel. Adhere to the following parameters (*see* **Note 17**): 0.5-mm-thick gels for 1 min, 0.75–1-mm gels for 5 min, and 1.5-mm gels for 15 min.

5. Cut the 0.45 μm nitrocellulose transfer membrane (*see* **Note 18**) to be slightly larger than the gel, then cut a piece of filter paper to be slightly larger than the membrane.
6. Slowly wet the nitrocellulose membrane from one edge into a glass dish of transfer buffer.
7. Equilibrate the membrane for 15 min.
8. To begin preparing the gel transfer sandwich, open the gel holder cassette on a disinfected, flat surface. Place a fiber pad that was submerged in 1× transfer buffer from **step 2** on the top surface of the cassette, then place a gel support sheet/filter paper over the fiber pad.
9. Place the gel face down onto the support, ensuring that the trimmed edge is on the right side. Pour 5–10 mL of 1× transfer buffer.
10. Remove the wet nitrocellulose membrane from the tray prepared in **steps 5–6**, carefully resubmerge the membrane in the transfer buffer, and then position it above the gel, lowering it from the center outward (*see* **Note 19**).
11. Wet the filter paper with transfer buffer for a few seconds and then slowly cover the nitrocellulose membrane to prevent air bubbles. Place a second fiber pad over the filter paper and close the gel cassette holder. This finalizes the preparation of the gel/membrane transfer sandwich.
12. Slide the assembled gel/membrane transfer sandwich on the transfer cassette into the tank transfer unit, with the membrane on the anode side (+) and the gel on the cathode side (−).
13. Pour 1× transfer buffer into the tank transfer unit until the electrode panels are completely covered but do not reach the electrical connectors (*see* **Note 20**).
14. Connect the anode and cathode leads to the appropriate power outputs.
15. Once the system is prepared, connect the system to the power supply and begin running the electrotransfer utilizing a 200–250 mA constant current and a voltage of 100 V. The duration is dependent on the gel thickness; for thicker gels (>0.75 mm in thickness), transfer times will be greater than 1 h.
16. After completion of electrotransfer, turn off the power supply and disconnect it. Remove the cassette holder from the tank and carefully disassemble the gel transfer sandwich with forceps.
17. Incubate the 0.45 μm nitrocellulose membrane with Starting-Block Blocking Buffer (5% non-fat dry milk in 1× TBS) for 1 h.

18. Apply 20 mL prepared dye reagent (50% (v/v) methanol, 0.05% (w/v) Coomassie Brilliant Blue R-250, 10% acetic acid, 40% Water) to the gel, and allow for gel to be incubated within a shaker for 1 h.
19. Following incubation, rinse and soak the gel under running distilled water.
20. The primary antibodies, such as Polyclonal Rabbit anti-Human FAR1 Antibody, were prepared overnight in 4 μL 1× TBST 5% non-fat dry milk. Retrieve and incubate the prepared primary antibodies at 4 °C overnight on the nitrocellulose membrane, gently rocking on a shaker.
21. Following incubation, wash the nitrocellulose membrane with 1× TBST twice for 3 min on the shaker, at room temperature.
22. Prepare 4 μL Goat anti-Rabbit IgG (H + L) Secondary Antibody, HRP, with 8 μL of 1× TBST 5% non-fat dry milk and incubate for 2 h, gently rocking on a shaker.
23. Wash the nitrocellulose membrane again with 1× TBST twice for 3 min on the shaker, at room temperature.
24. Prepare enhanced chemiluminescence ECL mixing components of Clarity™ in a 1:1 ratio (*see* **Note 21**). Incubate the ECL mix on the nitrocellulose membrane for 3–5 min.
25. Densitory analysis of protein bands of interest will be quantifiable utilizing imaging software, such as ImageJ Software.

3.5 Immunoblotting for Assessing AKT Phosphorylation

Immunoblotting procedure will be applied for TM cell samples of each condition described in Subheading 3.2.

1. Sample preparation (protein extraction and measurement of protein concentration) from TM cells. Follow the procedural steps outlined in Subheading 3.4.1.
2. For western blotting, follow the procedural steps outlined in Subheading 3.4.2. Use the appropriate antibodies detailed in Subheading 2.5 pertaining to AKT phosphorylation.

4 Notes

1. Gelatin coating on the well plate is utilized to ensure the migration of TM cells and their attachment to the well, while maintaining cells healthy and adequate in count. Gelling occurs rapidly at room temperature; therefore, perform the manipulations on ice bath/ice packs. Ideal temperature will be within the range of 2 °C to 8 °C. Ensure that coating is done under the hood [11].

2. Suitable corneal tissue samples are dependent upon storing procedures and the time period from death to culturing. Note the dates of expiration, death, enucleation, and preservation to determine optimal tissue for analysis; extended time periods will result in poor establishment and viability of primary cell cultures. Sterile environments also contribute to viability of corneal tissue samples. Reduce the potential contamination by sterilization of the sample's exterior prior to dissection and dissect/scrape off all conjunctiva. Further sterilization can be accomplished by rinsing corneal tissue samples with Betadine or 70% ethanol for 1–2 min, followed by extensive washing with sterile PBS, prior to handling [12].
3. An optimal TM cell media helps promote growth of TM cell cultures. 20% FBS supplementation or antifungal agents, such as 1% amphotericin B, may prove to be beneficial for stimulating and maintaining TM cell growth, particularly for more dated corneal tissue samples.
4. Passaging normal TM cells mobilizes the dissociation of cell-cell and cell-matrix adhesions, as well as migrates cells off from plates. It is important to note that TM cells have a finite number of doublings and begin to change in appearance after undergoing 6–8 passages. The maximum recommended number of passages for human TM cells to ensure viability of cell samples is seven passages. Given that initial number of cell isolates and the duration of cell strain passaging are directly proportional, culture monitoring is critical [12].
5. Dilution guidance is provided from the Horizon Discovery siRNA resuspension protocol. It is recommended to aliquot siRNA in a more concentrated manner. If siRNA is purchased from another manufacturer, follow associated dilution protocols. For example, by adding 50 μL 5× siRNA buffer to the 5 nmol siRNA, a 100 μM stock is obtained. The stock siRNA can then be stored in smaller 2 μL tubes. When siRNA is needed to be converted to 10 μM stock from 100 μM stock for transfection, an additional 18 μL 5× siRNA buffer is added to the 100 μM stock stored in the 2 μL tubes. This allows for a tenfold dilution.
6. An siRNA dilution is performed for each condition listed here: ATP8B2 gene knockdown, Negative Control (Scramble) and Positive Control (GAPDH). The mock condition is a control to mimic the transfection protocol without any siRNA. When performing the dilutions, utilize the respective Horizon Discovery siRNA pool or control pool ordered.
7. Dilutions and quantities of materials can be altered based on the number of wells being used for transfection to conserve

materials. Smaller pilot studies are also recommended to ensure that available siRNA is working properly against cells.

8. Protocol for RNA isolation developed using manufacturer's recommendations. Qiagen cat. #19560 can be used for microtubes and cat. #19566 as caps.
9. Buffer RLT Plus and ethanol quantities are dependent on cell count and dish size. They can be specified using manufacturer's recommendations.
10. At this step, it is optional to place the RNeasy spin column in another 2 mL collection tube. A further centrifuge can be performed for 1 min at full speed to dry the membrane more.
11. Protocol is developed with reference to manufacturer's recommendations.
12. Reverse transcriptase minus (RT-) negative control will contain all previous reagents used in reverse transcription except Maxima Enzyme Mix. The Reverse transcriptase minus (RT-) negative control is important to analyze the RNA sample for DNA contamination. No template control (NTC) will contain all previous reagents used in reverse transcription except for the RNA template. The no template control (NTC) is important for checking reagent contamination.
13. The product of cDNA synthesis can be stored for up to 7 days at −20 °C. If a longer period is needed, store at −70 °C. Repetitive freezing and thawing of cDNA should be avoided.
14. Recommended experimental setup for RT-qPCR includes having two control groups: (1) No Template Control (NTC) and (2) Reverse Transcriptase Minus (RT-) control. For RT-qPCR, it is important to have two sets of primer triplicates for each triplicate of the main conditions. For example, condition (1) "ATP8B2 gene knockdown" will have three triplicates (1, 2, and 3). During RT-qPCR, two primers will be used, such as an (a) ATP8B2 primer and a (b) housekeeping primer. Each triplicate condition (1, 2, & 3) will have three replicates of each primer (e.g., 1.a1, 1.a2, 1.a3 & 1.b1, 1.b2, 1.b3). This applies to the control groups as well.
15. Calculations, reaction setup recommendations, and thermocycler parameters are available from manufacturer's recommendations. If necessary, cDNA should be diluted to follow manufacturer recommendations. Similarly, primer design recommendations should be followed by manually designing primers or using commercially available primer designs software. Any primer designed will need a forward and reverse primer to complete RT-qPCR. Thermocycler should be set with parameters based on manufacturer's recommendations.

16. Protein extraction protocol has been adapted from Sigma-Aldrich Cell Lysis and Protein Extraction for Western Blotting protocol.
17. This step allows for the removal of excess SDS and prevents swelling of the gel during electrotransfer. This ultimately affects transfer yields if SDS concentration is unbalanced for the respective equilibration time [13].
18. A 0.45 μm nitrocellulose membrane is ideal for Western blotting techniques. The 0.45 μm pore size is optimal for most analytical blotting and transfers. The Bio-Rad nitrocellulose provides exceptional signal-to-noise results and exhibits versatility and efficiency with a variety of solutions [3, 13].
19. Air bubbles will prevent the electrotransfer of the desired proteins. To prevent the formation of air bubbles, slide the removed wet nitrocellulose membrane across the edge of the tray. When resubmerging the membrane in the transfer buffer, place the membrane on the gel from the center outward to mobilize air bubbles toward the edge of the gel [13].
20. Avoid leaving the transfer buffer in the covered transfer tank for more than 2 h prior to use to avoid evaporation of the methanol within the buffer, which can potentially change the buffer composition [13].
21. According to BioRad, for smaller-sized membranes (~ 7 × 8.5 cm), 7 mL of solution is sufficient. For mid-sized membranes (~8.5 × 13.5 cm), 12 mL of solution is sufficient for imaging.

References

1. Takatsu H et al (2014) Phospholipid flippase activities and substrate specificities of human type IV P-type ATPases localized to the plasma membrane. J Biol Chem 289(48): 33543–33556. https://doi.org/10.1074/jbc.M114.593012
2. Acar N et al (2009) Red blood cell plasmalogens and docosahexaenoic acid are independently reduced in primary open-angle glaucoma. Exp Eye Res 89(6):840–853. https://doi.org/10.1016/j.exer.2009.07.008
3. Honsho M, Mawatari S, Fujiki Y (2022) ATP8B2-mediated asymmetric distribution of plasmalogens regulates plasmalogen homeostasis and plays a role in intracellular signaling. Front Mol Biosci 9:915457. https://doi.org/10.3389/fmolb.2022.915457
4. Honsho M, Asaoku S, Fujiki Y (2010) Posttranslational regulation of fatty acyl-CoA reductase 1, Far1, controls ether glycerophospholipid synthesis. J Biol Chem 285(12): 8537–8542. https://doi.org/10.1074/jbc.M109.083311
5. Honsho M, Abe Y, Fujiki Y (2017) Plasmalogen biosynthesis is spatiotemporally regulated by sensing plasmalogens in the inner leaflet of plasma membranes. Sci Rep 7:43936. https://doi.org/10.1038/srep43936
6. Xu N et al (2012) Akt: a double-edged sword in cell proliferation and genome stability. J Oncol 2012:951724. https://doi.org/10.1155/2012/951724
7. Luna C et al (2009) Alterations in gene expression induced by cyclic mechanical stress in trabecular meshwork cells. Mol Vis 15:534–544
8. Hossain MS et al (2013) Plasmalogens rescue neuronal cell death through an activation of AKT and ERK survival signaling. PLoS One 8(12):e83508. https://doi.org/10.1371/journal.pone.0083508

9. Timmons S et al (2009) Akt signal transduction dysfunction in Parkinson's disease. Neurosci Lett 467(1):30–35. https://doi.org/10.1016/j.neulet.2009.09.055
10. Saab S et al (2014) Plasmalogens in the retina: from occurrence in retinal cell membranes to potential involvement in pathophysiology of retinal diseases. Biochimie 107(Pt A):58–65. https://doi.org/10.1016/j.biochi.2014.07.023
11. Waduthanthri KD, Montemagno C, Çetinel S (2019) Establishment of human trabecular meshwork cell cultures using nontransplantable corneoscleral rims. Turk J Biol 43:89–98. https://doi.org/10.3906/biy-1810-69
12. Keller KE et al (2018) Consensus recommendations for trabecular meshwork cell isolation, characterization and culture. Exp Eye Res 171: 164–173. https://doi.org/10.1016/j.exer.2018.03.001
13. Goldman A et al (2015) Electroblotting from polyacrylamide gels. Curr Protoc Protein Sci 82:10.17.11–10.17.16. https://doi.org/10.1002/0471140864.ps1007s82

Chapter 18

Diacylglycerol Signaling in Retinal Ganglion Cells

Ishan Chaplot, Isabella Moceri, Maria D. Cabrera Gonzalez, and Sanjoy K. Bhattacharya

Abstract

With impaired retinal ganglion cell (RGC) function and eventual RGC death, there is a heightened risk of experiencing glaucoma-induced blindness or other optic neuropathies. Poor RGC efficiency leads to limited transmission of visual signals between the retina and the brain by RGC axons. Increased focus on studying lipid messengers found in neurons such as endocannabinoids (eCBs) has importance due to their potential axonal pathway regenerative properties. 2-Arachidonoylglycerol (2-AG), a common eCB, is synthesized from an sn-1 hydrolysis reaction between diacylglycerol (DAG) and diacylglycerol lipase (DAGL). Examination of DAG production allows for future downstream analysis in relation to DAGL functionality. Here, we describe protocol guidelines for extracting RGCs from mouse retinas and subsequent mass spectrometry analysis of the DAG content present within the RGCs.

Key words Diacylglycerol signaling, Retinal ganglion cells, Mass spectrometry, Diacylglycerol, Axon regeneration

1 Introduction

Retinal ganglion cells (RGCs) mediate the transmission of visual signals between the retina of the eye and the brain. RGC axons compose the optic nerve, which allows for captured light information to be delivered to the visual cortex of the brain [1]. The risk of glaucoma-induced blindness or other optic neuropathies increases with impaired RGC function and death. With poor RGC axon development after injury, the optic pathway cannot function normally. Increased research in efforts to promote axon regeneration in RGCs has led to attention on the potential signaling pathways involved [2]. Endocannabinoids (eCBs) function as lipid messengers that can activate axonal developmental pathways [2]. 2-Arachidonoylglycerol (2-AG) is a prominent eCB involved in the binding with cannabinoid receptors such as cannabinoid receptor type 1 (CB1R) situated primarily in neurons. 2-AG is originally synthesized from an sn-1 hydrolysis of diacylglycerol (DAG)

Marco Brotto and Sanjoy K. Bhattacharya (eds.), *Lipid Signalling: Methods and Protocols*, Methods in Molecular Biology, vol. 2816, https://doi.org/10.1007/978-1-0716-3902-3_18,

through diacylglycerol lipase (DAGL) [3]. DAG is synthesized from phosphatidylinositol bisphosphate (PIP_2) through the progression of the phospholipase C pathway [4]. It is known that increased eCB localization during nerve injury exists, which allows for eCB regulation opportunities through enzyme inhibition in vitro. Monoacylglycerol lipase (MAGL) operates as the enzyme responsible for degrading 2-AG. Studies have aimed to inhibit eCB degrading enzymes such as MAGL to offset degradation of rapid eCB accumulation following injury [5]. As a result, the relationship between DAGL and MAGL becomes a topic of importance due to interconnectivity in the eCB system for future therapeutic applications. This chapter first outlines protocol guidelines for extracting and culturing RGCs from mouse retinas, followed by mass spectrometry analysis of DAG content in RGCs after lipid extraction. Analysis of DAG content allows for future conclusions regarding 2-AG production.

2 Materials

2.1 Anticipation of Retinal Ganglion Retrieval

1. 100 and 150 mm petri dishes.
2. TrisHCl: 50 mM, pH 8.5.
3. Goat anti-mouse IgM.
4. Goat Anti-rabbit IgG H + L (affinity purified).
5. Poly D-lysine (PDL): Add 5 mL of water to 5 mg PDL (Sigma-Aldrich P6407), filter with 0.22 μm filter. Make 200 mL aliquots and store at −20 °C.
6. Sterile water.
7. CO2 incubator.
8. Laminin (2 μg/mL diluted in Neurobasal).
9. Neurobasal (ThermoFisher 10888022).
10. Tris buffer solution (50 mM, pH 9.5).

2.2 Preparation of Retinal Ganglion Cell Extraction

1. Dulbecco's Phosphate-Buffered Saline (DPBS): 1 g/L glucose and 36 mg/L sodium pyruvate.
2. Earle's Buffered Salt Solution (EBSS), no magnesium and calcium.
3. 1 N NaOH.
4. Papain (60–100 units) (Worthington Biochemicals).
5. L-cysteine (Sigma-Aldrich).
6. 0.4% DNase solution: Add EBSS to DNAse (D4527) at 1 mL per 12,400 units. Make 200 μL aliquots and store at −20 °C.
7. 0.22 μm filters.

8. CD90 antibody (BioRad #MCA02R).
9. Bovine Serum Albumin (BSA) (Sigma-Aldrich A8806).
10. Low Amount of Papain Inhibitor Ovomucoid (Lo Ovo) Stock (10×): Add 600 mg BSA to 40 mL of DPBS and mix. Then, add 600 mg of trypsin inhibitor and dissolve by mixing. Adjust to pH 7.4 with approximately 500 μL of 1 N NaOH. Filter through 0.22 μm filter, and make 1 μL aliquots stored at −20 °C.
11. High Amount of Papain Inhibitor Ovomucoid (Hi Ovo) Stock (6×): Add 1200 mg BSA to 20 mL DPBS and mix. Add 1200 mg trypsin inhibitor and dissolve by mixing. Adjust to pH 7.4 with approximately 900 μL of 1 N NaOH, add more if necessary until solution is no longer acidic. Filter through 0.22 μm filter, and make 1 μL aliquots stored at −20 °C. Make two tubes of this for a total of 40 mL.
12. Rabbit anti-Mouse Macrophage antibody (Rb anti-Mo) (Accurate Chemical, AIAD31240).
13. 5 × 15 mL tubes labeled: 10-papain, 10-papain-filter, 9-Lo Ovo Ovo, Lo Ovo-Rb anti-Mo, 5-Hi Ovo Ovo
14. 4% BSA solution: Add 2 g BSA into 50 mL DPBS. Dissolved at 37 °C adjusted to pH 7.4 using 200 μL 1 N NaOH. Filter using 0.45 and 0.22 μm filters. Then store at −20 °C in 1 mL aliquots.
15. DPBS/0.2% BSA solution: Add 1 mL of 4% BSA to 19 mL of DBPS. Store at 4 °C.

2.3 Retina Dissection and Papain Digestion

1. Dissection equipment: scissors, #55 forceps, #5 forceps, #11 blade scalpel, dissecting scope, and a laboratory retina dissection spatula.
2. Petri dish with DPBS.
3. Retinas extracted from mice.
4. Papain solution.
5. CO_2 incubator (37 °C).

2.4 Trituration of Retinas

1. Lo Ovo solution.
2. Hi Ovo solution.
3. Centrifuge.

2.5 Immunopanning

1. Pre-equilibrated EBSS: Prepare 15 mL of EBSS in a centrifuge tube and place it in a 10% CO_2 incubator. The cap should be left loose. During equilibration. a media color change is expected.
2. Insulin 100× (0.5 mg/mL): Add 10 mg insulin (Sigma-Aldrich I6634) and 100 μL 1.0 N HCl into 20 mL of sterile water. Mix

well and filter through a 0.22 μm filter. Can store this for 4–6 weeks at 4 °C.

3. Panning buffer: Add 2 mL of 0.2% BSA solution to 18 mL of DPBS, then add 200 μL insulin.
4. 0.4% DNAse.
4. DPBS.
5. Phosphate buffer saline (PBS).
6. Sato Stock (100×): Prepare 40 mL neurobasal medium and add 400 mg BSA, 400 mg transferrin, 64 mg putrescine, 10 μL progesterone (2.5 mg/100 μL EtOH), 400 μL sodium selenite (4 mg/100 μL 0.1 N NaOH and 10 mL neurobasal). Make sure to mix well and filter through a 0.22 μm filter. Store at −20 °C in 200 μL aliquots.
7. T3 (100×): In 400 μL of N NaOH dissolve 3.2 mg of triiodothyronine (Sigma-Aldrich). Add 15 μL of this solution into 30 mL DPBS. Make sure to mix well and filter through a 0.22 μm filter, then discard the first 10 mL. Store at −20 °C in 200 μL aliquots.
8. NAC 1000× (5 mg/mL): In 10 mL neurobasal dissolve 50 mg N-acetyl cysteine (Sigma-Aldrich A8199). Store at −20 °C in 20 μL aliquots.
9. Forskolin 1000× (5 mM): To 10 mg forskolin add 4.8 mL DMSO for 2.1 mg/mL. Create 50 μL aliquots and store at −20 °C.
10. NB-Sato culture medium: Prepare 20 mL of Neurobasal and add 200 μL 100× of Sato stock, 200 μL of Insulin, 200 μL of T3, 20 μL of NAC, 200 μL of 100× Pen/Strep, 200 μL of 100× L-glutamine (1 mM), 400 μL of 50× B27 (Gibco 17,504,044), and 200 μL of sodium pyruvate (100 mM). Filter the solution with a 0.22 μm filter; Can store solution for up to 7 days at 4 °C. Upon use of NB-Sato culture medium add 20 μL of 1000× CNTF (10 μg/mL), 20 μL of 1000× BDNF, 20 μL of 1000× Forskolin (5 μM). Full medium can be stored for a maximum of 3 days at 4 °C. The NB-Sato culture medium will be referred to as growth media.
11. Trypsin solution (2.5%): Prepare 100 mg trypsin (Sigma-Aldrich T4665) and dissolve in 43.3 mL EBSS at a concentration of 30,000 units per 1 mL. Filter the solution with a 0.22 μm filter and store in 200 μL aliquots at −80 °C. It is important to make sure that trypsin does not lose activity.
12. Trypsin/EBSS solution: Add 200 μL of 2.5% trypsin to 4 mL of pre-equilibrated EBSS.

2.6 Release of RGCs

1. 30% Fetal Bovine Serum (FBS): Mix 6 mL FBS (heat inactivated serum) with 14 mL DPBS. Filter the solution with a 0.22 μm filter and store at 4 °C.
2. Trypsin/EBSS solution.
3. Trypan Blue.
4. Hemocytometer.
5. Pre-equilibrated NB-Sato culture medium.

2.7 Mass Spectrometry Analysis of DAG

1. 1:1 Ratio of LC-MS grade acetonitrile and isopropanol.
2. EquiSPLASH LIPIDOMIX Quantitative Mass Spec Internal Standard (Avanti Polar Lipids Inc.).
3. Thermo Scientific Vanquishb Horizon Binary UHPLC.
4. Accurcore Vanquish C18+ Column (150 mm × 2.1 mm, 1.5 um) Thermo Scientific.
5. Q Exactive mass spectrometer coupled to a heated electrospray ionization (HESI) source.
6. 200 mM ammonium formate.
7. Formic Acid.
8. HPLC-grade acetonitrile, isopropanol, and water.
9. Mobile phase A: 50:50 ACN: H2O with 5 mM ammonium formate and 0.1% formic acid.
10. Mobile phase B: 88:10:2 IPA:ACN:H2O with 5 mM ammonium formate and 0.1% formic acid.

3 Methods

3.1 Anticipation of Retinal Ganglion Retrieval

3.1.1 Secondary Antibody Coating to Panning Dishes

1. In a 100 mm petri dish, add 30 μL Goat anti-mouse IgM and 12 mL TrisHCl (50 mM, pH 8.5). This is for Thyl/CD90 panning.
2. In two 150 mm Petri dishes, add to each 20 mL TrisHCl (50 mM, pH 8.5) and 60 μL of affinity purified goat anti-rabbit IgG. These are anti-macrophage panning dishes.
3. After mixing and coating all dishes thoroughly, cover each plate and incubate on a flat surface at 4 °C overnight (*see* **Note 1**).

3.1.2 Poly D-Lysine and Laminin Coating for Culture Dishes

1. Add a 200 μL aliquot of Poly-D-Lysine (PDL) (1 mg/mL) to 20 mL sterile water (final concentration is 0.01 mg/mL in cell culture water). The amount of PDL solution added differs between well plate sizes and coverslip material (*see* **Note 2**).
2. Leave at room temperature for 30 min. Depending on the material, cover slips differ in the amount of time coated (*see* **Note 3**).

3. Shake off the PDL liquid, and wash with sterile water two times by applying and swirling 2–3 mL water and discarding (*see* **Note 4**).
4. For the laminin coating, use 2 μg/mL laminin derived from a 500× stock solution. The amount of laminin solution added differs between well plates and coverslips (*see* **Note 5**).
5. Incubate for a minimum of 3 h in a 10% CO_2 incubator at 37 ° C.

3.2 Preparation of Retinal Ganglion Cell Extraction

3.2.1 Papain Solution Preparation

1. Add 5 μL 1 N NaOH and 10 mL DPBS with 60–100 units of papain in a tube labeled "10-papain."
2. Dissolve the papain by placing the tube in a 37 °C water bath.
3. Add 2 mg of L-cysteine to the papain solution and mix properly.
4. Then, add a 100 μL aliquot of 0.4% DNase stock to the papain solution.
5. Mix well and filter the papain solution through a 0.22 μm filter that was pre-rinsed with DBPS into a tube labeled "10-papin-filtered."

3.2.2 Low Amount of Papain Inhibitor Ovomucoid (Lo Ovo)

1. Add 9 mL DPBS to a tube labeled "9-Lo Ovo". Then, add 100 μL of 0.4% DNase stock to the solution. Mix well.
2. Add 1.0 mL of 10× Lo Ovo Stock to the tube.
3. Then transfer 6 mL from the "9-Lo Ovo" tube to the tube labeled "Lo Ovo- Rb anti-Mo".
4. Add 80 μL rabbit anti-mouse macrophage to the "Lo Ovo- Rb anti-Mo" tube.

3.2.3 High Amount of Papain Inhibitor Ovomucoid (Hi Ovo)

1. Add 5.0 mL DPBS and 1.0 mL of 10× Hi Ovo to the tube labeled "5-Hi Ovo," then add approximately 6 μL NaOH to the same tube to adjust pH to an orange color.

3.2.4 Completion of Antibody Dishes

1. The 100 mm dish is to be rinsed with DPBS three times. In 10 mL of 0.2% BSA add 50 μL of CD90 (anti-Thy1), then add onto the dish and incubate for a minimum of 3 h at room temperature.
2. Leave the 150 mm dishes on bench at room temperature.

3.3 Retina Dissection and Papain Digestion

1. Retina dissection is to be done using proper materials and safety precautions (*see* **Note 6**).
2. Remove any leftover DPBS from the retina dissection using a pipette.
3. Move the retinas to a 50 mL tube containing papain solution (10-papain-filtered tube) by using a gentle suction with a cut pipette tip and incubate for 30 min at 37 °C while slowly agitating once in between.

3.4 Trituration of Retinas

1. Aspirate the papain-based solution from the retinas using a pipette after papain digestion is completed.
2. Use 4 mL of Lo Ovo- “9-Lo Ovo” solution to rinse the tissue (*see* **Note 7**).
3. Triturate the retinas with a pipette using 2 mL from the “Lo Ovo- Rb anti-Mo” solution (*see* **Note 8**).
4. Take 1 mL of the supernatant after the clumps have settled and place it in the empty “9-Lo Ovo” tube. Then, to break up possible cell clumps by pipetting repeatedly.
5. Triturate the retinas again by taking 1 mL from the “Lo Ovo-Rb anti-Mo” tube and adding it to the retinas. After the solution has settled, transfer 1 mL of the supernatant to the “9-Lo Ovo” tube. Keep repeating this process until all cells are in suspension and in the “9-Lo Ovo” tube (*see* **Note 9**).
6. Centrifuge the cells at 80*g* for 10 min at room temperature (25 °C).
7. Resuspend the pellet in 1 mL of Hi-Ovo solution after aspirating supernatant. Spin the cells again at 80*g* for 10 min after adding the remaining Hi Ovo solution.

3.5 Immunopanning

1. Resuspend the cells in 1 mL panning buffer after aspirating Hi Ovo solution.
2. Then add an additional 14 mL panning buffer for a total 15 mL volume along with 50 μL 0.4% DNase stock (*see* **Note 10**).
3. Rinse the first 150 mm coated dish three times with DPBS and then add cell suspension to the dish.
4. Incubate for 10 min then shake slowly and incubate for another 10 min to allow for some cell adhesion to the plate.
5. Rinse the other 150 mm coated dish twice with PBS and once with DPBS.
6. Take the first 150 mm dish and shake it on a flat surface, then transfer the cells to the second 150 mm dish. Use a pipette for any leftover solution.
7. Then incubate the second plate for 45 min while shaking every 15 min.
8. Locate the 100 mm CD90 plate and wash it three times with DPBS.
9. Pour the suspension from the second 150 mm dish onto the 100 mm dish. Then incubate for 50 min and shake gently every 10 min (*see* **Note 11**).
10. During the immunopanning process, finalize the cell culture well media quantities and prepare NB-Sato growth media. First, aspirate the laminin out of the wells and add NB-Sato

growth media. NB-Sato Growth media quantity will be different based on each well (*see* **Note 12**).

11. Toward the end of the immunopanning process, prepare the Trypsin/EBSS solution.

3.6 Release of RGCs

1. From the 100 mm CD90 dish, discard the cell suspension by slowly pouring.
2. Wash the dish with DPBS six times to remove any nonadherent cells and use a microscope to verify that all nonadherent cells have been removed. Wash two more times with DPBS if needed.
3. Rinse the dish by adding 6 mL of pre-equilibrated EBSS and stirring quickly.
4. Add 4 mL of the EBSS-Trypsin solution and incubate for 4 min at 37 °C.
5. After incubation, add 2 mL of 30% FBS to cells and then pipette toward the center of the dish and slowly move around the plate to remove the cells. Repeat this process by pipetting toward the edge of the plate and around the plate (*see* **Note 13**).
6. Transfer the removed RGC cells to a 30 mL tube.
7. Add 5 mL of 30% FBS to the plate again and use pipette to dissociate the cells by pipetting in the center and around the edges. Once all cells are dislodged, combine them together and rinse the plate with 4 mL of 30% FBS (*see* **Note 14**).
8. Mix the cells gently and prepare 50 μL in a separate tube for counting purposes. Centrifuge the cells at 80*g* for 15 min.
9. Add 50 μL of trypan blue to the 50 μL of solution in a separate tube, then use a hemocytometer to count cells.
10. After centrifugation, resuspend the cell pellet in pre-equilibrated growth media at about 1 μL per 1000 cells (*see* **Note 15**).
11. Feed the cells by removing half of the medium and replacing it with pre-equilibrated and pre-warmed growth medium. This can be done every other day.

3.7 Mass Spectrometry Analysis of DAG

Lipid samples are extracted from cells using Bligh and Dyer method or alternate method of choice, and speed-vacuumed to preserve their integrity. Run the lipid samples directly after extraction, but if needed, they can be kept at −80 °C for long-term storage.

3.7.1 Sample Preparation

1. Prepare Mobile Phase A in a solvent bottle. Prepare this by combining methanol/chloroform (60:40) with 0.2% formic acid and 10 mM ammonium acetate.

2. Prepare Mobile Phase B in a solvent bottle. Prepare this by combining methanol/chloroform (60:40) with 0.2% formic acid and 10 mM ammonium acetate.
3. Cure dry the lipid samples in 50 μL of chloroform/methanol (1:1).
4. Prepare EquiSPLASH standard by diluting stock solution to 100 μg/mL. Spike 1 μL of that into each sample.
5. Prepare an internal standard with 49 μL 1:1 chloroform/methanol and 1 μL of EquiSPLASH to run as a quality control.
6. Prepare a blank by adding 50 μL chloroform/methanol to a vial. Run blank every 8–10 samples.
   (a) Column temperature: 30 °C
   (b) Injection volume: 5 μL
   (c) Gradient time: 20 min, 15–65% solvent B
   (d) Flow rate: 0.260 mL/min

3.7.2 HESI Source Parameters

Positive Mode:

(a) Spray Voltage: 4.0 kV
(b) Capillary temperature: 325 °C
(c) Auxiliary heater: 275 °C
(d) Sheath gas flow rate: 35 units
(e) Aux gas flow rate: 15 units
(f) S-lens RF: 70

Negative Mode:

(a) Spray Voltage: 2.50 kV
(b) Capillary temperature: 325 °C
(c) Auxiliary heater: 275 °C
(d) Sheath gas flow rate: 30 units
(e) Aux gas flow rate: 15 units
(f) S-lens RF: 70

3.7.3 Q Exactive Positive Mode Parameters

Full Scan:

(a) Chromatogram Peak Width: 6 s
(b) Method Duration: 30 min
(c) Scan Range: 250–1200 m/z
(d) Resolution: 70,000
(e) AGC target: 1E6
(f) Maximum IT: 240 ms

DD/MS2

(a) Resolution: 17,500
(b) AGC target: 1E5
(c) Maximum IT: 50 ms
(d) Loop count: 10
(e) Isolation window: 1.0 m/z
(f) Collision energies: 20, 30, 40 eV

3.7.4 Q Exactive Negative Mode Parameters

Full Scan:

(a) Chromatogram Peak Width: 6 s
(b) Method Duration: 30 min
(c) Scan Range: 250–1500 m/z
(d) Resolution: 70,000
(e) AGC target: 1E6
(f) Maximum IT: 240 ms

DD/MS2:

(a) Resolution: 17,500
(b) AGC target: 1E5
(c) Maximum IT: 50 ms
(d) Loop Count: 0
(e) Isolation Window: 1.0 m/z
(f) Collision energies: 20, 30, 40 eV

Data Analysis

1. Upload the raw files from the Q Exactive to LipidSearch 5.0 and begin a new study.
2. Initialize the identification parameters as follows:
   (a) Parent and product search tolerance: 5 ppm
   (b) Filters: top rank, main isomer peak, FA priority
   (c) M-score threshold: 5.0
   (d) C-score threshold: 2.0
   (e) Quantification m/z tolerance: 5 ppm
   (f) Retention time tolerance: 0.5 min
   (g) Adducts in positive mode: +H, $+NH_4$, H, $+H-H_2O$, $+2H_2+$
3. Alignment parameters:
   (a) Retention time for alignment: 0.1 min
   (b) Filters: top ranked, main isomer peak

4. Run alignment.
5. Identify the internal standards found within all samples.
6. Under "Quantification," select the correct standards and indicate the concentration as 100 μg/mL.
7. Hit "normalize."
8. Export the normalized data to Excel Workbook.
9. Consolidate the data for further analysis.

4 Notes

1. The described methods in Subheading 3.1 should be done 24 h before the day of the full RGC panning and RGC extraction experiment. All other methods are to be done the day of RGC panning and RGC extraction. It is helpful to make solutions in a separate tube, then add to the dishes. Only Petri dishes should be used as the described coating does not work for tissue culture dishes. The coated plates should be used within 2 weeks if stored at 4 °C.
2. Depending on plate size: Add 50 μL per well in a 96-well plate, 200 μL per well in a 48-well plate, 400 μL per well in a 24-well plate, 750 μL per well in a 12-well plate, and 1.5 mL per well in a 6-well plate.
3. Plastic (Aclar) coverslips are coated for 10 min, and glass coverslips are coated for 30 min.
4. Plates coated in PDL should be used within 1 week if stored at 4 °C and wrapped with paraffin. Once laminin is applied, the culture dishes need to be used within a day.
5. Depending on plate size for addition of laminin: Add 50 μL per well in a 96-well plate, 200 μL per well in a 48-well plate, 350 μL per well in a 24-well plate, 750 μL per well in a 12-well plate, and 1.5 mL per well in a 6-well plate.
6. Mouse retina dissection should be completed with proper training and equipment. Use a Petri dish with DPBS to house the retinas during the extraction process. Make sure to use biohazard bags for accumulated waste. Clean the equipment with water and EtOH. During papain digestion, DPBS/0.2% BSA solution (Subheading 2.2), 30% FBS (Subheading 2.6), and panning buffer (Subheading 2.5) can be made in advance.
7. Make sure to gently rinse and let the cells settle for 30 s. Then aspirate the solution again.
8. Using a cut pipette tip may help for the retinas. Collect the retinas with the pipette carefully and then slowly release them. Repeat this process three times.

9. To allow the antibody and cells to bind properly, incubate the cells for approximately 10 min at room temperature. Transfer to preferred tube for centrifugation.
10. 50 μL of solution can be removed and used for counting cell count.
11. Extra RGCs can be extracted by moving the supernatant from the third 100 mm dish to another 100 mm dish containing CD90.
12. NB-Sato growth media quantities: For each well in a 96-well plate, add 150 μL, for each well in a 24 well plate, add 500 μL, for each well in a 6-well plate, add 2 mL.
13. It is beneficial to mark the plate to identify a starting point. Utilize a microscope to see if the cells have been dislodged from the plate. Incubate the cells again if the cells are still adherent.
14. It is important to use the microscope to check the edges of the plate for any leftover cells. Keep pipetting if there are more leftover cells.
15. For a 96-well plate, it is recommended to plate 5000 cells in 100 μL RGC growth media per well. For a 24-well plate, it is recommended to plate 15,000–20,000 cells in 400–500 μL RGC growth media per well. Avoid plating cells densely as this will cause cells to float from the dish. 50,000 cells are viable for qPCR procedures. Maximum cell recommendations per well are 25,000 cells for a 24-well plate, 100,000 for a 12-well plate, and 250,000 for a 6-well plate.

References

1. Pasquaré SJ, Chamorro-Aguirre E, Gaveglio VL (2023) The endocannabinoid system in the visual process. J Photochem Photobiol 13: 100159. https://doi.org/10.1016/j.jpap.2022.100159
2. Stark DT, Caprioli J (2016) Subcellular localization of a 2-arachidonoyl glycerol signaling cassette in retinal ganglion cell axonal growth in vitro. Invest Ophthalmol Vis Sci 57(15): 6885–6894. https://doi.org/10.1167/iovs.16-20748
3. Reisenberg M et al (2012) The diacylglycerol lipases: structure, regulation and roles in and beyond endocannabinoid signalling. Philos Trans R Soc Lond Ser B Biol Sci 367(1607): 3264–3275. https://doi.org/10.1098/rstb.2011.0387
4. Dong W et al (2012) Does diacylglycerol serve as a signaling molecule in plants? Plant Signal Behav 7(4):472–475. https://doi.org/10.4161/psb.19644
5. Martinez-Torres S et al (2023) Injury-induced activation of the endocannabinoid system promotes axon regeneration. iScience 26(6): 106814. https://doi.org/10.1016/j.isci.2023.106814

Chapter 19

Quantifying ABCA1/apoA-1 Signaling Pathways with AFM Imaging and Lipidomic Analysis

Carolina Cruz-Wegener, Jonah Kutell, Louay Aldaher, Noa deHaseth, Sanjoy K. Bhattacharya, and Noel Ziebarth

Abstract

The role of lipid metabolic pathways in the pathophysiology of primary open-angle glaucoma (POAG) has been thoroughly elucidated, with pathways involved in lipid-related disorders such as hypercholesterolemia and hyperlipoprotein accumulation being of particular interest. The ABCA1/apoA-1 transduction pathway moderates reverse cholesterol transport (RCT), facilitating the transport of free cholesterol (FC) and phospholipids (PL) and preventing intracellular lipid aggregates in retinal ganglion cells (RGCs) due to excess FCs and PLs. A deficiency of ABCA1 transporters, and thus, dysregulation of the ABCA1/apoA-1 transduction pathway, may potentiate cellular lipid accumulation, which affects the structural and mechanical features of the cholesterol-rich RGC membranes. Atomic force microscopy (AFM) is a cutting-edge imaging technique suitable for imaging topographical surfaces of a biological specimen and determining its mechanical properties and structural features. The versatility and precision of this technique may prove beneficial in understanding the effects of ABCA1/apoA-1 pathway downregulation and decreased cholesterol efflux in RGCs and their membranes. In this protocol, ABCA1$^{-/-}$ RGC mouse models are prepared over the course of 3 days and are then compared with non-knockout ABCA1 RGC mouse models through AFM imaging of topographical surfaces to examine the difference in membrane dynamics of knockout vs. non-knockout models. Intracellular and extracellular levels of lipids are quantified through high-performance liquid chromatography with tandem mass spectrometry (HPLC-MS/MS).

Key words ABCA1, apoA-1, Cholesterol efflux, Retinal ganglion cells, Lipid signaling, Atomic force microscopy, Tapping mode, Lipidomics, Mass spectrometry

1 Introduction

Primary open-angle glaucoma (POAG), a progressive optic neuropathy, is a predominant cause of irreversible blindness globally, but it remains largely undiagnosed in the affected population at an early stage [1]. An extensive body of research has supported the potential implication of lipid metabolism disorders, including hypercholesterolemia and hyperlipoprotein accumulation, on intraocular pressure (IOP) and, ultimately, the assessment and prognosis

Marco Brotto and Sanjoy K. Bhattacharya (eds.), *Lipid Signalling: Methods and Protocols*, Methods in Molecular Biology, vol. 2816, https://doi.org/10.1007/978-1-0716-3902-3_19,

of POAG [2]. Studies have demonstrated a significant difference in plasma total cholesterol concentrations between individuals with primary open-angle glaucoma (POAG) and non-POAG individuals [3]. As such, this relationship bears the question of understanding the influence of cholesterol and lipoprotein levels on retinal ganglion cells (RGCs), the primary glaucomatous neuronal cells whose axonal degeneration is responsible for progressive vision loss [4].

Lipidomic analysis at the cellular level has identified several POAG-linked gene polymorphisms that play critical roles in determining membrane surface tension and strength, including ABCA1, an ATP-binding cassette (ABC) transporter [5, 6]. ABCA1 is an essential protein in the lipid signaling transduction pathway of reverse cholesterol transport (RCT) in RGCs, mediating unidirectional cellular phospholipids and cholesterol efflux to lipid-free apolipoprotein A-I (apoA-I), and is responsible for forming high-density lipoprotein (HDL) [5, 7, 8]. ApoA-1 is a major component of HDL that interacts with ABCA1 to facilitate the transport of free cholesterol (FC) and phospholipids (PL) and plays a role in apoptotic pathways in RGCs [7]. The mechanisms of the ABCA1/apoA-1 interaction are largely unknown; however, studies have verified that the ABCA1/apoA-1 pathway aids in the prevention of intracellular lipid aggregates effectively by ridding RGCs of excess FC and PL [9].

A deficiency or absence of ABCA1 transporters disrupts the transduction pathway, leading to intracellular lipid accumulation, may potentially impact the fluidity, permeability, and rigidity of these increasingly cholesterol-rich membranes [10]. Ultimately, hypercholesteremia- and hyperlipoprotein-induced accumulation within RGC plasma membranes expedites the retinal damage characteristic of POAG [5]. The available literature has yet to address the role of atomic force microscopy (AFM) in outlining the changes to RGC plasma membrane properties because of deficient ABCA1/apoA-1 interactions. However, AFM is promising in quantifying and imaging these structural and mechanical changes to establish the effects of the ABCA1/apoA-1 pathway. Further comprehension of the transduction pathway can be achieved with a lipid concentration analysis in the extracellular and intracellular environments via liquid chromatography with tandem mass spectrometry.

Atomic force microscopy is an extension of scanning force microscopy (SFM) that provides versatile and high-spatiotemporal resolution imaging of a biological specimen's topographical surfaces, mechanical properties, and structural features. With its numerous modalities, AFM has proven to be a potent tool for studying dynamic biological processes, even at the unimolecular level, via single-molecule force spectroscopy (SMFS) [11, 12]. The baseline mechanism involves the atomic force microscope, which has a cantilever with a molecularly sharp probe that is raster scanned over the membrane simultaneously as cantilever deflection is measured. To maintain constant deflection, the height of the cantilever is changed using the piezoelectric mechanism. The quantifiable height for the scanned area then allows for the development of the membrane's three-dimensional topography [13]. AFM may be a suitable approach for a comparative study of the effect of ABCA1 knockout vs. non-knockout on the topographical surfaces of RGCs, as well as the affected signaling pathway involving cholesterol efflux from the inner to the outer membrane of RGCs. More precisely, a prediction regarding impacted cholesterol levels can be deduced by the rigidity of the membrane prior to ABC1 knockout and single-molecule detection of cholesterol [14].

Here, we demonstrate how imaging and lipidomic analysis can be applied to better understand the mechanisms of ABCA1/apo-A1 signaling transduction pathways involving cholesterol efflux and its implications on cellular cholesterol levels, IOP, and POAG assessment. Topographical and single-molecule imaging and analysis are achieved using AFM, while quantification of the extracellular and intracellular levels of lipids is accomplished utilizing high-performance liquid chromatography with tandem mass spectrometry (HPLC-MS/MS). Each procedure will be done accordingly for both ABCA1-knockout mice and wild-type mice to compare the structural differences resulting from a deficiency of the ABCA-1 transporter.

2 Materials

2.1 RGC Cell Culturing

1. DBA/1-Abca1 tm1Jdm/J KO Mice, Strain #: 003897, RRID: IMSR_JAX:003897 (The Jackson Laboratory); Preparation of Transgenic ABCA1$^{-/-}$ Mice (*see* **Note 1**).
2. C57BL/6J Wild-Type Mice, Strain #: 000664, RRID: IMSR_JAX:000664 (The Jackson Laboratory).
3. PBS 1%FBS: 99 mL PBS, 1 mL FBS, chilled at 4 °C.
4. Poly D lysine (PDL), 1 mg/mL.
5. RGC growth medium (37 °C).
6. Neurobasal medium: 1 M NaOH.
7. Laminin Mouse Protein, Natural from Gibcoc™, Cat. No. 23017015 (ThermoFisher).
8. 25 g Papain Enzyme, CAS No. 9001-73-4 (Carl Roth).
9. Eppendorf tube.
10. Dissection microscope.
11. 40 mm Petri dish.
12. 70 μm strainer.
13. 10 mL syringe.
14. Polypropylene tube for cell transfer.
15. P1000 pipette.

16. Hemocytometer.
17. 70% Alcohol.
18. 0.4% Trypan blue, fluorescent dye.

2.2 Atomic Force Microscopy

2.2.1 Functionalization of Probe Tip

1. Gold Coated Soft Tapping Mode AFM Probe, Cat. No. HQ: NSC18/Cr-Au BS-15 (*see* **Note 2**).
   (a) Cantilever Specifications (* designates the ranges for each specification).
      (i) Force Constant: 2.8 N/m (1.2–5.5 N/m)*
      (ii) Resonance Frequency: 75 kHz (60–90 kHz)*
      (iii) Length: 225 μm (1–230 μm)*
      (iv) Width: 27.5 μm (24.5–30.5 μm)*
      (v) Thickness: 3 μm (2.5–3.5 μm)*
2. Ar (g) 5.0 contained in pressure-reducing cylinder.
3. HPLC-grade chloroform.
4. Aminopropyltriethoxysilane (APTES).
5. Triethylamine.
6. 5 L glass desiccator with large silicone O-ring seal.
7. Removed lids from 1.5 mL Eppendorf reaction vials.
8. 40 mm glass crystallization dishes.
9. 38 mm polystyrene Petri dishes.
10. Maleimide-PEG(27)-NHS, 100 mg, MW: 1570.76 g/mol, SKU: PEG1209.0100 (Iris-Biotech).
11. Reaction chamber.
12. Singular strip of Parafilm™ (16 mm × 16 mm).
13. Sensor molecule (may be any of the following: 1–2 μM thiol-protein, or 10–100 μM thiol-DNA, or 1–2 mM of a small molecule with an SH group), in HBS-EP Buffer.
14. HBS-EP Buffer, 0.01 M HEPES pH 7.4, 0.15 M NaCl, 3 mM EDTA 0.005% v/v Surfactant P20, Cat. No. BR100188 (Cytiva).
15. PBS.
16. Sodium azide (0.2–0.3 mg/mL).
17. HEPES acid stock: 5 μL of 1 M HEPES acid, add 20% NaOH until pH = 7.5, store in frozen aliquots.
18. Tris(2-carboxyethyl)phosphine hydrochloride (TCEP) in water: 2 μL of 100 mM stock, unbuffered, freshly thawed aliquot.
19. HEPES basic stock solution: 2 μL of 1 M stock solution HEPES acid, add 20% NaOH until the pH = 9.6.

2.2.2 Sample Preparation for AFM

1. Micro-Tec boro silicate glass coverslips #1 18 × 18 mm.
2. KOH pellets, SKU: P5958-250G (Sigma Aldrich).
3. Ethanol, Absolute (200 Proof), Molecular Biology Grade, Fisher BioReagents™ (Sigma Aldrich).
4. Magnetic stirrer.
5. PTFE baskets (Teflon).
6. Sonicator.
7. Deionized water.
8. 95% ethanol.
9. High purity nitrogen gas.
10. Custom cut square coverslips, approx. ~13 × 13 mm.
11. Steel Mounting Disc for AFM Specimens, 12 mm Diameter (SPI Supplies).
12. Epoxy Adhesive, Jar, 1 lb., 5:1 Mix Ratio, 16 h Functional Cure (Devcon, part #: 20845).
13. Expanded Plasma Cleaner, PDC-001 (115 V) (Harrick Plasma).
14. Recording buffer: 10 mM HEPES pH 8.0, 200 mM KAc, 5 mM $MgAc_2$.

2.2.3 Adjusting Measurements for Scanning and Imaging in Liquid

1. Computer System for recording.
2. Inverted microscope w/motorized stage controlled by Bio-Scope stage controller.
3. Video Camera.
4. Newport Stabilizer Vibration Control System.
5. AFM hardware and software.
6. Functionalized Cantilever Tip(s).

2.2.4 Data Analysis and Arithmetic Equations

1. Most imaging AFM systems come with analysis software, typically based in Igor Pro, which can perform the surface analyses. Additional software that is also suitable include Mountains9® or Gwyddion.
2. Optional: numerical computing software package, such as MATLAB (MathWorks).

2.3 Lipidomic Analysis via HPLC-MS/MS

2.3.1 Sample Preparation with HPLC

1. HPLC-MS/MS grade acetonitrile and isopropanol: 1:1 ratio, freshly prepared in a glass vial and stored at room temperature.
2. EquiSPLASH™ LIPIDOMIX® Quantitative Mass Spectrometry Internal Standard (Avanti Polar Lipids Inc.).
3. Thermo Scientific™ Vanquish™ Horizon Binary UHPLC.

4. Accucore™ Vanquish™ C18+ Column (150 mm × 2.1 mm × 1.5 μm) Thermo Scientific™.
5. 200 mM ammonium formate.
6. Formic acid.
7. HPLC grade acetonitrile, ispropanol, water.
8. Mobile Phase A: 50:50 ACN:H2O with 5 mM ammonium formate and 0.1% formic acid.
9. Mobile Phase B: 88:10:2 IPA:ACN:H2O w/5 mM ammonium formate and 0.1% formic acid.
10. Chloroform/methanol 2/1 v/v.
11. Standard mix.
12. Glass beads.
13. Heindolph multi reax test tube shaker.
14. Isopropanol.
15. Methyl tert-butyl ether.

2.3.2 Tandem Mass Spectrometry

1. Q Exactive™ Plus Hybrid Quadrupole-Orbitrap™ Mass Spectrometer (Thermo Scientific™).
2. Heated Electrospray Ionization (HESI-II) Probe (Thermo Scientific™), coupled to the Q Exactive™.
3. LipidSearch 5.0 Software for cellular lipid identification and quantification.

3 Methods

3.1 RGC Cell Culturing [15–18]

1. Following euthanization (*see* **Note 3**), place retina into a PBS/1% FBS solution in the 40 mm Petri dish, then wash the tissues with fresh PBS before being placed onto a 70 μm nylon strainer, which has been dampened with PBS/1%FBS, agitate circularly with the back of a 10 mL syringe to release the cells (*see* **Note 4**).
2. Filter the cells through the 70 μm strainer into the collection tube with a P1000 pipette, then rinse the strainer.
3. Dilute with PBS solution up to 1 mL/retina, the centrifuge at 200×*g* for 7 min. Discard the supernatant and resuspend the pellet in PBS, 1 mL/5 retinae. Incubate overnight at 4 °C.
4. Following incubation, fill the Petri dish with neural cell medium and store at 4 °C horizontally until cell counting.
5. To quantify the cell count with hemocytometer, wipe down the hemocytometer and coverslip with alcohol, swirl the cell suspension, and place 20 μL 0.4% trypan blue with 20 μL of the cell suspension in the microcentrifuge.

6. Add 10 μL of the blue/cell suspension to hemocytometer and count the trypan blue negative cells.
7. To plate sample, resuspend in RGC growth medium on a sterile Petri dish with PDL and laminin (50 μg/mL in 5 mL neurobasal medium) (*see* **Note 5**).

3.2 Atomic Force Microscopy

3.2.1 Amino functionalization Through Esterification and Silanization [19, 20]

1. Submerge the silicon cantilevers (*see* **Note 6**) in a glass crystallization dish containing HPLC-grade chloroform for 5 min. Repeat twice with new sterile dishes and fresh HPLC-grade chloroform.
2. Dry wash the cantilevers in argon stream, then store in a polystyrene Petri dish.
3. Flush the 5 L desiccator with argon gas through top valve opening for 5 min to sterilize and remove unwanted substances. Reinsert the valve to desiccator cover.
4. Remove the desiccator cover and place the two Eppendorf vial lids inside desiccator with approximately 10 cm of distance between the lids.
5. Transfer 60 μL APTES into one lid and 20 μL triethylamine into the other lid.
6. Place the cantilevers on glass tray inside the desiccator, positioned equidistant from the lid with triethylamine and the lid with APTES.
7. After placing cover on the desiccator, flush desiccator with argon gas via the top valvular opening for 30 s. Reinstate the top valve and place the desiccator in 5% CO_2 incubator for 2 h.
8. Retrieve the Eppendorf vials lids containing APTES and triethylamine from desiccator.
9. Place cover over desiccator and remove top valve, repeating the argon gas flush for a minimum of 5 min to ensure thorough removal of triethylamine traces.
10. Re-incubate aminofunctionalized cantilevers within desiccator for a minimum of 2 days to ensure optimum cross-linking among APTES molecules and between APTES and surface.

3.2.2 Immobilization and Attachment of PEG-Maleimide Linker to Cantilever Tip [19, 20]

1. Dissolve 1 mg of maleimide-PEG-NHS in 0.5 mL of HPLC grade chloroform (*see* **Notes** 7 and **8**). Transfer solution to the reaction chamber, then add 30 μL of triethylamine.
2. Immediately following the addition of triethylamine, submerge the aminofunctionalized cantilever tips into the mixture.

Securely seal the chamber with the lid, then incubate for 2 h at room temperature.

3. Triple wash the cantilevers in glass crystallization dishes with chloroform, then incubate the cantilevers for 10 min per wash. After triple washing, dry the cantilevers with argon gas.
4. Place Parafilm™ clean-side up in a polystyrene Petri dish, then place the cantilever onto the Parafilm, with its tips pointing toward the center.
5. Prepare a solution containing 100 μL of sensor molecule in HBS with 1–2 mM EDTA (*see* **Note 9**).
6. Transfer the mixture containing the sensor molecule onto the cantilever(s) with a pipette, then cover the Petri dish, and incubate for 2–4 h.
7. Triple wash the cantilever(s) in PBS for 5 minutes per wash in polystyrene Petri dashes.
8. Set aside prepared cantilever for usage. Alternatively, store the cantilever(s) in another polystyrene Petri dish at 4 °C in buffer solution containing sodium azide (0.2–0.3 mg/mL).

3.2.3 Sample Preparation for AFM

1. Prepare a saturated KOH solution by adding 90 g of KOH pellets dissolved in 350 mL absolute ethanol in a 1 L beaker. Stir solution with magnetic stirrer until solution turns dark orange in color (approx. ~4 h).
2. Place glass coverslips (*see* **Note 10**) on the edges of a Teflon basket(s). Douse the coverslips with the saturated KOH solution and immerse in a sonicator for 3 min.
3. Rinse the coverslips with deionized water, then transfer into a beaker to be sonicated with distilled deionized water for 3 min. Repeat step once, with rinsing in between.
4. Rinse the cover slips with 95% ethanol, then dry with nitrogen gas. Store in a desiccator.
5. To prepare the support assembly, manually cut square coverslips with measurements of ~13 × 13 mm, then attach coverslips to 12 mm diameter AFM specimen discs with epoxy. Maintain equal distribution of epoxy between the specimen disc and glass coverslips to prevent air bubble.
6. Leave the support assembly to solidify overnight.
7. Prior to use, plasma cleanse the glass coverslips in oxygen for 10 min, 250 mTorr, using ~30 W forward RF power.
8. Prepare a dilution of 45 μg/mL RGC cell sample in equal amounts of 10 mM Tris and 200 mM KCl buffer adsorption.
9. Deposit the solution onto a glass support system and incubate for 1 h.

10. Following incubation, rinse the samples with 10 mL of recording buffer (20 mM Tris pH ~ 8.5, 200 mM Kac, 5 mM $MgAc_2$). Samples are now ready for imaging.

3.2.4 Calibration of Atomic Force Microscope

1. Load the cantilever into the cantilever holder of the AFM system.
2. Turn on the AFM controller hardware and open the software. On the software, select "Contact Mode," which will trigger several windows to open, including the Sum and Deflection Meter window.
3. Turn on the laser. Adjust the laser and the mirror by turning the knobs on the sides of the AFM until the laser beam is aligned with the cantilever tip. This is confirmed using the Sum and Deflection Meter, as signal will increase as the laser becomes aligned with the tip.
4. Lower the cantilever so that is a few microns above a hard surface, such as a Petri dish.
5. On the software, navigate to the operation to perform a single force scan, which is used to calculate the inverse optical lever sensitivity (InvOLS). The InvOLS describes how much photodiode signal is given per cantilever deflection and is important for converting the cantilever deflection from volts to meters.
6. On the resulting force graph, add markers to the beginning and end of the contact region of the force graph. Select the operation to calculate InvOLS.
7. To get the mean InvOLS value from multiple force scans, navigate to the operation to perform "continuous" force scans. Once about 20 scans are acquired, stop acquisition. Calculate the InvOLS from the scans (*see* **step 6**), and then calculate the Gaussian mean of the scans.
8. For acquisition of the Thermal Power Spectral Density (TPSD) of the cantilever in fluid, place the cantilever in fluid, but far from the surface of the Petri dish. If the cantilever is too close, there will be interactions between the cantilever and the surface that will interfere with the power spectrum, and the resulting spring constant value may be affected.
9. Navigate on the software to perform a thermal scan. The TPSD graph will form. When this graph has reached equilibrium, stop acquisition.
10. The TPSD graph will have a clearly defined peak, which is the resonant frequency of the cantilever. The software can fit this graph and calculate the spring constant.

3.2.5 Adjusting Measurements for Scanning and Imaging in Liquid [21–24]

AFM may be performed in contact mode or tapping/non-contact mode in the liquid medium; our study will utilize tapping mode (*see* **Notes 11** and **12**). Measurements are applicable for both contact and tapping modes. For further specifications of operating and imaging in a specific AFM system, refer to the system's operation manual for a more detailed set of guidelines.

1. Place the prepared sample onto the inverted microscope stage.
2. Turn on the control software and AFM system hardware. On the software, select "AC Mode," which will trigger several windows to open, including the Sum and Deflection Meter window.
3. Carefully mount the functionalized cantilever onto its holder using tweezers to gently hold the cantilever chip on its side. Ensure that the functionalized tip is aligned on the surface that will come in contact with the sample. Take care when mounting the cantilever into the holder, as the tip is easily damaged (*see* **Note 13**).
4. Mount the cantilever holder system onto the measurement head of the AFM system.
5. To align the laser optics, first focus the optical microscope on the AFM cantilever and adjust the incident laser beam onto the cantilever tip.
6. Make small adjustments to the alignment of the laser beam on the cantilever tip until the reflected beam produces an optimal/maximal signal. This is monitored via the Sum and Deflection signal.
7. Additional alignment of the photodetector is dependent upon the AFM system being used.
8. Set the drive frequency to a value that is nearly equivalent to the resonance frequency of the AFM cantilever (from calibration measurements). Adjust the drive amplitude as necessary.
9. Adjust the oscillation damping for tapping mode. Oscillation damping will determine the interaction forces experienced between the cantilever tip and the sample surface.
10. Optimization of AFM Scan parameters (*see* **Note 14**). Applicable for nearly all AFM probes to ensure that **Trace** and **Retrace** height is contouring in the height channel. For a visual representation of the following optimization settings, please refer to NanoWorld in Supplemental Materials 1.
    (a) Proper AFM tip tracking of surface topography is attainable when the Trace and Retrace lines closely follow each other and exhibit overlap. If mismatch between their lines occurs (no overlapping of lines), gradually reduce the AFM tip velocity via reduction of the **Scan Rate**, **Tip Velocity**, or **Scan Size**.

(b) Proper AFM tip tracking of surface topography is attainable when the Proportional Gain (PG) and Integral Gain (IG) are set properly and does not demonstrate any visible noise, allowing for Trace and Retrace lines closely follow each other. If mismatch of the lines occurs, gradually increase the **Proportional Gain** and **Integral Gain** functions to reconverge the Trace and Retrace Lines. Mismatch of lines may also occur when the PG and IG functions are increased far beyond this equilibrium point and will lead to "noise" in the lines resulting from feedback oscillations. To reestablish this equilibrium point, gradually reduce the PG and IG functions until the "noise" is no longer visible.

(c) Proper AFM tip tracking of surface topography is attainable when the Setpoint parameter achieves an appropriate value that allows for the Trace and Retrace lines closely follow each other. If mismatch occurs, gradually decrease the **Setpoint** parameter until the Trace and Retrace Lines reconverge. Avoid reduction of the Amplitude Setpoint beyond the appropriate point, as it increases the chances of tip wearing and reduced tip functionality.

11. Engagement of the probe via Course vs. Fine approach—refer to specific AFM system being used (*see* **Note 15**).

    (a) **Course Approach:** Involves careful, manual lowering of the AFM probe, keeping a safe distance of parts of a millimeter from the surface to avoid damaging the topographical surface of the sample. Sample damage will occur if not proceeded with precaution.

    (b) **Fine Approach:** Involves setting the engage parameters to then engage the cantilever tip onto the sample surface. Refer to the AFM system's manual for further instruction about engage parameters. The traditional dual piezoelectric scanner-stepper motor engagement can cause structural damage with improper parameters set in place.

12. Begin scanning of RGC cell sample, consistently monitoring the parameters and interpreting the AFM images. Read the AFM-specific manuals for further details regarding imaging formation and interpretation.

13. During the imaging process, the most critical parameters to monitor are the Setpoint and the Feedback Gain. If the Setpoint parameter repeatedly increases without being manually decreased to maintain stability, it will result in movement of the laser point from the photodetector, consequently decreasing the amplitude signal. The Feedback Gain must remain at a reasonable range of 0.5–0.7.

14. Successive force-distance (FD) curves are recorded with a scanning area of 2 × 2 μm containing 32 × 32 (1024) FD curves. Sample each FD curve with 256 points.

3.2.6 Data Analysis and Arithmetic Equations

Data analysis techniques have been modified for the purpose of the protocol [23, 25–28]. More extensive guidelines for analyzing AFM data and performing calculations are available in current literature and respective AFM manuals.

1. Use the Z-sensor signal channel rather than the direct height signal for analysis. The height signal, which comes directly from the piezoelectric actuator, is not accurate when sample features are greater than 0.1 μm because piezoelectric materials exhibit mechanical hysteresis. In contrast, the Z-sensor signal is obtained from the linear variable differential transformer (LVDT) position sensors attached to the piezoelectric actuator and therefore, correspond to the true topography of the sample surface.
2. Consider possible modifications of the acquired data via the following data processing parameters:
   (a) **Line-by-line leveling** acquires all horizontal or vertical lines produces in AFM image and then fitting into a polynomial equation.
   (b) **Inclusion/exclusion leveling** allows users to designate areas of interest on sample and removal of isolated features on flat surface.
   (c) **Matrix filtration** averages adjacent points together within an image, which is then used to sharpen/blur the image.
3. Use the analysis software included with the AFM system used to calculate quantitative parameters that characterize surface roughness: root-mean-square (RMS) deviation, average deviation, skewness (asymmetry around the mean), and kurtosis (peakedness or flatness compared to the normal distribution) [28]. Calculations are based on 100 × 100 nm^2 non-overlapping, uniform areas and with equivalent pixel density (1.9 nm/pixel) (*see* **Note 17**).
4. MATLAB may also be utilized to calculate the primary surface parameters from the AFM dataset. *See* [25, 27] for further instruction on how to approach these arithmetic calculations.
5. Represent the group averaged data as means ± S.E.M. Statistical evaluations are performed using repeated-measures analysis of variance (ANOVA). Follow up ANOVA with either *post-hoc* tests or independent two-tail *t*-tests.

3.3 Lipidomic Analysis via HPLC-MS/MS

3.3.1 High-Performance Liquid Chromatography (HPLC) with Autosampler

1. To prepare Mobile Phase A, combine 487 mL acetonitrile, 487 mL water, 25 mL 200 mM ammonium formate, and 1 mL formic acid in a 1 L solvent bottle.
2. To prepare Mobile Phase B, combine 857 mL isopropanol, 97.5 mL acetonitrile, 19.5 mL water, 25 mL 200 mM ammonium nitrate, and 1 mL formic acid in a separate 1 L solvent bottle.
3. Reconstitute dried lipid samples in 49 μL of 1:1 acetonitrile/isopropanol.
4. To prepare the EquiSPLASH standard, dilute stock solution to 100 μg/mL. For each sample, add 1 μL of diluted EquiSPLASH standard.
5. To prepare an internal standard (IS), combine 49 μL 1:1 acetonitrile/isopropanol and 1 μL EquiSPLASH.
6. Prepare a blank with 50 μL acetonitrile/isopropanol to a vial. The blank will be run every 8–10 samples.
7. The following parameters for Vanquish Horizon will be utilized for mass spectrometric analysis (*see* **Note 18**):
   (a) Column temperature: 55 °C
   (b) Injection volume: 8 μL
   (c) Gradient time: 30 min, 10–100%, solvent B
   (d) Flow rate: 0.260 mL/min
8. Alternatively, traditional HPLC may be performed if autosampler is inaccessible.

3.3.2 Q Exactive™ MS and Electrospray Ionization (ESI) Interface Parameters

Apply the appropriate parameters for Q Exactive and ESI Interface (*see* **Note 18**). Each lipid sample will be run on both positive and negative modes for the Q Exactive™ MS and ESI Interface.

3.3.3 LipidSearch 5.0 Parameters for FC Quantification

1. Transfer Q Exactive files to LipidSearch 5.0 Software (ThermoFisher) and begin a new study.
2. Apply the appropriate identification parameters (*see* **Note 18**). Pay particular attention the following lipid classifications: Cholesterol (d7), lysoPC (PC), lysoPE (PE), lysoPI (PI), lysoPS (PS), lysoPA (PA), triglycerides, sphingomyelin, sphingoid base, Acyl-CoA (*see* **Note 18**).
3. Apply the following alignment parameters:
   (a) Click on the sample and assign a number.
   (b) Align the positive and negative mode data together, as well as any technical replicates together.
   (c) Retention time for alignment: 0.1 min.
   (d) Filters: top ranked, main isomer peak.

4. Run alignment and find internal standards within table demonstrating identified FCs.
5. Select correct standards under "Quantification" with each standard present across each individual sample being compared before running algorithmic quantification. Indicate the concentration as 100 μg/mL and press "Normalize."
6. Export normalized data to an Excel Workbook and consolidate data for further analysis.

4 Notes

1. For a simplified procedure, wild-type mice and ABCA1-knockout mice will be utilized for retinal extractions and observation. However, experimental knockout of ABCA1 in mice is a plausible approach for observation. For detailed procedures of different routes of creating ABCA1 lines in transgenic mice, *see* Supplemental Materials 1.
2. HQ:NSC18/Cr-Au BS AFM cantilevers are optimal for tapping mode in liquid given their high stability and sensitivity to magnetic and electric forces. The gold coating of the cantilever tip further enhances its sensitivity and specificity, permitted by gold's high affinity for thiol PEG linkers. Furthermore, gold coating is needed to attain proper signal during imaging.
3. Isoflurane gas can be used as an alternative to CO_2 euthanization.
4. Avoid using NaN3 during the preparation of the PBS solution. NaN3 is a toxic substance that can be harmful for live RGC cells.
5. Alternatively, use a pestle or 15 IU/mL papain, 5 mM L cysteine, and 200 U/mL DNase I for 15 min at 37 °C, followed by PBS/10% FBS for enzymatic digestion of the sample to yield the retina. An equivalent yield may be accomplished through this route.
6. Cantilever functionalization is dependent upon material and coating of cantilever tip. The ligand molecule must be attached to the tip to ensure that binding strength between the tip and molecule (PEG-linker) overpowers the interaction between the ligand molecule and surface receptors.
7. During molecular recognition measurements with functionalized tips, PEG-linkers overcome the difficulties of distinguishing between desired specific interactions and undesirable interactions by providing mobility to the ligand molecule. MAL-PEG-NHS, a heterobifunctional PEG derivative that is reactive with amine binding targets, was selected given that

ABCA1 has two identified binding sites: (1) GHNGAGKT and (2) GVNGAGKS (*see* Supplemental Materials 2). With these binding sites exhibiting more positively charged residues with nitrogen-containing ends, we would expect MAL-PEG-NHS to interact with binding sites of ABCA1 to produce greater detailed depictions of the transmembrane structure.

8. Due to the toxicity of chloroform, pretreatment of the cantilever(s) must be performed within a well-ventilated fume hood to avoid inhaling of chloroform vapors. Always carry out procedures with proper protective equipment.
9. If native conformation of the sensor molecule does not contain an essential disulfide, add 5 μL HEPES (1 M, pH 7.5), 2 μL TCEP hydrochloride (100 mM in water, unbuffered, thawed), and 2 μL HEPES (1 M, pH 9.6).
10. Glass coverslips have become of increasing interest for AFM imaging of live cells both in solid- and liquid-mode AFM analyses. Although mica is often elected as the substrate, imaging with AFM at the molecular level is less feasible given the difficulty of mounting biological specimen onto a solid flat substrate that provides consistent stability. As such, glass coverslips may increase affinity interactions.
11. Given the fragility and size of cells, imaging must be conducted in liquid medium, with extra precaution being taken to prevent drying and structural damage of the RGC cells.
12. To visualize the fluidity and mechanical properties of the RGC cell membranes with and without ABCA1 knockout, tapping mode may prove to be more resourceful. Tapping mode will also prevent structural damage of the RGC cell samples so that they may be used and prepared for HPLC-MS/MS analysis.
13. If the cantilever flips over due to improper handling while the mounting of the cantilever onto its holder, assume that the tip may be damaged. Acquire a new functionalized cantilever and proceed.
14. For proper selection of peaks for the resonance frequency parameter, elect a peak that is higher than most peaks seen.
15. All AFM probe parameters help ensure that Trace and Retrace height is contouring in the height channel. For a visual representation of the described optimization settings, please refer to NanoWorld's guide (*see* Supplemental Materials 3). Given the natural wearing of AFM probe over time through usage, ensuring that optical scan parameters are in place will slow the process of wearing. Carefully observe the scan progress and correct the scan parameters as necessary.

16. During the engagement process, the amplitude gradually increases until a certain point at which the cantilever tip reaches the liquid surrounding the sample, and no direct contact with the sample. The amplitude then decreases and drops as it encounters the sample. Retract the cantilever after engagement with the surface.
17. MATLAB may also be utilized to calculate the primary surface parameters from the AFM dataset, such as arithmetic mean height (R_a), root mean square height (R_q), maximum height above the mean surface plane ($p_{\max}$), and maximum depth below the mean surface plane ($v_{\max}$) [25, 27].
18. Vanquish Horizon, Q Exactive™ MS Parameters, Electrospray Ionization (ESI) Interface Parameters, and LipidSearch 5.0 Parameters are derived from prior lipidomic analyses [29]. The lipid classes of interest identified in the LipidSearch 5.0 parameters all have been researched as potential effectors of ABCA1 expression [29]. More specifically, decreased expression or knockout of ABCA1 is expected to exhibit lower extracellular levels of free cholesterol (FC), high-density cholesterol (HDL), and phospholipids particularly phosphatidylcholine (PC), which are secreted by ABCA1. Sphingomyelins and Lysophospholipids may also be detected at lower levels (*see* Supplemental Materials 4–6).

Additional Resources

1. Joyce C et al (2003) Study of ABCA1 function in transgenic mice. Arteriosclr Thromb Vasc Bio 23:965–971. https://doi.org/10.1161/01.ATV.0000055194.85073.FF
2. UniProt (n.d.) O95477 ABCA1_HUMAN. https://www.uniprot.org/uniprotkb/O95477/entry#subcellular_location
3. NanoWorld. Choose your AFM settings properly! https://www.nanoworld.com/optimize-AFM-scan-parameters.pdf
4. Wang Y, Oram JF (2005) Unsaturated fatty acids phosphorylate and destabilize ABCA1 through a phospholipase D2 pathway. J Biol Chem 280(43):35896–35903. https://doi.org/10.1074/jbc.M506210200
5. Kobayashi A et al (2006) Efflux of sphingomyelin, cholesterol, and phosphatidylcholine by ABCG1. J Lipid Res 47(8): 1791–1802. https://doi.org/10.1194/jlr.M500546-JLR200
6. Liu X et al (2012) ABCA1, ABCG1, and SR-bi: transit of HDL-associated sphingosine-1-phosphate. Clin Chim Acta 413(3–4, 384):–390. https://doi.org/10.1016/j.cca.2011.11.002

References

1. Weinreb RN, Khaw PT (2004) Primary open-angle glaucoma. Lancet 363(9422): 1711–1720. https://doi.org/10.1016/S0140-6736(04)16257-0
2. Pavljašević S, Ašćerić M (2009) Primary open-angle glaucoma and serum lipids. Bosn J Basic Med Sci 9(1):85–88. https://doi.org/10.17305/bjbms.2009.2863
3. Xu M et al (2020) Plasma lipid levels and risk of primary open angle glaucoma; a genetic study using Mendelian randomization. BMC Opthalmol 20:390. https://doi.org/10.1186/s12886-020-01661-0
4. Munemasa Y, Kitaoka Y (2012) Molecular mechanisms of retinal ganglion cell degeneration in glaucoma and future prospects for cell body and axonal protection. Front Cell Neurosci 6:60. https://doi.org/10.33892/fncel.2012.00060
5. Kim Y-K et al (2021) AICAR upregulates ABCA1/ABCG1 expression in the retinal pigment epithelium and reduces Bruch's membrane lipid deposit in ApoE deficient mice. Exp Eye Res. https://doi.org/10.1016/j.exer.2021.108854
6. Petty HR (2018) Frontiers of complex disease mechanisms: membrane surface tension may link genotype to phenotype in glaucoma. Front Cell Dev Biol 6. https://doi.org/10.3389/fcell.2018.00032
7. Zhao G-J et al (2021) The interaction of ApoA-1 and ABCA1 triggers signal transduction pathways to mediate efflux of cellular lipids. Mol Med 18(1):149–158. https://doi.org/10.2119/molmed.2011.00183
8. Storti et al (2019) Impaired ABCA1/ABCG1-mediated lipid efflux in the mouse retinal pigment epithelium (RPE) leads to retinal degeneration. elife 13:8e45100. https://doi.org/10.7554/eLife.4510
9. Oram JF (2003) HDL apolipoproteins and ABCA1: partners in the removal of excess cellular cholesterol. Arterioscler Thromb Vasc Biol 23(5):720–727. https://doi.org/10.1161/01.ATV.0000054662.44688.9A
10. Khadka NK et al (2021) Mehcanical properties of the high cholesterol-containing membrane: an AFM study. Biochim Biophys Acta Biomembr 1863(8):183625. https://doi.org/10.1016/j.bbamem.2021.183625
11. Xia F, Youcef-Toumi K (2022) Review: advanced atomic force microscopy modes for biomedical research. Biosensors (Basel) 12(12):1116. https://doi.org/10.3390/bios1212111
12. Zlatanova J, Lindsay SM, Leuba SH (2000) Single molecule force spectroscopy in biology using the atomic force microscope. Prog Biophys Mol Biol 74(1–2):37–61. https://doi.org/10.1016/s0079-6107(00)00014-6
13. Muller DJ (2008) AFM: a Nanotool in membrane biology. Biochemistry 47(31): 7986–7998. https://doi.org/10.1021/bi800753x
14. Yang J et al (2022) Cholesterol homeostasis regulated by ABCA1 is critical for retinal ganglion cell survival. Sci China Life Sci 66:211–225. https://doi.org/10.1007/s11427-021-2126-2
15. Chintalapudi SR et al (2016) Isolation and molecular profiling of primary mouse retinal ganglion cells: comparison of phenotypes from healthy and glaucomatous retinas. Front Aging Neurosci 8:93. https://doi.org/10.3389/fnagi.2016.00093
16. Gao F et al (2016) Comparative analysis of three purification protocols for retinal ganglion cells. Mol Vis 22:387–400. PMID: 27122968
17. Chintalapudi SR et al (2017) Isolation of primary murine retinal ganglion cells (RGCs) by flow cryometry. J Vis Exp 125:55785. https://doi.org/10.3791/55785
18. Winzeler A, Wang JT (2013) Purification and culture of retinal ganglion cells from rodents. Cold Spring Harb Protoc. https://doi.org/10.1101/pdb.prot074906
19. Ebner et al (2019) Functionalization of AFM tips and supports for molecular recognition force spectroscopy and recognition imaging. At Force Microsc:117–151. https://doi.org/10.1007/978-1-4939-8894-5_7
20. Gruber HJ (2016) Functionalization of AFM tips with maleimide linkers. Institute of Biophysics, JKU. https://www.jku.at/fileadmin/gruppen/216/07_AFM_tip_with_maleimide_2016_05_06.pdf
21. Sumbul F et al (2020) One-step calibration of AFM in liquid. Front Phys 8. https://doi.org/10.3389/fphy.2020.00301
22. NanoAndMore USA (n.d.) AFM measurement basics. https://www.nanoandmore.com/how-to/afm-measurement-basics
23. Wu et al (2016) Atomic force microscopy protocol for measurement of membrane plasticity and extracellular interactions in single neurons in epilepsy. Front Aging Neurosci 8:88. https://doi.org/10.3389/fnagi.2016.00088
24. Füzik T et al (2017) Imaging of biological samples in liquid environment. Technicka 5, Application Note 080, Prague, Czech

Republic. 166:28. https://www.ntmdt-si.com/resources/applications/imaging-of-biological-samples-in-liquid-environment

25. McDonough M et al (2020) Data processing & analysis for atomic force microscopy (AFM). Undergraduate theses and capstone projects. 18. https://dc.suffolk.edu/undergrad/18
26. Roduit C et al (2009) Stiffness topography by atomic force microscopy. Biophys J 97(2): 674–677. https://doi.org/10.1016/j.bpj.2009.05.010
27. Sarney WL (2022) Using MATLAB to calculate parameters from data collected by atomic force microscopy. DEVCON Army Research Laboratory. ARL-TN-1127. https://apps.dtic.mil/sti/pdfs/AD1174507.pdf
28. Sueiras V, Moy VT, Ziebarth NM (2015) Lens capsule structure assessed by atomic force microscopy. Mol Vis 21:316–323
29. Harvey FC, Collao V, Bhattacharya SK (2023) High-resolution liquid chromatography–mass spectrometry for lipidomics. In: Bhattacharya SK (ed) Lipidomics, Methods in molecular biology. Springer Nature, p 2625. https://doi.org/10.1007/978-1-0716-2966-6_4

Chapter 20

Oxidized Low-Density Lipoprotein: Preparation, Validation, and Use in Cell Models

Narayanasamy Angayarkanni and Kannadasan Anand Babu

Abstract

Lipoproteins in plasma are constituted by the least dense chylomicron, very-low-density lipoprotein (VLDL), low-density lipoprotein (LDL), and high-density lipoprotein (HDL) that can be separated using commercially available medium such as iodixanol. Iodixanol constitutes the self-generated density gradient to fractionate lipoproteins by rapid ultracentrifugation method, replacing time-consuming protocols. Filling the centrifuge tubes is technically easier and faster than layering salt gradients and is reproducible. The separated lipoproteins by this method are closest to the native state with 80 to 100% recovery possible. Low-density lipoprotein is the major carrier of cholesterol in systemic circulation. The plasma isolated LDL is purified to be used as native LDL and for the preparation of oxidized LDL (oxLDL). The oxLDL is characterized for its oxidation, by various methods based on assay of the lipid and protein oxidation products such as TBARS, conjugated diene formation, and by other methods such as agarose gel electrophoresis. Rapid isolation of LDL particles from human plasma is useful for lipid peroxidation studies, characterization of subclass for functional studies and clinical correlation especially in cardiovascular diseases apart from lipidomic, and proteomic studies. OxLDL preparations are done in vitro chiefly based on copper-induced oxidation; glucose and other prooxidants. Which are used for various studies using animal model and in vitro cell models especially to understand macrophage-mediated atheroma formation, vascular endothelial cell dysfunction, cell signaling studies has scope for extensive research in metabolic dysfunction of various cells. This chapter deals with one of the applications in the in vitro cell models using macrophage (THP-1 cell line) and human retinal pigment epithelial cell (ARPE-19 cell line) to study the oxLDL uptake using fluorescently labeled oxidized LDL (DiI-oxLDL).

Key words Low-density lipoprotein, OxLDL, DiI-oxLDL, Iodixanol, Cholesterol, Ultracentrifugation, Density gradient

1 Introduction

Low-density lipoprotein is the major carrier of cholesterol in systemic circulation. Oxidized LDL (oxLDL) denotes "oxidatively modified LDL" [1]. Oxidative stress induces oxidation of LDL that results in lipid peroxidation and the apo B protein, the major protein component of LDL as well as the lipids. The oxLDL acts as the key atherogenic metabolite that promotes atherosclerosis

Marco Brotto and Sanjoy K. Bhattacharya (eds.), *Lipid Signalling: Methods and Protocols*, Methods in Molecular Biology, vol. 2816,
https://doi.org/10.1007/978-1-0716-3902-3_20,

causing cardiovascular dysfunction (CVD) through inflammatory and immunologic mechanisms [2, 3]. The tissue resident macrophages internalize the oxLDL through their scavenger receptors, CD36, SRA-I/II, and SR-B1, and subsequently, oxLDL accumulates, causing endothelial dysfunction and foam cell formation, thereby promoting atherosclerotic lesions in the blood vessels [4]. The implications of oxLDL mediated retinal pigment epithelial (RPE) cell dysfunction is associated with the pathogenic mechanism of Age related macular degeneration (AMD), RPE being yet another phagocytic cell.

Many studies focus on lipid oxidation of LDL in vitro that involves isolation of LDL to elucidate oxLDL-mediated disease mechanisms such as in CVD and stroke. However, measuring LDL oxidation in vivo is highly challenging, and hence, the majority of investigations on LDL oxidation have been carried out using isolated LDL. LDL preparations are unstable and prone to oxidation during the process of isolation, handling, and storage [5]. Therefore, lipoprotein fractionation and extraction from human plasma for the in vitro experiments are of utmost significance. The laboratory procedures based on ultracentrifugation required a minimum of 56 h of ultracentrifugation involving multiple steps and are prone to get inclusion of lipid hydroperoxides during processing. In the later years, various methods were developed on the basis of one-step isopycnic density gradient ultracentrifugation [6]. In recent years, several rapid methods have been employed on the basis of ultracentrifugation techniques using various substances such as Iodixanol, Ammonium sulphate and Polyethylene glycol for the density gradient [7, 8].

This chapter elaborates the rapid method of LDL isolation based on one-step isopycnic density gradient ultracentrifugation using iodixanol as a gradient medium, followed by protocol for in vitro oxidation to form oxLDL and methods of validating the oxidation in vitro.

2 Materials

Prepare all solutions using sterile MilliQ water and analytical grade reagents and store at room temperature (unless mentioned). Cautiously follow all waste disposal regulations when disposing of waste materials.

2.1 LDL Isolation

1. *Iodixanol:* Optiprep^{TM}: 60% (W/V) Iodixanol solution (Axis-Shield, Norway)

   Iodixanol is a medium for self-generated density gradients to fractionate lipoproteins close to their native state with recovery close to 100%. It is a non-ionic, non-toxic, and metabolically inert compound, iso-osmotic at all densities.

2. *1× Phosphate buffered saline (PBS)*: 137 mM NaCl, 2.7 mM KCl, 10 mM Na_2HPO_4, and 1.8 mM KH_2PO_4 (pH 7.4). Add about 800 mL water to a glass beaker. Weigh 8 g NaCl, 0.2 g KCl, 1.44 g Na_2HPO_4, 0.24 g KH_2PO_4 and transfer to the glass beaker. Stir well until dissolved. Adjust pH with HCl (*Refer* **Note 1**). Make up to 1 L with water (*Refer* **Note 2**).
3. *12% iodixanol-plasma solution*: Mix 2.6 mL of plasma, 0.7 mL of 60% (w/v) iodixanol solution (Optiprep™), and 0.2 mL of PBS to get 12% (w/v) iodixanol-plasma solution (final density = 1.067 kg/L).
4. *9% iodixanol solution*: Mix 7.5 mL of 60% (w/v) iodixanol solution (Optiprep™), and 42.5 mL PBS to get 9% (w/v) iodixanol solution (final density = 1.050 kg/L).
5. Optiseal centrifuge tubes (Beckman).
6. Plastic syringe (2 and 5 mL) and Metal cannula (18 or 21 G).
7. Vertical or near-vertical rotor, NVT 65 (11.2 mL), capable of achieving at least 350,000*g* (Beckman).

2.2 Purification of LDL

1. *10 mM sodium phosphate buffer (pH 7.4)*: Mix 423 mL of Solution A and 577 mL of Solution B. Autoclave and store at room temperature.

   Solution A: Dissolve 1.38 g NaH_2PO_4-H_2O in 1 L water.

   Solution B: Dissolve 1.42 g Na_2HPO_4 in 1 L of water.
2. *Dialysis buffer*: 154 mM NaCl and 300 μM ethylenediaminetetraacetic acid (EDTA) in 10 mM sodium phosphate buffer (pH 7.4). Weigh 9 g NaCl, and 111.67 mg disodium EDTA•$2H_2O$ and transfer into 950 mL of 10 mM sodium phosphate buffer (pH 7.4) containing the beaker. Stir well until it dissolves. Make up to 1 L with the buffer.

2.3 Oxidation of LDL

1. *10 mM Copper chloride*: Weigh 1.35 mg $CuCl_2$ and dissolve in 1 mL of sterile water and filter through a 0.45 μm filter. (*Refer* **Note 3**).
2. *500 mM EDTA (pH 8.0)*: Weigh 1.861 g of disodium EDTA•$2H_2O$ and transfer to 8 mL of water in the beaker. Stir vigorously on a magnetic stirrer. Adjust the pH to 8.0 with NaOH (*Refer* **Note 4**). Make up to 10 mL with water.
3. *500 mM glucose*: Dissolve 900 mg glucose in 10 mL of 1× PBS.
4. *10 mM Diethylenetriaminepentaacetic acid (DTPA)*: Dissolve 39.33 mg DTPA in 10 mL of 1× PBS.
5. *1 mM Homocysteine (Hcy)*: Dissolve 1.35 mg Hcy in 10 mL of 1× PBS.

2.4 Validation of oxLDL

A. **Thiobarbituric Acid Reactie Substances (TBARS) ASSAY:**

1. *150 mM NaCl*: Dissolve 0.9 g NaCl in 100 mL of water.
2. *20% (w/v) trichloroacetic acid (TCA)*: Dissolve 20 g TCA in 75 mL of water and make up to 100 mL. Store at 4 °C.
3. *0.67% (w/v) thiobarbituric acid (TBA) solution*: Dissolve 670 mg TBA in 100 mL of water.
4. *Thiobarbituric acid (TBA) reagent*: Mix 1:1 v/v of 0.67% TBA aqueous solution and glacial acetic acid. Freshly prepare for each assay. Store in the dark.

B. **Reactive amino group determination:**

1. *4% w/v Sodium bicarbonate*: Dissolve 4 g $NaHCO_3$ in 100 mL of water.
2. *0.1% (w/v) picrylsulfonic acid (PSA)*: Add 20 μL of 5% (w/v) PSA solution into 980 μL of water (*Refer* **Note 5**).
3. *1N HCl:* Add 827 μL of concentrated HCl (12.1 M) to a final volume of 10 mL of water.
4. *10% (w/v) sodium dodecyl sulfate (SDS)*: Dissolve 1 g SDS in 10 mL of water. Store at 4 °C.

C. **Agarose gel electrophoresis:**

1. *10× Tris acetate EDTA (TAE) buffer*: 400 mM Tris base, 200 mM glacial acetic acid (GAA) and 10 mM EDTA (pH 8.5). Weigh 48.44 g Tris base, 11.42 mL GAA and 3.72 g disodium EDTA and transfer into 950 mL of water containing beaker. Stir well until it dissolves. Dilute 10 times with water to get 1× TAE buffer.
2. *0.6% Agarose gel*: Add 0.3 g agarose and transfer into 50 mL of 1× TAE buffer. Boil and add 2.5 μL of 10 mg/mL ethidium bromide. Pour into the gel tray with the appropriate comb.

D. **Electron Microscopy:**

1. *2% sodium phosphotungstate*: Dissolve 1 g of sodium phosphotungstate in 50 mL of water. Filter with Whatman filter paper to remove debris and store at room temperature.

3 Methods

Carry out all procedures at room temperature unless otherwise specified.

3.1 LDL Isolation by Ultracentrifugation [7]

1. Blood collection: Each LDL preparation should be prepared from 30 mL of venous blood collected into EDTA (4.2 mM final concentration) containing tubes from a normolipemic,

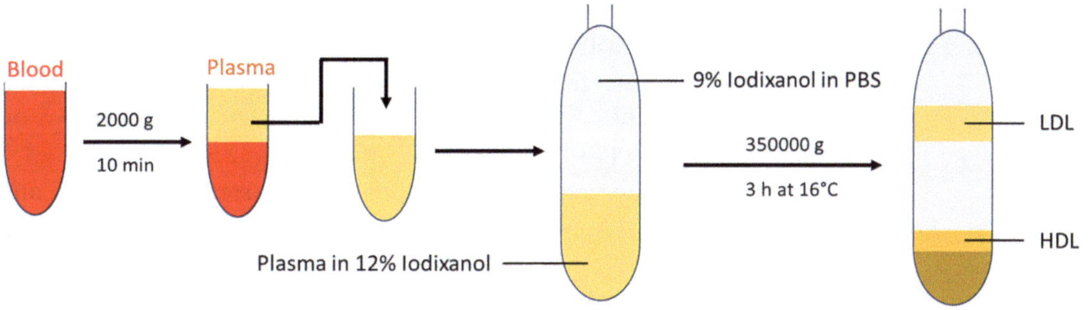

Fig. 1 Flowchart for LDL isolation
Overall protocol for the isolation of LDL by iodixanol gradient using ultracentrifugation method

nondiabetic, healthy human donor after 12 h of fasting. For bulk LDL preparation, pooled plasma can be used.

2. Overall protocol of LDL isolation is given in the flowchart (Fig. 1).
3. Plasma should be separated within 30 min by centrifugation at 2000 rpm for 10 min at 4 °C.
4. Add 7.9 mL of 9% (w/v) iodixanol solution into Beckman Optiseal centrifuge tubes (11.2 mL) and carefully underlayer 3 mL of 12% (w/v) iodixanol-plasma solution with a syringe and metal cannula.
5. The centrifuge tubes should be kept in a near-vertical rotor with proper balancing and centrifuge at 341,000*g* for 3 h at 16 °C in a Beckman Optima XL-100 ultracentrifuge with acceleration program 5 and deceleration program 5.
6. After the completion of the centrifugation, carefully take the centrifuge tubes without disturbing samples.
7. Remove the interphase yellow LDL layer from the centrifuge tubes with a syringe and metal cannula.

3.2 Purification of LDL

1. Transfer the collected crude LDL into a dialysis bag and subject to dialysis for 3 h with a dialysis buffer at 4 °C.
2. Change the dialysis buffer to a fresh one and subject to extensive dialysis overnight at 4 °C (*Refer* **Note 6**).
3. Store LDL under dark at 4 °C after purging with nitrogen to prevent oxidation.
4. LDL characterization is done by analysis of total cholesterol, fatty acid, and protein content.

3.3 Preparation of oxLDL

A. **Copper-induced oxidation** [9] (*Refer* **Note 7**)

1. Incubate 3 mg/mL protein of LDL with 10 μM $CuCL_2$ at 37 °C for 24 h.

2. Terminate oxidation by adding 100 μM EDTA and subject to extensive dialysis (*Refer* Subheading 3.2).
3. Measure free total cholesterol, triglycerides, and protein content based on kit methods based on spectrophotometry [10].
4. LDL preparations should be filter-sterilized (0.22 μm, Millipore GS), stored at 4 °C in the dark after purging with nitrogen, and used within 1 month of preparation.

B. **Glucose-induced oxidation** [11] (*Refer* **Note 8**)

1. G-LDL (Glycated): Incubate 1 mg/mL protein of LDL with 50 mM glucose at 37 °C for 72 h under antioxidant conditions (1 mM DTPA with 270 μM EDTA under nitrogen).
2. HOG-LDL (Glycoxidation): Incubate G-LDL with 10 μM $CuCL_2$ at 37 °C for 24 h under air followed by repeated dialysis.

C. **Homocysteine-induced oxidation** [12] (*Refer* **Note 9**)

1. Hcy-LDL: Incubate 1 mg/mL protein of LDL with 100 μM Hcy at 37 °C for 24 h.
2. After incubation, load the Hcy-LDL solution into Sephadex PD-10 column equilibrated with 1× PBS to remove excess Hcy.

3.4 Validation of oxLDL

A. **Oxidation measurement by spectrophotometry** [9]:

1. Oxidation of LDL results in the formation of conjugated dienes,which can be measured by spectrophotometry at 234 nm.
2. Transfer the native LDL (unoxidized) and oxLDL (oxidized) in a quartz cuvette to measure the degree of oxidation.
3. Measure the absorbance of native LDL and oxLDL in the spectrophotometer at 234 nm.
4. Shift in absorbance indicates the degree of oxidation (Fig. 2).

B. **TBARS assay for peroxidation measurement** [13]: **(*Refer* Note 10)**

1. Add LDL or oxLDL (50 μg protein) to 1.5 mL of 150 mM NaCl in a 15 mL boiling tube.
2. Add 500 μL of 20% TCA solution and 0.5 mL of TBA reagent into the above LDL solution.
3. Boil at 95 °C for 1 h.
4. Cool the mixture with tap water.

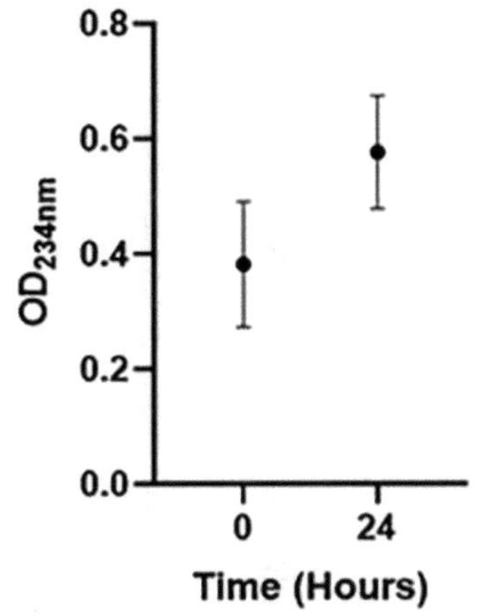

Fig. 2 Oxidation measurement as conjugated dienes by spectrophotometry An increase in absorbance over the time indicates the formation of conjugated dienes results in oxidation of LDL

5. Add 2 mL of n-butanol and vortex mix vigorously for at least 15 s.
6. Centrifuge at 4000*g* for 10 min.
7. Carefully remove the top n-butanol layer and measure the fluorescence in a fluorescence spectrophotometer with excitation at 515 nm and emission at 550 nm.
8. Use 1,1,3,3-Tetraethoxypropane [Malonaldehyde bis (diethyl acetal)] as standard and the results are expressed as malondialdehyde (MDA) equivalents.
9. Increase in fluorescence indicates the degree of oxidation (Fig. 3).

C. **Determination of oxidation by reactive amine group measurement** [14]: (*Refer* **Note 11**)

1. Add LDL or oxLDL (50 μg protein) to 1 mL of 4% $NaHCO_3$ and 50 μL of 0.1% PSA.
2. Incubate at 37 °C for 1 h.
3. After incubation, add 100 μL of 1N HCl and 100 μL of 10% SDS.
4. Measure the absorbance at 340 nm in a UV spectrophotometer.
5. Use valine as standard in the range of 5–50 nmol for linear range.
6. Decrease in absorbance indicates degree of oxidation (Fig. 4).

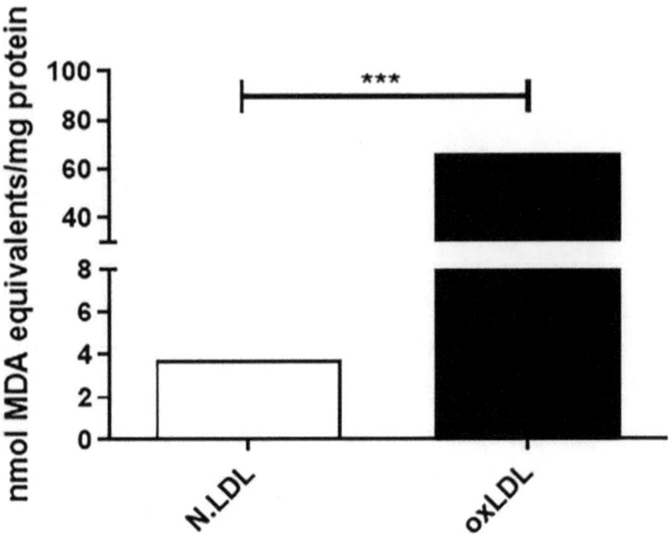

Fig. 3 TBARS assay. There was an increase in MDA equivalents in the copper treated LDL indicating the oxidation of LDL

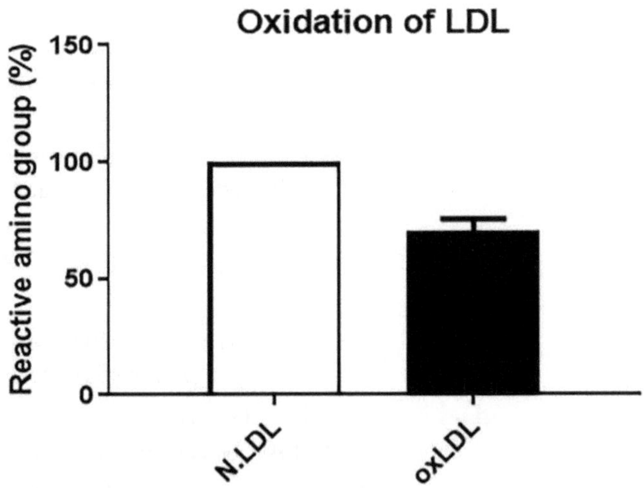

Fig. 4 Reactive amine group measurement. There was a decrease in the reactive amine group indicating the oxidation of LDL. The data represented a relative percentage to native LDL (N.LDL)

D. **Agarose gel electrophoresis** [15]:
   1. Mix LDL or oxLDL (50 μg protein) with sample loading dye (bromophenol blue).
   2. Add dye mixed LDL or oxLDL into wells of 0.6% agarose gel and subject to electrophoresis.

3. After electrophoresis, the gel should be stained with Coomassie brilliant blue for the protein component of LDL.
4. The other gel can be stained with Sudan black B for the lipid component of LDL.
5. Faster movement of LDL indicates the degree of oxidation.

E. **Electron Microscopy** [13]:

1. Mix 100 μL of 500 μg/mL LDL or oxLDL with equal volume of 2% sodium phosphotungstate.
2. Overlay a small aliquot of the mixture on a Formvar-carbon-coated grid and examine under electron microscopy.
3. The particle size ranges of LDL and HDL are 16.0–31.3 nm and 7.6–16.0 nm based on lipoprotein subclasses [16].

F. **Carotenoid distribution**

1. Carotenes are preferentially distributed in large, less dense LDL and VLDL particles, and xanthophylls are preferentially distributed in small, more dense HDL particles [17].

3.5 Experimental Cell Based In Vitro Studies Using Native and oxLDL Such as in Macrophages

A. **DiI-labeled oxLDL uptake assay: (*Refer* Note 12)**

1. Labeling of LDL and oxLDL with fluorescent compound 1, 1′-dioctadecyl-3,3,3′,3′-tetramethyl-indocarbocyanine perchlorate (DiI) can be done as per the standard protocol [18].
2. Accordingly, incubate 1 mg LDL (native) or oxLDL with 50 μL of DiI (3 mg/mL in DMSO) at 37 °C for 16 h followed by repeated dialysis.
3. Human retinal pigment epithelial cell line (ARPE-19, ATCC-CRL-2302; ATCC, USA) and human acute monocytic leukaemia cell line (THP-1, RCB3686; Riken, Japan) are used for the study.
4. Maintain THP-1 cell line in culture media (Roswell Park Memorial Institute (RPMI) 1640 Medium containing 10% fetal bovine serum, 100 IU/mL penicillin, and 100 μg/mL streptomycin) at 37 °C in 5% CO_2.
5. Differentiate the monocyte cell line into macrophages by 50 ng/mL phorbol 12-myristate-13-acetate (PMA) for 48 h under RPMI 1640 medium, supplemented with 10% (v/v) FBS.
6. Give serum starvation overnight with RPMI 1640 + 1% FBS.

7. Expose the macrophage to 200 μM H_2O_2 for 3 h, and 5 mM N-acetyl cysteine pretreatment for 30 min followed by 200 μM H_2O_2 for 3 h.
8. Subject to oxLDL uptake assay by DiI labelled oxLDL (DiI-oxLDL).
9. Similarly, culture ARPE-19 with Dulbecco's modified Eagle's medium/Nutrient Mixture F-12 (DMEM/F-12) and exposure to treatment conditions followed by oxLDL uptake assay.
10. A differential uptake was seen by both AREP19 cells and the THP-1 macrophage cell line, with oxLDL treatment showing higher uptake than the native LDL treatment (Fig. 5).
11. Effect of insults/compounds: Incubate DiI-oxLDL (50 μg/mL for spectrofluorimetry and 10 μg/mL for fluorescent microscopy) with pretreated ARPE-19 and THP-1 macrophage cells (experimental insults) such as H_2O_2 as pro-oxidant and with N-acetyl cysteine pretreatment as antioxidant control.
12. Wash the cells thrice with 1XPBS.
13. Lyse the cells for spectrofluorimetry measurement and fix the cells for fluorescent microscopy measurement (Figs. 6 and 7).

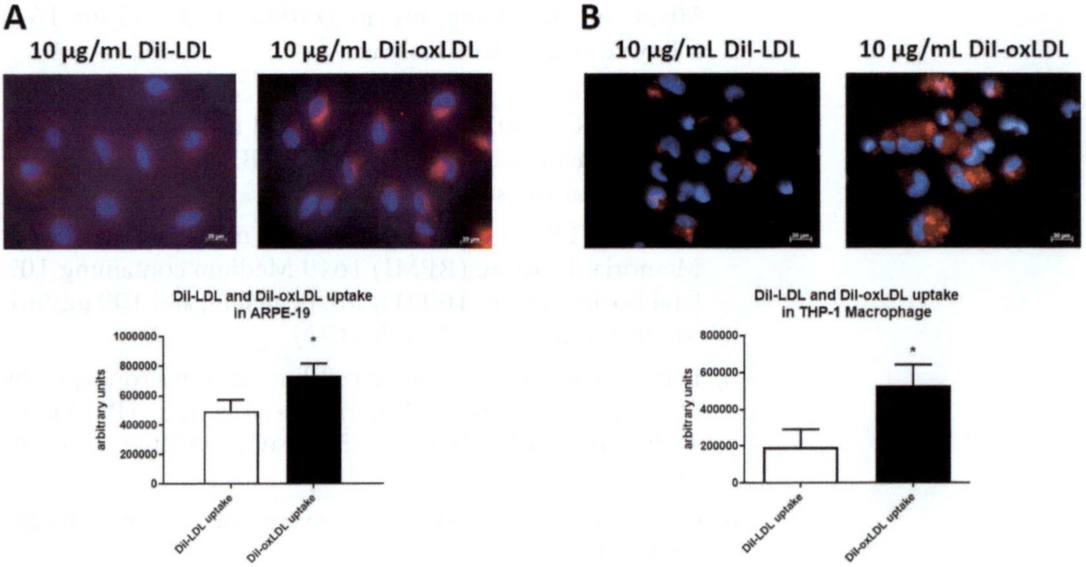

Fig. 5 Comparison of DiI-LDL & DiI-oxLDL uptake. (**a**) ARPE-19. (**b**) THP-1 macrophage
The oxLDL uptake was significantly higher than the native LDL uptake in both ARPE-19 and THP-1 macrophage cells

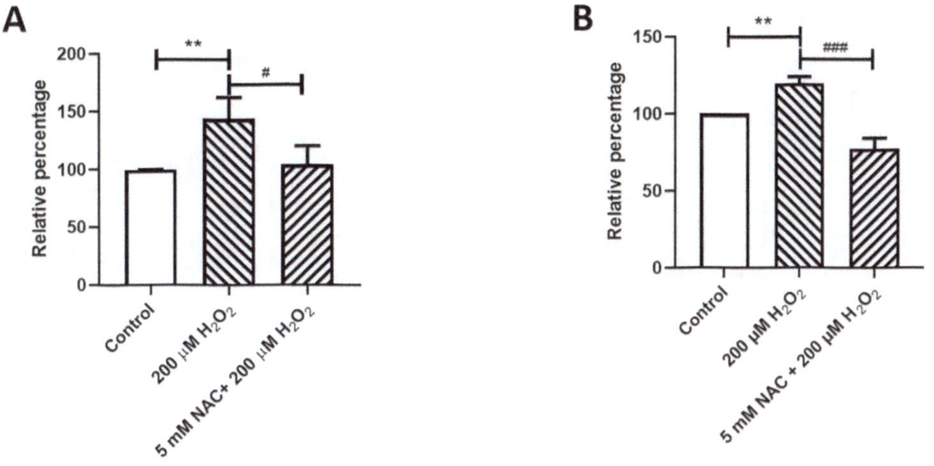

Fig. 6 DiI-oxLDL uptake assay by spectrofluorimetry. (**a**) DiI-oxLDL uptake assay in ARPE-19. (**b**) DiI-oxLDL uptake assay in THP-1 macrophage

The oxLDL uptake was significantly increased under prooxidant (H_2O_2) condition, and it was reversed by antioxidant (N-acetyl cysteine) condition.

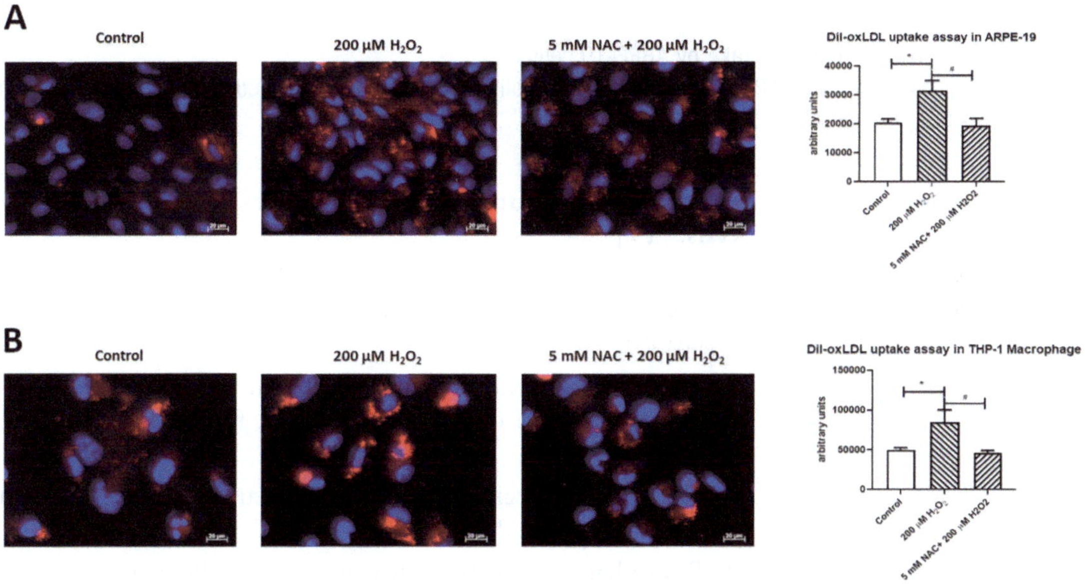

Fig. 7 DiI-oxLDL uptake assay by fluorescent microscopy. (**a**) ARPE-19. (**b**) THP-1 macrophage

To substantiate the results of fluorimetry, fluorescent microscopy imaging also revealed that oxLDL uptake was significantly increased under H_2O_2 treated condition, and it was reversed by N-acetyl cysteine condition

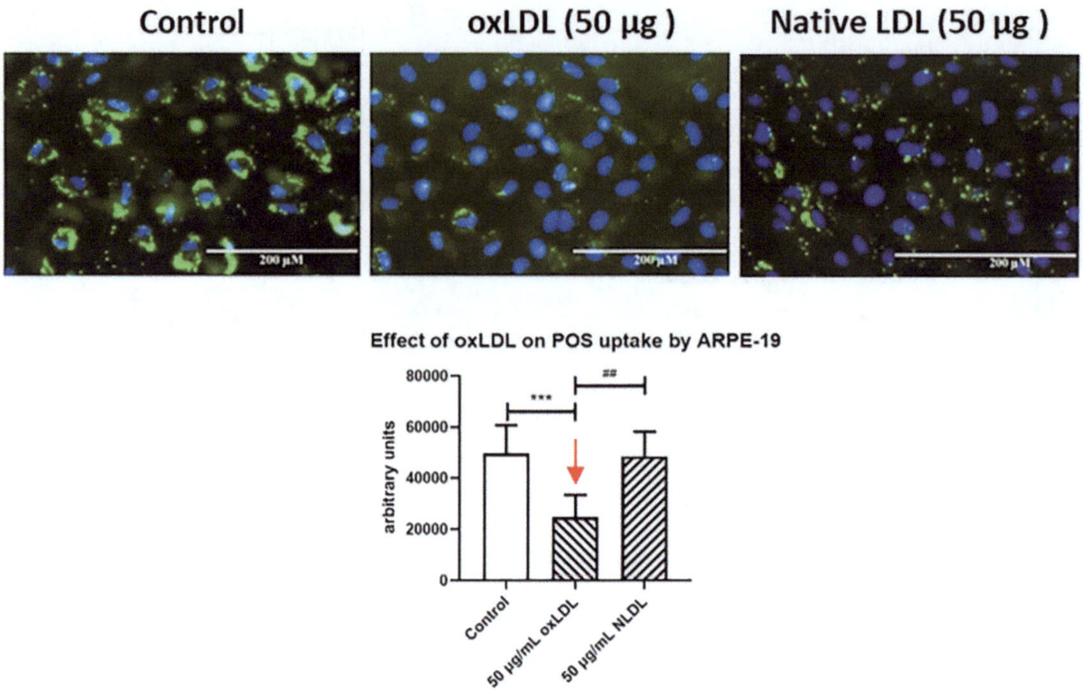

Fig. 8 Effect of oxLDL on POS uptake by ARPE-19 cells
Photoreceptor outer segments (POS) uptake was drastically reduced under oxLDL condition compared to the native LDL(NLDL) treated ARPE19 cells

B. **Photoreceptor outer segment uptake assay in ARPE-19 cells:** [19]

1. The retinal pigment epithelium (RPE) plays a crucial role in phagocytosis of shed photoreceptor rod outer segments (ROS).
2. The proximal end of the outer segment constantly forms new membrane discs to compensate for those lost at the distal tip.
3. The RPE has been studying the molecular mechanisms used to phagocytose POS for several decades.
4. POS phagocytosis was measured by the POS uptake assay with fluorescently labeled isolated POS particles.
5. POS can be isolated from the retina by sucrose density gradient ultracentrifugation method.
6. After POS isolation, it has to be labeled with a fluorescent probe (FITC) to measure the uptake by spectrofluorimetry or by imaging with fluorescent microscopy (Fig. 8).

3.6 Other Methods of LDL Isolation

A. **LDL isolation by immunoaffinity chromatography (IAC):** Among the most valuable methods for the isolation of specific biomolecules is immunoaffinity chromatography (IAC), which involves the principle of specific interactions between an affinity ligand, such as monoclonal antibody, and its respective target antigen [20]. Earlier, IAC has been widely used to isolate LDL particles from plasma and also co-isolated with VLDL and other anti-apoB-containing lipoproteins using anti-apoB antibodies immobilized on cross-linked agarose [21]. But an agarose matrix inherently lacks mechanical strength and swells in aqueous solution. In recent years, monolithic stationary supports in IAC have become increasingly popular for the rapid isolation of low-abundance biomolecules from complex biological matrices. The definition of a monolithic stationary phase is a single piece of interconnected porous material with through-pore networks.

B. The anti-apoB-100 mAb has been used for the isolation of LDL. Ligand, anti-apoB-100 mAb, which selectively recognizes human apoB-100 species, which is a protein component of LDL, immobilized onto monolithic disks [22]. (*Refer* **Note 13**)

LDL particles are classified into subclasses based on the density and size by various methods such as large (LDL-I), intermediate (LDL-II), small (LDL-III), and very small (LDL-IV). However, the classification of LDL varies based on different analytical methods, and it is important to take care when comparing the results of clinical studies using different methods.

LDL analysis is widely performed using ultracentrifugation density gradient and gradient gel electrophoresis (GGE) with their modifications. LDL subclass phenotypes are used for the classification of an atherogenic lipoprotein phenotype [7]. Predominance of small, dense LDL (sdLDL) that includes LDL-III or LDL-IV, diameter (258 Å) is known to confer additional atherogenicity and therefore increase the risk for coronary heart disease.

4 LDL Subfractionation by Different Methods: [23]

1. **Ultracentrifugation salt density gradient (density):** [24]
   - (a) LDL-I: 1.019–1.023 g/mL
   - (b) LDL-II: 1.023–1.034 g/mL
   - (c) LDL-III: 1.034–1.044 g/mL
   - (d) LDL-IV: 1.044–1.060 g/mL
2. **Ultracentrifugation iodixanol density gradient (density):** [25]
   - (a) LDL-I: 1.016–1.028 g/mL
   - (b) LDL-III: 1.028–1.043 g/mL

3. **Non-denaturing gradient gel electrophoresis (GGE) (peak diameter):** [26]
   (a) LDL-I: 26.0–28.5 nm
   (b) LDL-II: 25.5–26.4 nm
   (c) LDL-III: 24.7–25.5 nm
   (d) LDL-IV: 22.0–23.2 nm
4. **Ion mobility (peak size):** [27]
   (a) LDL-I: 21.9–23.8 nm
   (b) LDL-II: 21.1–21.9 nm
   (c) LDL-III: 20.17–21.1 nm
   (d) LDL-IV: 18.0–20.17 nm
5. **NMR (peak size):** [28]
   (a) LDL-I: 21.3–22.7 nm
   (b) LDL-II: 19.8–21.2 nm
   (c) LDL-III: 18.3–19.7 nm
6. **Homogenous assay:** Low-density lipoprotein (LDL) particles are heterogeneous with respect to size and density. The homogeneous assay is the measurement of the sd-LDL-Cholesterol. Homogeneous assay eliminates the non-LDL lipoproteins with specific surfactant A. Sphingomyelinase specifically reacts with large buoyant LDL (lb-LDL) and eliminates it. Polyoxyethylene alkyl ether (Surfactant B) dissociates the sd-LDL-C and the cholesterol released from sd-LDL, which is subsequently measured by using cholesterol oxidase/esterase kit method [29].

4.1 Brief Method on the Ultracentrifugation Iodixanol Density Gradient (Density) for LDL Subfraction [7]

1. Plasma is pre-stained by Coomassie Blue R250.
2. 3 mL Plasma is adjusted to 12% iodixanol and is under-layered beneath 9% iodixanol and subject to ultracentrifugation (3 h, 65,000 rpm 16 °C, (341,000*g*)) in a near vertical rotor (Beckman NVT65).
3. A digital photograph of separated LDL bands can be downloaded onto a PC and analysed using Total Lab 1D gel-scan software (Pharmacia, UK).
4. LDL subclass phenotypes can be characterized by Rf values.

5 Notes

1. Diluted HCL (e.g., concentrated HCl (12 N) can be used at first to narrow the gap from the starting pH to the required pH. From then on, it would be better to use a series of HCl

(e.g., 6 and 1 N) with lower ionic strengths to avoid a sudden drop in pH below the required pH.

2. By preparing 10× instead of 1× PBS solution, you can dilute 100 mL of the 10× solution to 900 mL of distilled water to obtain 1 L of 1× PBS buffer. Store at 4 °C.
3. The presence of a suitable ligand such as chloride can allow Cu (II) to act as an efficient aerobic redox catalyst.
4. EDTA will completely dissolve in pH 8.0 only. Hence, pH should be adjusted with NaOH until it gets dissolved completely.
5. 0.1% (w/v) PSA should be freshly prepared for each assay. Use it in the dark. Stock 5% (w/v) PSA store at 4 °C.
6. Purification/concentration of the isolated LDL can be done by ultrafiltration through a collodium bag under nitrogen. This dialysis/concentration method is better than filtration through short gel columns in preventing oxidation and dilution of the sample [30].
7. Metal ions such as Cu2+ or Fe2+ are used for LDL oxidation in vitro, as the resulting oxLDL has similar biological activities to that oxidized in vivo. Hydroperoxides are produced from the polyunsaturated fatty acyl chains present in LDL, which in turn gives rise to aldehydes. These aldehydes form Schiff bases with the lysine residues of apoB of LDL.
8. The glycation process is initiated by a chemical reaction between the reactive carbonyl group of a sugar or an aldehyde with a nucleophilic free amino group of a protein, leading to the rapid formation of an unstable Schiff base. Hyperglycaemia induces non-enzymatic glycation, followed by glycoxidation of LDL lipids and proteins enhancing uptake by macrophages thereby promoting atherosclerosis.
9. Homocysteine binds to thiol group of cysteine residues of apoB, leads to S-homocysteinylated LDL which increases the atherogenicity of LDL.
10. Lipid peroxidation (LPO) results in a mixture of complex secondary products such as malondialdehyde (MDA) and 4-hydroxynonenal that are measured to monitor LPO. MDA reacts with TBA to form a conjugate and these levels measured spectrophotometrically at 532 nm (pink colour) or fluorometrically (Ex530nm /Em540nm) and is indicative of extent of oxidative stress. Pre-analytical handling and storage of biological specimen if improper can lead to oxidation and so the TBARS levels artificially elevated. Kits are available for MDA assay.

11. The aldehydes formed due to oxidation, Schiff bases with lysine residues of apoB. Hence, the reactive amino group of lysine residues are reduced after oxidation. These reactive amino group can be measured as indicator of the degree of oxidation.
12. LDL preparations for in vitro oxidation studies should be evaluated for plasma levels of ascorbate, urate, and vitamin E levels.
13. ELISA for the detection and quantification of apoB-100 can be done by using a human Human apolipoprotein BELISAPRO kit (Code 3715-1HP-2, Mabtech AB, Sweden).

References

1. Lara-Guzmán OJ, Gil-Izquierdo Á, Medina S, Osorio E, Álvarez-Quintero R, Zuluaga N, Oger C, Galano J-M, Durand T, Muñoz-Durango K (2018) Oxidized LDL triggers changes in oxidative stress and inflammatory biomarkers in human macrophages. Redox Biol 15: 1–11
2. Esterbauer H, Gebicki J, Puhl H, Jürgens G (1992) The role of lipid peroxidation and antioxidants in oxidative modification of LDL. Free Radic Biol Med 13:341–390
3. Boullier A, Bird DA, Chang M-K et al (2006) Scavenger receptors, oxidized LDL, and atherosclerosis. Ann N Y Acad Sci 947:214–223
4. Brown MS, Goldstein JL (1983) Lipoprotein metabolism in the macrophage: implications for cholesterol deposition in atherosclerosis. Annu Rev Biochem 52:223–261
5. Kleinveld HA, HL H-L, Stalenhoef AF, Demacker PN (1992) Improved measurement of low-density-lipoprotein susceptibility to copper-induced oxidation: application of a short procedure for isolating low-density lipoprotein. Clin Chem 38:2066–2072
6. Kelley JL, Kruski AW (1986) Density gradient ultracentrifugation of serum lipoproteins in a swinging bucket rotor. Methods Enzymol 128: 170–181
7. Davies IG, Graham JM, Griffin BA (2003) Rapid separation of LDL subclasses by iodixanol gradient ultracentrifugation. Clin Chem 49:1865–1872
8. Wang N, Xu Q, Liu Y, Jin Y, Harlina PW, Ma M (2018) Highly efficient extraction and purification of low-density lipoprotein from hen egg yolk. Poult Sci 97:2230–2238
9. Jenkins AJ, Velarde V, Klein RL, Joyce KC, Phillips KD, Mayfield RK, Lyons TJ, Jaffa AA (2000) Native and modified LDL activate extracellular signal-regulated kinases in mesangial cells. Diabetes 49:2160–2169
10. Sawle A, Higgins MK, Olivant MP, Higgins JA (2002) A rapid single-step centrifugation method for determination of HDL, LDL, and VLDL cholesterol, and TG, and identification of predominant LDL subclass. J Lipid Res 43: 335–343
11. Fu D, Wu M, Zhang J, Du M, Yang S (2012) Mechanisms of modified LDL-induced pericyte loss and retinal injury in diabetic retinopathy. Diabetologia 6:3128–3140
12. Zinellu A, Sotgia S, Scanu B, Pintus G, Maria A, Cossu A, Deiana L, Sengupta S, Carru C (2009) S-homocysteinylated LDL apolipoprotein B adversely affects human endothelial cells in vitro. Atherosclerosis 206: 40–46
13. Yokode M, Kita T, Kikawa Y, Ogorochi T, Narumiya S, Kawai C (1988) Stimulated arachidonate metabolism during foam cell transformation of mouse peritoneal macrophages with oxidized low density lipoprotein electron microscopy oflipoproteins assayfor lipid peroxides. J Clin Invest 81:720–729
14. Steinbrecher UP (1987) Oxidation of human low density lipoprotein results in derivatization of lysine residues of apolipoprotein B by lipid peroxide decomposition products. J Biol Chem 262:3603–3608
15. Wang K, Li Y, Luo C, Chen Y (2020) Dynamic AFM detection of the oxidation-induced changes in size, stiffness, and stickiness of low-density lipoprotein. J Nanobiotechnol 18: 167
16. Okazaki M, Yamashita S (2016) Recent advances in analytical methods on lipoprotein subclasses: calculation of particle numbers from lipid levels by gel permeation HPLC using "spherical particle model". J Oleo Sci 65: 265–282
17. Lowe GM, Bilton RF, Davies IG, Ford TC, Billington D, Young AJ (1999) Carotenoid

composition and antioxidant potential in subfractions of human low-density lipoprotein. Ann Clin Biochem 36:323–332

18. Dehouck B, Fenart L, Dehouck MP, Pierce A, Torpier G, Cecchelli R (1997) A new function for the LDL receptor: transcytosis of LDL across the blood-brain barrier. J Cell Biol 138: 877–889
19. Mazzoni F, Safa H, Finnemann SC (2014) Understanding photoreceptor outer segment phagocytosis: use and utility of RPE cells in culture. Exp Eye Res 126:51–60
20. Moser AC, Hage DS (2010) Immunoaffinity chromatography: an introduction to applications and recent developments. Bioanalysis 2: 769–790
21. McConathy WJ, Koren E, Wieland H, Campos EM, Lee DM, Kloer HU, Alaupovic P (1985) Evaluation of immunoaffinity chromatography for isolating human lipoproteins containing apolipoprotein B. J Chromatogr 342:46–66
22. Liangsupree T, Multia E, Metso J, Jauhiainen M, Forssén P, Fornstedt T, Öörni K, Podgornik A, Riekkola M-L (2019) Rapid affinity chromatographic isolation method for LDL in human plasma by immobilized chondroitin-6-sulfate and anti-apoB-100 antibody monolithic disks in tandem. Sci Rep 9:11235
23. Ivanova EA, Myasoedova VA, Melnichenko AA, Grechko AV, Orekhov AN (2017) Small dense low-density lipoprotein as biomarker for atherosclerotic diseases. Oxidative Med Cell Longev 2017:1273042
24. Berneis KK, Krauss RM (2002) Metabolic origins and clinical significance of LDL heterogeneity. J Lipid Res 43:1363–1379
25. Yee MS, Pavitt DV, Tan T, Venkatesan S, Godsland IF, Richmond W, Johnston DG (2008) Lipoprotein separation in a novel iodixanol density gradient, for composition, density, and phenotype analysis. J Lipid Res 49:1364–1371
26. Krauss RM, Burke DJ (1982) Identification of multiple subclasses of plasma low density lipoproteins in normal humans. J Lipid Res 23:97–104
27. Caulfield MP, Li S, Lee G, Blanche PJ, Salameh WA, Benner WH, Reitz RE, Krauss RM (2008) Direct determination of lipoprotein particle sizes and concentrations by ion mobility analysis. Clin Chem 54:1307–1316
28. Otvos JD, Jeyarajah EJ, Bennett DW, Krauss RM (1992) Development of a proton nuclear magnetic resonance spectroscopic method for determining plasma lipoprotein concentrations and subspecies distributions from a single, rapid measurement. Clin Chem 38:1632–1638
29. Ito Y, Fujimura M, Ohta M, Hirano T (2011) Development of a homogeneous assay for measurement of small dense LDL cholesterol. Clin Chem 57:57–65
30. Vieira OV, Laranjinha JA, Madeira VM, Almeida LM (1996) Rapid isolation of low density lipoproteins in a concentrated fraction free from water-soluble plasma antioxidants. J Lipid Res 37:2715–2721

Chapter 21

Lipidomics Profiling of the Linoleic Acid Metabolites After Whole-Body Vibration in Humans

Kamal Awad, Stephen Leroy Newhart, Leticia Brotto, and Marco Brotto

Abstract

Bioactive lipids have been identified as dynamic signaling lipid mediators (LMs). These fats have the ability to activate responses and control bodily functions either directly or indirectly. Linoleic Acid (LA) and Alpha Linoleic Acid (ALA) are types of omega 3 fatty acids that possess inflammatory properties and promote resolution of inflammation either through their own actions or through their metabolites known as oxylipins. In this chapter, we provide an explanation of a method that combines chromatography with tandem mass spectroscopy (LC MS/MS) to identify and measure all the metabolites derived from LA and ALA. Additionally, we employed the described methodology to analyze human serum samples obtained before and after whole-body vibration exercise training. The results indicated an increase in some of the LA and ALA LMs that have beneficial effects in regulating the cardiovascular system.

Key words Linoleic acid, Whole-body vibration, Lipidomics, Alpha linoleic acid, LC-MS/MS

1 Introduction

The study of lipidomics has become more widely used in scientific research to better our understanding of the role lipids play in daily bodily function and metabolism. Lipids have a substantial function in the production of multiple hormones, they allow for the absorption of vitamins, provide energy, cushion joints, they make up the cell membrane of all cells, and have many other functions [1–3]. The goal of our research study was to assess the effects of oscillatory Whole-Body Vibration (OWBV) exercise on lipidomics and lipid signaling in the body. OWBV has been shown through previous research to improve physical characteristics like dynamic balance, muscle strength, muscle endurance, and flexibility [4–6]. It has additionally been shown to normalize high levels of blood pressure when performed twice a week for 30 min. Whole-body vibration exercise has shown multiple times in current

Marco Brotto and Sanjoy K. Bhattacharya (eds.), *Lipid Signalling: Methods and Protocols*, Methods in Molecular Biology, vol. 2816, https://doi.org/10.1007/978-1-0716-3902-3_21,

research to improve physical function and characteristics, therefore providing solid evidence of muscular activation and metabolic increase [7, 8].

Recently, lipidomics and mapping the lipid signaling mediators have emerged in the molecular biology and biomedical research fields due to their potential in revealing the intricate roles of lipid signaling in various biological processes and diseases. In a study by Jiang et al., it was found in mice that WBV improved lipid metabolism pathways (such as linoleic acid, α-linolenic acid metabolism, glycerophospholipid metabolism pathways) in skeletal muscle. Furthermore, WBV decreased the levels of triglycerides, total cholesterol, low-density lipoprotein cholesterol, and very-low-density lipoprotein in blood; decreased the lipid deposition in skeletal muscle; decreased the protein expression of monocyte chemoattractant protein-1and interleukin-6; improved the protein levels of phosphorylated insulin receptor substrate-1, phosphate phosphoinositide 3-kinase, and p-AKT; improved the protein levels of klotho; and decreased the protein expression of p53 [9]. In a meta-analysis study observing the effects of Whole-Body Vibration on lipid metabolism, the first random effects ($I^2 = 70\%$, p-value $= 0.02$) revealed a statistically significant mean change for percent fat mass (−2.56, 95% CI: −3.81, −1.31) [8]. The second fixed effects revealed a significant mean change (−1.91, 95% CI: −2.64, −1.18) for fat mass/kg [10].

In light of this information, it is justified to believe that the lipid metabolism and lipid utilization are increased during exercise bouts on a whole-body vibration platform. Our research team already showed improvement of WBV against a control group in a previous study [6]. The literature is widely underdeveloped in assessing the effects of whole-body vibration exercise on lipid function and utilization in humans. Some hypotheses suggested that the whole-body mechanical vibration could enhance the blood circulation and metabolism leading to improvement in the lipid metabolism. Yet, the underlying mechanism and a clear cause-and-effect relationship still unrevealed. Thus, the purpose of our research study was to assess the effects of OWBV exercise on lipidomics and lipid signaling in the body. This chapter will detail the protocols used in WBV exercise, specimen collection, as well as sample preparation, and lipidomics analysis. The lipidomics analysis will focus on the profiling and quantification of linoleic acid (LA), alpha-linoleic acid (ALA), and various signaling lipid mediators derived from using a Liquid Chromatography-Tandem Mass Spectrometry (LC-MS/MS) technique. These metabolites play important roles in various physiological processes, including inflammation regulation, cell signaling, and immune response. Further, these metabolites can have both pro-inflammatory and anti-inflammatory effects depending on the specific context and the presence of other related compounds.

2 Materials

2.1 Subjects

In this protocol, subjects were humans classified as sedentary adults, including males and females, who self-reported as sedentary adults from two local medical clinics. All subjects volunteered and qualified for the inclusion criteria set forth by the research team. Inclusion criteria required the participant to be older than 35 years of age, have no surgeries within the last year, and identify as sedentary. Subjects were excluded if they reported an acute musculoskeletal injury, severe pain, or pregnancy. Subjects who possessed a contraindication mentioned in the WBV platform's operations manual (e.g., epilepsy or active migraines) or indicated another medical condition (e.g., high blood pressure, cancer) on their health history form could participate only with their physician's clearance. There were no subjects who presented a contraindication for the equipment. All subjects participated in the experimental group, and pre/post measures were collected on everyone.

2.2 WBV Protocol

1. Dr. Fuji® FJ-700 vibration platform
2. 21 g collection needle
3. 6 mL lithium heparin VACUETTE tubes
4. PTEF/silicon screw cap

2.3 LCMS Analysis

1. LCMS instrumentation: Shimadzu LCMS-8060 system (Shimadzu Scientific Instruments, Inc., Columbia, MD, USA).
2. Mobile Phase PumpA: 0.1% Formic acid in water.
3. Mobile Phase Pump B: 100% Acetonitrile.
4. LC Separation: C8 column (Ultra C8, 150 × 2.1 mm, 3 μm, RESTEK, Manchaca, TX, USA).
5. Software: Shimadzu LabSolutions software V5.99 SP2.

3 Methods

The Dr. Fuji® FJ-700 vibration platform (Fremont, CA) (Fig. 1a) was used in all intervention sessions. Dr. Fuji® FJ-700 has an oscillation amplitude range between 10 and 18 mm and a frequency range from 3 to 10 Hz (Levels 1–10). The platform oscillates rather than using a vertical displacement pattern (Fig. 1b, c). The oscillating (pivoting) amplitude pattern allows for greater magnitude of displacement when compared to linear or vertical vibration plates. An oscillating machine can support more body weight and have a more positive physical impact on the body because it is not creating an up and down piston-like vibration but rather a swing-like vibration that encourages the body to actively contract muscles on

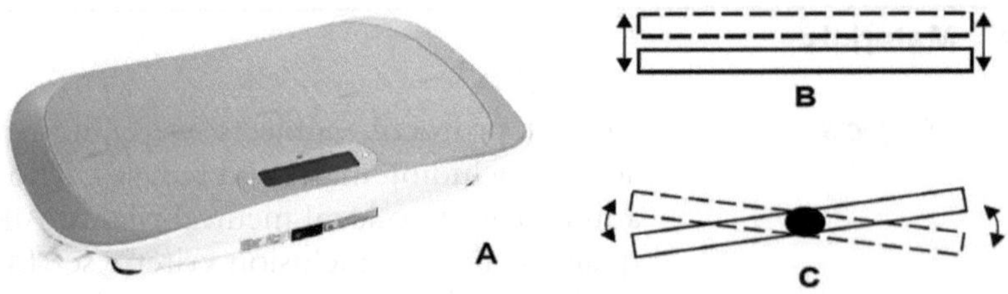

Fig. 1 (**a**) Dr. Fuji-700 vibration plate and oscillation patterns of vibration plates (**b**) Vertical displacement and (**c**) Oscillation

alternating sides to maintain equilibrium. The Dr. Fuji® FJ-700 amplitude range is based on the width of foot placement on the vibration plate. The amplitude experienced will be greater with a wider foot stance. The oscillation pattern and greater amplitude range provide gravity force amplification or muscular overload during body weight exercises.

3.1 WBV Exercise Protocol

The experimental group performed static standing on the vibration platform for 20 min a day, two times a week for 8 weeks (16 sessions). Every subject wore shoes. The participants stood with knees slightly flexed and/or maintained a neutral back position to prevent any possible injuries. All subjects began at frequency level 2 (4.76 Hz) on the platform on their first week, progressed to level 4 (5.55 Hz) on their second week, level 6 (6.67 Hz) on their third week, level 8 (7.41 Hz) on their fourth week, then level 10 (9.5 Hz) for weeks 5–8, as to gradually overload the body. All subjects were able to comfortably acclimate and tolerate this prescription. The research study was conducted ethically according to international standards. To investigate the effects of WBV on bone and muscle function, and physical activity measures in sedentary adults over 35 years of age, we used a 1 × 2 (group × time) mixed model repeated measures design. Independent variables included intervention group and time (pre and post). Dependent variables collected were lipid profile and other biomarkers pre and post the WBV exercise.

3.2 Blood Draw Procedure and Transportation

Blood samples were collected from participants at baseline and following the 8-week period of physical vibration standing. We drew the standard amount of blood, which is 30 mL for analysis per timepoint, 30 mL was collected at the pre-test, and 30 mL at the post test. When drawing the blood for serum, patients were seated, and samples were drawn through left AC. The area was disinfected with 70% isopropyl alcohol, phlebotomy was performed with a 21 g collection needle. Blood samples were collected into 6 mL lithium heparin VACUETTE tubes. Samples were appropriately labeled by date and the subject's de-identified assigned number and allowed to clot for a minimum of 30 min before being centrifuged, separated, and immediately packed in dry ice at −80° for courier pickup. Samples were then transported the same day to a deep freezer at the Brotto laboratory facility in the SEIR Building at the University of Texas at Arlington. Procedures were blinded, and the subject number was identified at random. The Brotto lab is certified for these analyses. The same process took place upon completion of the post-test blood draws. The samples will be stored in the Brotto lab for up to 24 months after the study completion for any further analysis or repeats. Part of the blood was used for targeted real-time PCR gene arrays to reveal the expression levels of genes associated to musculoskeletal health, and the rest was used for targeted lipidomics to profile the bioactive lipids known to be involved in inflammation, oxidative stress, tissue injury, and degeneration. This lipidomics approach was developed at the Brotto Lab in collaboration with the Shimadzu Research Institute and allows for the specific measurement of up to 180 lipid mediators in blood. We applied these methods for the first time to the study of WBV.

3.3 Lipidomics Protocol

In this section, solid-phase extraction (SPE) pretreatment for liquid samples (Serum, Plasma, or body fluids) followed by an LC/ESI-MS/MS method using the Shimadzu LCMS-8060 system (Shimadzu Scientific Instruments, Inc., Columbia, MD, USA) will be described for the profiling and quantification of LA and ALA derived LMs. This method provides a simultaneous quantification for LA and ALA derived LMs with on-column detection at femtomolar level without the necessity of chemical derivatization. In this protocol, we mainly focused on the quantification and profiling of LA and ALA metabolites. Linoleic acid has several metabolites, derived through the cytochrome p-450, that have significant roles as lipid mediators (LMs). These LMs include 12,13-dihydroxy-9Z-octadecenoic acid (12,13-DiHOME), 9,10-dihydroxy-12Z-octadecenoic acid (9,10-DiHOME),13-hydroxyoctadecadienoic acid (13-HODE), 9-hydroxyoctadecadienoic acid (9-HODE), 9-hydroperoxyoctadecadienoic acid (9-HpODE), 13-Oxo-9E,11E-octadecadienoic acid (13-OxoODE), 13-hydroperoxyoctadecadienoic acid (13-HpODE), 9-Oxooctadeca-10,12-dienoic acid (9-Oxo-ODE),12,13-epoxyoctadecenoic acid (12,13-EpOME), and 9,10-epoxyoctadecenoic acid (9,10-EpOME). The ALA includes 9- and 13-hydroxyoctadecatrienoic acid (9-HOTrE and 13-HOTrE) and 13-hydroperoxyoctadecatrienoic acid (13-HpOTrE).

3.3.1 Solid-Phase Extraction

SPE is an analytical sample preparation technique that is used to isolate, concentrate, and purify specific compounds for further analysis. In this method, we have used our developed and detailed SPE protocol published in this book in the chapter entitled "Detailed Protocol for Solid Phase Extraction for Lipidomic Analysis." This protocol is adapted for different sample types (such as cells, serum, plasma, body fluids, soft tissues, and bone marrow); please see the mentioned chapter for full details on the SPE.

3.3.2 Standards of LA and ALA Lipid Mediators

Standards of the targeted LMs and their related deuterated (Internal Standards (IS) such as (±) 9, 10-DiHOME-d4, (±)9(10)-EpOME-d4 dissolved in methyl acetate, and 13(S)-HODE-d4 dissolved in acetonitrile) can be purchased from any company, but the purity must exceed 98%. Methanol, acetonitrile, water, and other solvents used for the LC mobile phase must be of HPLC-MS analytical grade. Reagents such as formic acid used during the protocol could be a reagent grade (>95%). The following steps are taken to prepare the IS stock solution and the standards working solution:

1. *Internal Standards (IS) stock solution*: The IS stock solution is prepared by mixing the specific volume of each isotope labeled internal standard with methanol for a final IS stock according to the following table. After preparing the IS stock, it is stored as aliquots in small tubes and frozen at −80 °C for future use.
2. *Standards Stock Solution (Std)*: The standard compounds are mixed in methanol to prepare the stock solution of standards mixture. Here, we used 9-HOTrE, 13-HODE, and 9-HODE as standard compounds with final concentration of 2.5 μg/mL of each compound in the final stock volume of 1 mL in methanol as described in Table 1. After preparing the standards, they are stored as aliquots in small tubes and frozen at −80 °C for future use.

Table 1
Preparation of IS and standards stock solutions (Std)

Internal standards (IS)	Stock conc. (μg/μL)	Stock volume (μL)	Final conc. in mixture (μg/mL)	Volume (μL) for 1 mL IS stock
(±) 9, 10-DiHOME-d4, 50 μg in 500 μL methyl acetate	0.1	500	0.5	5
(±)9(10)-EpOME-d4, 50 μg in 500 μL methyl acetate	0.1	500	0.5	5
13(S)-HODE-d4, 50 μg in 500 μL acetonitrile	0.1	500	0.5	5
Methanol				985
Final volume of the prepared IS stock				1000
Standards stock solution (Std)	**Stock conc. (μg/μL)**	**Stock volume (μL)**	**Final conc. in mixture (μg/mL)**	**Volume (μL) for 1 mL stock**
9(S)-HOTrE in ethanol	1	100	2.5	2.5
13(S)-HODE in ethanol	1	100	2.5	2.5
9(S)-HODE in ethanol	1	100	2.5	2.5
Methanol				992.5

Fig. 2 Autosampler glass vials with target insert and screw cap

3. *Working Solution and Calibration Curve:* Using the prepared Std stock, the working solution can be prepared by serial dilutions of eight concentrations ranging from 1 μg/mL to 1 ng/mL in methanol. At least 50 μL of the highest concentration (1 μg/mL) is prepared to be used for the lower concentrations and the calibration curve as well. To prepare the final mixture for the calibration curve, 5 μL of each working solution +5 μL of the IS solution +90 μL methanol is mixed in 300 μL of the target polyspring insert, placed in clear autosampler glass vials, then closed using the PTEF/silicon screw cap (Fig. 2) and vortexed for 10 s to ensure proper mixing.

3.3.3 LC-MS/MS Analysis

1. *Instrument:* Shimadzu LCMS-8060 system with the UF-Qarray ion guide that increases LC/MS/MS sensitivity by enhancing ion signal intensity and reducing noise. By improving the ion sampling device, the ion guide, and the vacuum efficiency, the LCMS-8060 delivers a new vision of sensitivity and makes a real difference to working better and faster. Multiple reaction monitoring (MRM) is successfully employed to provide sensitive and specific analysis of lipidomics, metabolomics, and proteomics.
2. *Mobile phase:* Pump A: 0.1% Formic acid in water, and Pump B: 100% Acetonitrile.
   (a) The LC system is equipped with system controller: CBM-20A, three LC-30 AD pumps, an SIL-30AC autosampler, and a CTO-30A column oven. The LC separation is conducted on a C8 column (Ultra C8, 150 × 2.1 mm, 3 μm, RESTEK, Manchaca, TX, USA). The oven temperature for the analytical column is set to 40 °C.
   (b) The triple-quadrupole mass spectrometer (LCMS-8060 system) is operated and optimized under electrospray ionization in negative mode. MS parameters are optimized as follows: interface: DUIS (ESI) and (APCI), voltage,

Table 2
Mobile phase and LC program parameters

Time (min)	Mobile phase A	Mobile phase B	Flow rate (mL/min)
0.00–5.00	75	25	0.4
5.00–10.00	65	35	0.4
10.00–20.00	25	75	0.4
20.00–20.10	5	95	0.4
20.10–25.00	5	95	0.4
25.00–25.10	90	10	0.4
25.10–30.00	90	10	0.4

4.0 kV; interface temperature, 280 °C; desolvation line (DL) temperature, 275 °C; heating block temperature, 400 °C; drying gas (N2), 10 L/min; nebulizing gas (N2), 3 L/min; heating gas (air), 10 L/min; and collision-induced dissociation (CID) gas (Ar), 230 kPa.

(c) Considering the MRM technique, the acquisition is divided into multiple segments. Start with MRM optimization method for lipidomics to obtain the m/z transitions and their MS parameters (Q1, CE, and Q3 voltages) based on the best MRM responses. These parameters are presented in Table 2 for the LA and ALA LMs.

(d) Once all parameters have been optimized, the optimized method is run with full range to scan and monitor all IS and Std LMs in a concentrated sample of IS + Std mixture.

(e) Once all LMs are identified and detected, the exact retention time in the method is recorded and updated if needed.

(f) The optimized method is saved, and the calibration curve samples are running; chromatograms are obtained from the calibration curve and shown in Fig. 3.

(g) The browser tab in the LabSolutions software is used to analyze and check the calibration curve accuracy with at least $R^2 > 99.99\%$.

(h) Once all parameters are optimized and the calibration curve results met the standards, samples will begin preparation. The samples prepared by SPE are reconstituted with the dried extracts in the 1.5 mL tubes by adding 50 μL methanol, vertexing, and then transferring 20 μL of each sample directly into autosampler glass vials (Fig. 2) for auto injection of 10 μL.

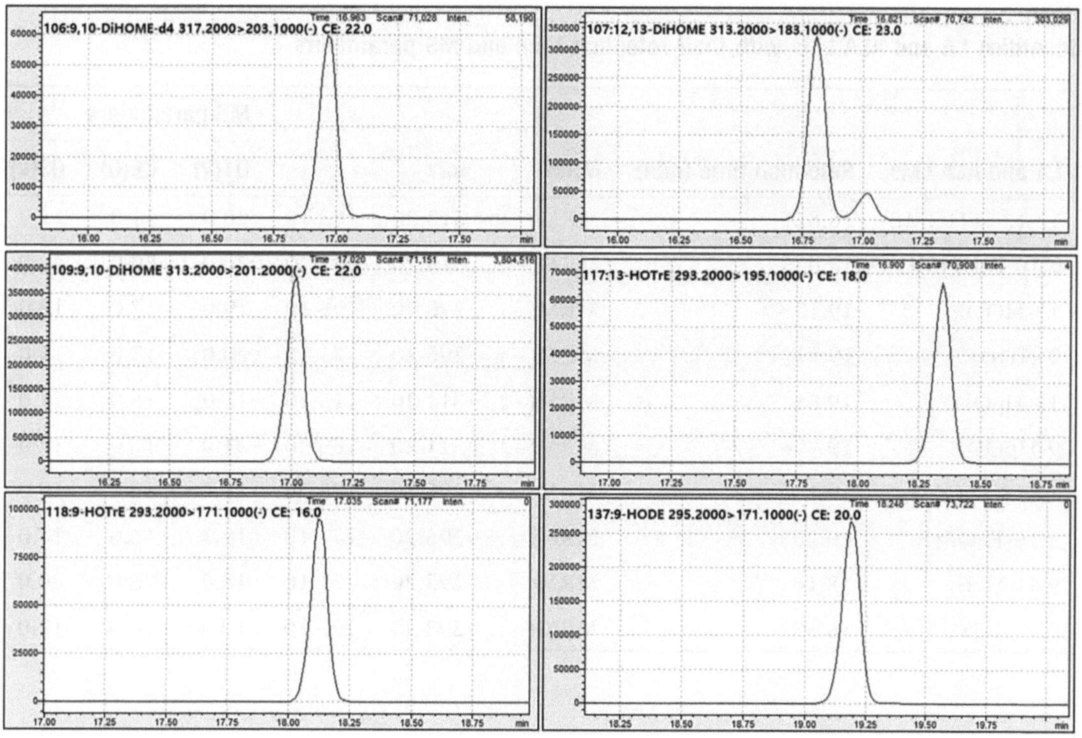

Fig. 3 MS/MS chromatograms of some of the detected linoleic acid and alpha linoleic acid metabolites and their deuterated standards

(i) After all samples are ran successfully, the raw data are uploaded into the browser with the calibration curve results to quantify the detected LMs in the tested samples (Table 3).

(j) An example of the quantified LMs from the WBV serum samples is presented in Fig. 4.

(k) External calibration curves are performed daily for this method, with the limits of quantitation (LOQ) of approximately 0.1 nM for each analyte under the present LC-MS/MS conditions.

(l) All analyses and data processing are performed using the Shimadzu LabSolutions software V5.99 SP2.

4 Example of Detected LA and ALA LMs in Serum Samples After the WBV Exercise

As described above, serum samples were used to profile and quantify the lipid mediators derived from the LA and ALA. As shown in Fig. 4, the concentration of 9-HOTrE LM increased in the post-WBV training group compared to the pre-test results of the same

Table 3
Identified LA and ALA LMs with their retention time and MS parameters

LA and ALA LMs	Retention time (min)	MRM	m/z	MS parameters		
				Q1(V)	CE(Q)	Q3(V)
12,13-DiHOME	16.821	MRM(−)	313.20 > 183.10	20.0	21.0	11.0
9,10-DiHOME	17.025	MRM(−)	313.20 > 201.20	20.0	20.0	20.0
13-HODE	19.17	MRM(−)	295.20 > 195.10	19.0	17.0	12.0
9-HODE	19.11	MRM(−)	295.20 > 171.10	20.0	17.0	17.0
13-HpODE	19.51	MRM(−)	311.20 > 113.10	11.0	16.0	24.0
9-HpODE	19.496	MRM(−)	311.20 > 185.20	21.0	15.0	12.0
12,13-EpOME	19.1	MRM(−)	295.20 > 195.20	18.0	16.0	19.0
9,10-EpOME	19.25	MRM(−)	295.20 > 171.10	18.0	16.0	17.0
9-HOTrE	18.16	MRM(−)	293.20 > 171.10	18.0	13.0	30.0
13-HOTrE	18.335	MRM(−)	293.20 > 195.10	19.0	16.0	12.0

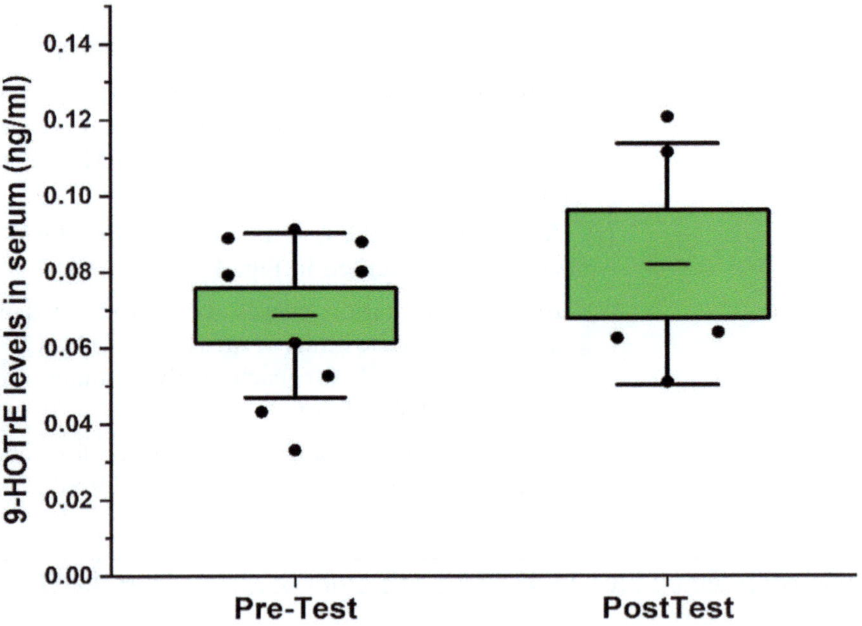

Fig. 4 9-HOTrE LM concentration increases with WBV training

group. Although no statistically significant difference was observed, there is a notable increase in the 9-HOTrE concentration. The 9-HOTrE is an omega-3 polyunsaturated fatty acid, which is known to have anti-inflammatory effects by inactivating the nod-like receptor family, the pyrin domain containing 3 (NLRP3) inflammasome [11–13]. Further, it has been reported that LA and ALA metabolites play an essential role in whole-body physiology and function including maintenance of cardiac health and vascular homeostasis [14, 15]. Thus, the results from our lipidomic profile confirm that WBV increases specific LA and ALA metabolites that have beneficial effects on the cardiovascular system and resolution of inflammation.

Acknowledgments

This project was funded by the National Institutes of Health Grants: NIA 2-PO1AG039355, NIA-R01AG056504; National Institute of Diabetes and Digestive and Kidney Diseases R01DK119066; and National Institutes of Neurological Disorders and Stroke (NINDS) 2-R01NS105621. The authors are thankful for the generous support from the George W. and Hazel M. Jay and Evanston Research Endowments and the UTA Shimadzu Institute for Research Technologies (https://www.uta.edu/sirt/).

References

1. Segatto M, Pallottini V (2020) Facts about fats: new insights into the role of lipids in metabolism, disease and therapy. Int J Mol Sci 21:6651
2. Su Z et al (2022) Lipid metabolism in cartilage development, degeneration, and regeneration. Nutrients 14:3984
3. Stevens SL (2021) Fat-soluble vitamins. Nurs Clin N Am 56:33–45. https://doi.org/10.1016/j.cnur.2020.10.003
4. Ritzmann R, Kramer A, Bernhardt S, Gollhofer A (2014) Whole body vibration training – improving balance control and muscle endurance. PLoS One 9:e89905. https://doi.org/10.1371/journal.pone.0089905
5. Merriman H, Jackson K (2009) The effects of whole-body vibration training in aging adults: a systematic review. J Geriatr Phys Ther 32:134
6. Newhart SL, Trowbridge CA (2021) Short-term training program using whole body vibration with body weight exercises improves physical functioning. J Athl Enhanc 10:8
7. Park S-Y, Son W-M, Kwon O-S (2015) Effects of whole body vibration training on body composition, skeletal muscle strength, and cardiovascular health. J Exerc Rehabil 11:289–295. https://doi.org/10.12965/jer.150254
8. Omidvar M, Alavinia SM, Craven BC (2019) The effects of whole body vibration therapy on reducing fat mass in the adult general population: a systematic review and meta-analyses. J Musculoskelet Neuronal Interact 19:455–464
9. Jiang D et al (2020) Metabolomics study of whole-body vibration on lipid metabolism of skeletal muscle in aging mice. Int J Sports Med 42:464–477. https://doi.org/10.1055/a-1268-8458
10. Alavinia SM, Omidvar M, Craven BC (2021) Does whole body vibration therapy assist in reducing fat mass or treating obesity in healthy overweight and obese adults? A systematic review and meta-analyses. Disabil Rehabil 43:1935–1947. https://doi.org/10.1080/09638288.2019.1688871
11. Ren J, Chung SH (2007) Anti-inflammatory effect of α-linolenic acid and its mode of action through the inhibition of nitric oxide production and inducible nitric oxide synthase gene expression via NF-κB and mitogen-activated

protein kinase pathways. J Agric Food Chem 55:5073–5080. https://doi.org/10.1021/jf0702693

12. Cappello V et al (2016) Ultrastructural characterization of the lower motor system in a mouse model of Krabbe disease. Sci Rep 6:1. https://doi.org/10.1038/s41598-016-0001-8
13. Zahradka P, Neumann S, Aukema HM, Taylor CG (2017) Adipocyte lipid storage and adipokine production are modulated by lipoxygenase-derived oxylipins generated from 18-carbon fatty acids. Int J Biochem Cell Biol 88:23–30. https://doi.org/10.1016/j.biocel.2017.04.009
14. Cambiaggi L, Chakravarty A, Noureddine N, Hersberger M (2023) The role of α-linolenic acid and its oxylipins in human cardiovascular diseases. Int J Mol Sci 24:6110
15. Hernandez-Saavedra D, Stanford KI (2022) In: Parinandi NL, Hund TJ (eds) Cardiovascular signaling in health and disease. Springer, Cham, pp 235–258

Chapter 22

Analysis of Lipoprotein Signaling in Iris Melanocytes

Molly Kennedy, K'Mani Blyden, Samyak Jain, Isabella Moceri, and Sanjoy K. Bhattacharya

Abstract

Lipids are compounds involved in many biologic functions including cell structure, metabolism, energy storage and are involved in signaling. A prominent lipid in these functions is cholesterol. Cholesterol also plays a part in the signaling of melanocytes, which contain melanosomes. The maturation of these melanosomes happens during melanocyte growth. The deficit of melanogenesis or melanosome maturation is associated with ocular albinism in the eye. Aberrations of melanosome maturation are also associated with pigment dispersion syndrome. Albinism and pigment dispersion manifestations are systemic. Both melanogenesis and melanocyte maturation are affected by cholesterol metabolism. Cholesterol signaling is a part of many pathways in the body, and evaluating these signals can have implications in systemic disease processes of melanogenesis and melanosome maturation, like ocular albinism and pigment dispersion. Cholesterol is carried by lipoprotein particles. Low-density lipoprotein (LDL) is usually the transport vehicle for cholesterol to reach tissues and organelles. The LDL uptake on cells often sends out a cascade of internal signaling within the cells. We describe here LDL signaling related to lipase activity changes using enzymatic methods with a kit. We describe analyses of cholesterol esters and free cholesterol with liquid chromatography and gas chromatography with or in tandem with mass spectrometry (GC-MS and LC-MS/MS). These analyses will provide insight into melanosome maturation and melanogenesis. The methods described here are applicable to all melanocytes within the body of a model mammalian organism.

Key words Cholesterol, Gas chromatography, Liquid chromatography, Mass spectrometry

1 Introduction

Cholesterol is a pivotal molecule that plays a vital role in numerous physiologic functions. As a lipid molecule, cholesterol contributes to the fluidity and stability of cell membranes. Moreover, it serves as a precursor for other essential biomolecules such as steroid hormones, bile acids, and vitamin D. However, cholesterol also emerges as a potent signaling molecule in melanogenesis within melanocytes and the iris pigment epithelium (IPE) [1]. It first enters the body through the absorption of dietary fats or through a de novo synthesis pathway involving hydroxyl-methyl-glutaryl

Marco Brotto and Sanjoy K. Bhattacharya (eds.), *Lipid Signalling: Methods and Protocols*, Methods in Molecular Biology, vol. 2816, https://doi.org/10.1007/978-1-0716-3902-3_22,

CoA reductase (HMG-CoA reductase). Due to its lipophilic nature, cholesterol relies on low-density lipoprotein (LDL) for transport in the blood.

LDL cholesterol has been shown to heighten melanogenesis in epidermal melanocytes and human melanoma cells [2]. This relationship involves the interaction of LDL with melanocyte LDL receptors, triggering a signal cascade that involves cAMP-activated lipases that cleave cholesterol esters, ultimately culminating in the increased transcription of key proteins vital to melanogenesis. Additionally, cholesterol may enhance the effects of melanin synthesis through its stabilizing effects of the melanocyte and melanosome membrane [2]. However, most studies elucidating melanocytes and melanogenesis predominantly focus on the skin not the other tissues including ocular tissues. Albinism and pigment dispersion result in significant diseases related to the eye, namely ocular albinism and pigment dispersion glaucoma. There remains a gap in investigation of role of cholesterol in both melanogenesis and melanocyte maturation. Bridging the gap in current knowledge and exploration of cholesterol's role in cell signaling within melanocytes and melanogenesis, particularly in the context of pathologies characterized by aberrant melanogenesis, is important. An in-depth understanding of lipoprotein signaling in iris melanocytes holds to inform mechanism-specific potential intervention possibilities.

We present a protocol designed for the isolation of iris stromal melanocytes [3–5], as well as the detection of lipase activity. We describe free cholesterol and esterified cholesterol analyses using gas and liquid chromatography coupled with tandem mass spectrometry, respectively [6, 7]. The significance of LDL internalization-led signaling extends beyond melanocytes, impacting various cell types where enhanced lipase activity induces substantial changes. A comprehensive evaluation of these steps becomes imperative for quantifying and comprehending lipid signaling and alterations in lipid composition in multiple cell types. This protocol, detailed in the subsequent sections, forms the basis for unraveling the intricate web of lipoprotein signaling in melanocytes and contributes to advancing our understanding of melanogenesis and melanocyte maturation.

2 Materials

2.1 Isolation of Iris Stromal Melanocytes

1. Culture dishes
2. Buffered solution (Ca++ and Mg++− free Hank's solution)
3. 0.25% trypsin solution
4. D-Hank's solution

5. Collagenase (400 U/mL) in F12 medium with 10% FBS
6. Geneticin
7. Antibodies:
   (a) Monoclonal mouse antibodies to human cytokeratin
   (b) Rabbit immunoglobulin to cow S-100 antigen
   (c) Monoclonal mouse antibody to human vimentin
8. Electron microscopy supplies:
   (a) 2% paraformaldehyde
   (b) 1% glutaraldehyde
   (c) 0.2 M phosphate buffer (PB), pH 7.4
   (d) 1% OsO4 supplemented with 1.5% ferrocyanide
   (e) Ethanol
   (f) Epon
   (g) Uranyl acetate and lead citrate for staining
   (h) Electron microscope
9. Ultracryomicrotome (Leica)
10. Methylcellulose
11. Antibodies and protein A coupled to 5, 10, or 15 nm gold (PAG 5, PAG 10, PAG 15)
12. Activity-based probe (ABP)—hexynyl-4-nitrophenyl-hexyl-phosphonate (ABP C6)
13. Tubes
14. Sonication equipment
15. Centrifuge
16. Protein concentration assay kit
17. Biotin click reaction components
18. Beads for enrichment
19. Quantitative proteomics tools

2.2 LDL Exposure

1. Low-density lipoprotein (LDL) (Thermo fisher #L3486) at 2.5 mg/mL stock
2. PC13:0/13:0 (1,2-ditridecanoyl-sn-glycero-3-phosphocholine) as a negative control (Avanti Polar Lipids #850340).
3. 1× Phosphate-buffered solution (PBS)
4. Pipettes
5. Eppendorf tube
6. Incubator
7. Heating block
8. 23-gauge needle

9. Centrifuge
10. Polyvinylidene difluoride membrane (PVDF)
11. Bovine Serum Albumin (BSA) (1 mg/mL stock solution)
12. MeOH (methanol)
13. Gelcode blue stain
14. Distilled water
15. ImageJ software

2.3 Lipase Activity Assay

1. Lipoprotein Lipase Assay Kit (Abcam, #ab204721)
2. LPL Assay buffer—provided in kit
3. LPL substrate in DMSO—provided in kit
4. LPL positive control—provided in kit
5. LPL inhibitor—provided in kit
6. Microplate reader capable of reading fluorescence at ex/em = 482/515 nm
7. Phosphate buffered saline (PBS)
8. Double distilled water
9. Pipettes with tips
10. Glassware
11. Tubes
12. 96-well plate
13. Incubator

2.4 Esterified Cholesterol Analysis with LC-MS/MS

1. 1 L solvent bottles (2)
2. Acetonitrile
3. Ammonium Formate (200 mM)
4. Formic acid
5. Water
6. Isopropanol
7. Acetonitrile
8. EquiSPLASH (Avanti Polar Lipids)
9. Acetonitrile/isopropanol 1:1 mix
10. Q Exactive Mass spectrometer with heated spray ionization (HESI) source
11. Thermo Scientific Vanquish Horizon Binary UHPLC
12. Thermo Scientific Accucore Vanquish C18+ Column

2.5 Cholesterol Analysis with GCMS

1. Cyrovials
2. Electric balance
3. Methanol/chloroform 2:1 mix
4. Methanol
5. Chloroform
6. Sodium Hydroxide methanol (0.5 mol/L)
7. Water
8. Centrifuge
9. Dounce homogenizer
10. Boron Trifluoride ($BF_{3)}$ in 14% methanol
11. 40 μL Methyl tert-butyl ether (MTBE)/Methanol (MeOH) (9:1)
12. Deuterated 3ß-hydroxy-5-cholestenoic acid-d5
13. Pasteur Pipettes
14. Airgas Helium
15. Agilent 798-A GC system
16. Agilent 7693 autosampler
17. Agilent CP-Sil 8 CB Column: 0.12 μm thick, 0.25 mm diameter
18. LipidSearch Software

3 Methods

3.1 Isolation of Iris Stromal Melanocytes

3.1.1 Preparation of Uveal Stromal Tissues

1. Create a circumferential scleral incision 8 mm posterior to the limbus, separating the anterior globe portions (anterior sclera, lens, iris, and ciliary body) and place into a culture dish
2. In the culture dish, excise the iris at the root with posterior portion facing up
3. Gently wash the iris with buffered solution (i.e., Ca++ and Mg+ +– free Hank's solution) and cover with 0.25% trypsin solution for 1 h at 37 °C
4. Separate the iris pigment epithelium from the stroma under a dissecting microscope.

3.1.2 Disaggregation of Uveal Stroma Using Trypsin: Collagenase

1. Immerse the isolated uveal stromal segment in 0.25% trypsin solution and refrigerate for 18 h at 4 °C and then incubate for 1 h at 37 °C.
2. Add the culture medium with 10% fetal bovine serum (FBS).
3. Collect the released cells into centrifuge tubes, centrifugate, resuspend with culture medium, and plate in culture dishes.

4. Wash with D-Hank's solution and incubate with collagenase (400 U/mL) in F12 medium with 10% FBS at 37 °C.
5. Replace the collagenase solution and collect the released cells and plate. Each hour for 3 h. (*See* **Note 1**).

3.1.3 Purification and Removal of Contaminants

1. If necessary, geneticin can be added to the culture medium to eliminate contaminating cells. Add geneticin to the culture medium to a final concentration of 100 μg/mL.
2. Subject the cells to this treatment for 3–7 days, replacing the medium twice a week.

3.1.4 Characterization Immunocytohistochemical Staining

1. Cells will be fixed in 4% paraformaldehyde for 30 min on an Eppendorf tube in a total 100 μL volume.
2. Next, the cells are subjected to 2000 rpm centrifugation for 10 min. About 10 μL of solution containing cells is left while pipetting out most of the solution. The cell suspension will be placed on a slide followed by application of antibodies and DAPI (4′,6-diamidino-2-phenylindole; to stain cellular nuclei) to confirm identity of isolated melanocytes using detection for the presence of bonafide markers.
   (a) Monoclonal mouse antibodies to ***human cytokeratin***
   (b) Rabbit immunoglobulin to cow ***S-100 antigen***
   (c) Monoclonal mouse antibody to ***human vimentin***
3. Exam morphology under microscopy.

3.1.5 Electron Microscopy of Premelanosome Protein (PMEL) and Tyrosinase

1. Fix the stromal melanocytes with a mixture of 2% paraformaldehyde, 1% glutaraldehyde in 0.2 M phosphate buffer (PBS), pH 7.4.
2. Postfix the cells with 1% OsO_4 supplemented with 1.5% ferrocyanide, dehydrated in ethanol, and embedded in epon.
3. Prepare ultrathin sections, counterstain with uranyl acetate and lead citrate for observation by electron microscope.

3.1.6 Immunogold Labeling and Quantification

1. Fix the primary melanocytes (or MNT-1 cells) using 2% paraformaldehyde.
2. Wash the cell pellets with PB, embed in 10% gelatin, and infuse in 2.3 M sucrose.
3. Freeze the mounted gelatin blocks in liquid nitrogen.
4. Prepare and collect ultrathin cryosections with 2% methylcellulose, 2.3 M sucrose.
5. Perform single, double, or triple immunogold labeling with antibodies and protein A coupled to 5, 10, or 15 nm gold (PAG 5, PAG 10, and PAG 15).

6. Quantify the distribution of Pmel17 and TRP1, Pmel17 and EEA1, or Pmel17, TRP1, and LAMP1 on double or triple immunogold-labeled cryosections.
7. Count the number of gold particles labeling each protein.

3.2 LDL Exposure

1. LDL was procured from a manufacturer and prepared as directed.
2. Plate about 10,000 melanocyte cells. (*See* **Note 2**).
3. Add 10 μL of LDL to 90 μL of 1× PBS. Total volume is 100 μL.
4. Two negative controls will be made. The first negative control is made by adding 10 μL of PC13:0/13:0 to 90 μL of PBS. The second will be 100 μL of plain PBS.
5. Pipette the mixtures and add to the cell cultures. Add negative controls to their own cell culture.
6. Incubate the cell cultures at 37 °C for 3 h.
7. After incubation, centrifuge the cells to obtain a cell pellet by centrifuging at 3000 rpm for 15 min.
8. Lyse the cells by alternative cycling at −80 °C and 25 °C (on a heating block) for 15 min each. Do this step (cycling) five times. (*See* **Note 3**).
9. Pass the lysate through a 23-gauge needle three times. Collect the lysate in an Eppendorf tube.
10. Measure the lysate protein concentrations using polyvinylidene difluoride (PVDF) membrane densitometry with a BSA standard.
    (a) To activate the PVDF membrane, saturate with MeOH, swish, and then discard. Rinse the membrane again with 20% MeOH and PBS, then discard the solution. Rinse a third time with distilled water and wash three times. Dry the membrane to ready it for application of sample proteins and standards.
    (b) Apply a BSA standard (1 mg/mL) to spots with 0.2, 0.3, 0.4, 0.5, and 1 μL (containing 0.2–1 μg) proteins to prepare a calibration curve.
    (c) Apply at least three concentrations of each protein sample (for example, 0.2, 0.3, 0.4, 0.5 μL) to spots on the same PVDF membrane.
    (d) After the membrane has dried, stain with Gelcode blue (ThermoFisher Scientific). After 2–10 min of staining, wash the membrane three times with distilled water for 10 min each, and let it dry.

(e) Take a picture of the membrane and upload to ImageJ to analyze the sample spots.

(f) From the standard curve estimates, determine the sample protein amounts to normalize for all subsequent analysis (lipase activity, cholesterol estimations). This will provide quantitative estimates per cell through the protein amount estimations. That is, the amount of protein obtained per cell and amount of activity or cholesterol amount per protein amount.

3.3 Lipase Activity Assay

1. Warm the LPL substrate in DMSO to 37 °C by placing it in an incubator.
2. Take the warmed LPL substrate (2.5 mg/mL) in DMSO and dilute 2 μL in 1 mL of LPL Assay buffer.
3. Take 220 μL of double distilled water and mix with LPL positive control. A negative control will be 220 μL of PBS alone or PBS with PC13:0/13:0.
4. Standard dilutions were made according to the table provided by the manufacturer.
5. Melanocyte cells were harvested and washed with PBS twice, then suspended in 200 μL of cold PBS. Pipette and then centrifuge this mix.
6. Transfer the centrifuged mix to a tube.
7. Wells were set up according to the tables provided by the manufacturer.
8. Reaction mixes were made according to manufacturer's provided table, with each mix containing 10 μL of solution.
9. Add each of the mixes into a well.
10. Add 10 μL of LPL substrate into each well, except the negative controls, which will have PBS or PC13:0/13:0.
11. Incubate for an hour at 37 °C.
12. During the hour of incubation, output at Excitation/Emission = 482/415 nm will be measured every 10 min.
13. The standard lipase is part of the kit. The activity using the standard will be measured identically, and the emission reading determined identically will be compared, and lipase activity will be tabulated as percent activity compared to the standard.

3.4 Esterified Cholesterol Analysis with LCMS/MS

1. Label each of the 1 L solvent bottles, with A and B.
2. A and B will contain the mobile phases.

   (a) For bottle A, add 1 mL formic acid, 487 mL of water, and 25 mL of ammonium formate together.

(b) For bottle B, add 1 mL formic acid, 19.5 mL of water, 25 mL of ammonium formate, 857 mL of isopropanol, and 97.5 mL of acetonitrile.

3. Add the esterified cholesterol lipid sample to a 49 μL 1:1 mix of acetonitrile/isopropanol.
4. Spike 1 μL of EquiSPLASH into each sample.
5. Begin Mass Spectrometry analysis:
   1. Vanquish Horizon Parameters
      (a) 8 μL of the sample should be injected. The column temperature for the sample should be 55 °C. Gradient time is 30 min for solvent b. Flow rate should be set at 0.260 mL/min.
   2. HESI Source Parameters:
      (a) Positive mode parameters: Spray voltage 4.0 kV, Capillary temperature 325 °C, Auxiliary heater 275 °C, Sheath gas flow rate 35 units, Aux gas flow rate 15 units, S-lens RF 70.
      (b) Negative mode parameters: Spray voltage 2.50 kV, Capillary temperature 325 °C, Auxiliary heater 275 °C, Sheath gas flow rate 30 units, Aux gas flow rate 15 units, S-lens RF.
   3. Q Exactive Parameters:
      (a) Positive Mode/Full scan: Chromatogram peak width 6 s, Method duration 30 min, Resolution 70,000, AGC target 1E6, Maximum injection time 240 ms, Scan range 250–1200 m/z.
      (b) DD/MS2: Resolution 17,500, AGC target 1E5, Maximum injection time 50 ms, Loop count 10, Isolation window 1.0 m/z, Collision energies 20, 25, 30, 35, 40 eV.
      (c) Negative mode/Full scan: Chromatogram peak width 6 s, Method duration 30 min, Resolution 70,000, AGC target 1E6, Maximum injection time 240 ms, Scan range 250–1500 m/z [6].
6. Upload the files to LipidSearch software to analyze.

3.5 Cholesterol Analysis with GCMS

1. After weighing an empty cryovial on an electric balance, add the lysed cell solution.
2. Add 300 μL methanol/chloroform (2:1) mix to the lysate cell solution. Vortex and homogenize this sample for 3 min.
3. Add 300 μL chloroform to the cell sample and homogenize again for about 2 min.
4. Centrifuge the sample at 10,000*g* for about 30 min.

5. After centrifugation, separate the top layer in the cryovial from the bottom layer, which will contain the cholesterol. Once separated, add each layer to its own labeled cryovial (weigh the cyrovials before adding the layers).
6. Centrifuge the cryovials until a pellet appears.
7. To confirm efficiency, an internal standard of deuterated 3-ß-hydroxy-5-cholestenoic acid-d5 (catalog 700151 Avanti Polar Lipids, Alabaster, AL) prepared in methanol/chloroform is used.
8. Take the cryovials with pellets and add 100 μL of sodium hydroxide methanol.
9. Heat the cryovials in an oven at 90 °C for 15 min.
10. Add 400 μL of BF_3 solution to cryovials.
11. Heat the cryovials in the oven again at 90 °C for 30 min, then vortex.
12. Centrifuge the cryovials at 5000*g* for 5 min.
13. Separate the layer with cholesterol with a Pasteur pipette.
14. Add the layer with cholesterol to a mix of MTBE/MeOH and inject 2 μL to splitless autosampler.
15. The sample is injected with an Agilent 7693 autosampler (Agilent Technologies) in splitless mode on the GC-MS (Agilent 7890A Plus GC-MS) and separated through a CP-Sil 8 CB column (non-polar phase: 5% phenyl methylpoly-siloxane) using the following temperature gradient: 100 °C for 4 min; increase to 300 C at steps of 10 C/min; hold for 6 min; decrease to 120 °C at steps of 15 C/min. Helium at a rate of 1 mL/min is used as the carrier gas [7].
16. LipidSearch software will be used to analyze data.

4 Notes

1. The disaggregation of uveal stromal cells, achieved through a sequential process involving trypsin followed by collagenase, yields superior outcomes in terms of cell quantity, viability, and growth rates compared to using trypsin or collagenase alone.
2. If using this method for mass spectrometric lipid analysis, use one million cells.
3. If no heating block is available for the lysing process, a good replacement is incubating the cells at room temperature instead.

References

1. Mueller A, Lam I, Kishor K et al (2024) Secondary glaucoma: toward interventions based on molecular underpinnings. WIREs Mech Dis 16: e1628. https://doi.org/10.1002/wsbm.1628
2. Schallreuter KU, Hasse S, Rokos H et al (2009) Cholesterol regulates melanogenesis in human epidermal melanocytes and melanoma cells. Exp Dermatol 18:680–688. https://doi.org/10.1111/j.1600-0625.2009.00850.x
3. Hu DN, McCormick SA, Ritch R, Pelton-Henrion K (1993) Studies of human uveal melanocytes in vitro: isolation, purification and cultivation of human uveal melanocytes. Invest Ophthalmol Vis Sci 34:2210–2219
4. Raposo G, Tenza D, Murphy DM et al (2001) Distinct protein sorting and localization to premelanosomes, melanosomes, and lysosomes in pigmented melanocytic cells. J Cell Biol 152: 809–824. https://doi.org/10.1083/jcb.152.4.809
5. Viertler M, Schittmayer M, Birner-Gruenberger R (2012) Activity based subcellular resolution imaging of lipases. Bioorg Med Chem 20:628–632. https://doi.org/10.1016/j.bmc.2011.04.018
6. Harvey FC, Collao V, Bhattacharya SK (2023) High-resolution liquid chromatography-mass spectrometry for lipidomics. Methods Mol Biol 2625:57–63. https://doi.org/10.1007/978-1-0716-2966-6_4
7. Neag EJ, Collao V, Bhattacharya SK (2023) Analysis of cholesterol lipids using gas chromatography mass spectrometry. Methods Mol Biol 2625:141–148. https://doi.org/10.1007/978-1-0716-2966-6_13

Index

Marco Brotto and Sanjoy K. Bhattacharya (eds.), *Lipid Signalling: Methods and Protocols*, Methods in Molecular Biology, vol. 2816, https://doi.org/10.1007/978-1-0716-3902-3,

U

W

Z

MIX
Papier aus verantwortungsvollen Quellen
Paper from responsible sources
FSC® C105338
FSC
www.fsc.org

If you have any concerns about our products,
you can contact us on
ProductSafety@springernature.com

In case Publisher is established outside the EU,
the EU authorized representative is:
Springer Nature Customer Service Center GmbH
Europaplatz 3, 69115 Heidelberg, Germany

Printed by Libri Plureos GmbH
in Hamburg, Germany